PROTON-EMITTING NUCLEI

Related Titles from AIP Conference Proceedings

513 Nuclear and Condensed Matter Physics: VI Regional CRRNSM Conference
Edited by Antonino Messina, April 2000, 1-56396-929-7

495 Experimental Nuclear Physics in Europe: ENPE 99, Facing the Next Millennium
Edited by Berta Rubio, Manuel Lozano, and William Gelletly, November 1999, 1-56396-907-6

481 Nuclear Structure 98
Edited by C. Baktash, September 1999, 1-56396-858-4

455 ENAM 98: Exotic Nuclei and Atomic Masses
Edited by B. M. Sherrill, D. J. Morrissey, and C. N. Davids, December 1998, 1-56396-804-5

447 Nuclear Fission and Fission-Product Spectroscopy: Second International Workshop
Edited by G. Fioni, H. Faust, S. Oberstedt, F.-J. Hambsch, October 1998, 1-56396-823-1

425 Tours Symposium on Nuclear Physics III
Edited by M. Arnould, M. Lewitowicz, Yu. Ts. Oganessian, M. Ohta, H. Utsunomiya, and T. Wada, April 1998, 1-56396-749-9

412 Intersections Between Particle and Nuclear Physics: 6th Conference
Edited by T. W. Donnelly, December 1997, 1-56396-712-X

To learn more about these titles, or the AIP Conference Proceedings Series, please visit the webpage **http://www.aip.org/catalog/aboutconf.html**

PROTON-EMITTING NUCLEI

PROCON '99
First International Symposium

Oak Ridge, TN   7–9 October 1999

EDITOR
Jon C. Batchelder
UNIRIB/Oak Ridge Associated Universities, Tennessee

Melville, New York
AIP CONFERENCE PROCEEDINGS ■ 518

Editor:

Jon C. Batchelder
JIHIR/ORNL
P.O. Box 2008, Bldg. 6008
Oak Ridge, TN 37831-6374
USA

E-mail: batcheld@mail.phy.ornl.gov

The articles on pp. 3–13, 59–67, 112–122, and 200–208 were authored by U.S. Government employees and are not covered by the below mentioned copyright.

Authorization to photocopy items for internal or personal use, beyond the free copying permitted under the 1978 U.S. Copyright Law (see statement below), is granted by the American Institute of Physics for users registered with the Copyright Clearance Center (CCC) Transactional Reporting Service, provided that the base fee of $17.00 per copy is paid directly to CCC, 222 Rosewood Drive, Danvers, MA 01923. For those organizations that have been granted a photocopy license by CCC, a separate system of payment has been arranged. The fee code for users of the Transactional Reporting Service is: 1-56396-937-8/00/$17.00.

© 2000 American Institute of Physics

Individual readers of this volume and nonprofit libraries, acting for them, are permitted to make fair use of the material in it, such as copying an article for use in teaching or research. Permission is granted to quote from this volume in scientific work with the customary acknowledgment of the source. To reprint a figure, table, or other excerpt requires the consent of one of the original authors and notification to AIP. Republication or systematic or multiple reproduction of any material in this volume is permitted only under license from AIP. Address inquiries to Office of Rights and Permissions, Suite 1NO1, 2 Huntington Quadrangle, Melville, N.Y. 11747-4502; phone: 516-576-2268; fax: 516-576-2450; e-mail: rights@aip.org.

L.C. Catalog Card No. 00-102142
ISBN 1-56396-937-8
ISSN 0094-243X
Printed in the United States of America

CONTENTS

Preface ix
Organizing Committees xi
Sponsors xiii
Conference Photo xiv

OVERVIEW AND PIONEERING EXPERIMENTS

Delayed Proton Emission from Nuclei: A Historical Perspective 3
   J. Cerny
Proton Emission Studies at GSI in the 1980s 14
   S. Hofmann
Munich Efforts to Search for Proton Radioactivity in Beam 24
   T. Faestermann
Proton Decay Rates and Nuclear Structure 34
   P. J. Woods

EXPERIMENTAL DIRECT PROTON RADIOACTIVITY MEASUREMENTS

Proton Drip-Line Studies at HRIBF 49
   K. Rykaczewski, J. C. Batchelder, C. R. Bingham, R. E. Bryan,
   T. Davinson, T. N. Ginter, C. J. Gross, R. Grzywacz, J. H. Hamilton,
   Z. Janas, M. Karny, B. D. MacDonald, J. W. McConnell, A. Piechaczek,
   J. Szerypo, K. S. Toth, W. B. Walters, P. J. Woods, and E. F. Zganjar
Fine Structure in Deformed Proton Emitters 59
   A. A. Sonzogni, C. N. Davids, P. J. Woods, D. Seweryniak,
   M. P. Carpenter, J. J. Ressler, J. Schwartz, J. Uusitalo,
   and W. B. Walters
First Observation of Proton Emission From ^{117}La 68
   F. Soramel, A. Guglielmetti, L. Stroe, L. Müller, R. Bonetti,
   F. Malerba, G. L. Poli, C. Boiano, A. Andrighetto, Z. C. Li,
   F. Scarlassara, C. Signorini, A. Dal Bello, R. Isocrate, Z. H. Liu,
   M. Ruan, M. Ivascu, P. Bednarczyk, and C. Broude
Nuclear Structure Effects in the Proton Decay of Odd-Odd Nuclides 74
   W. B. Walters
A Search for Neutron Single-Particle States Populated Via Proton
Emission from ^{146}Tm 83
   T. N. Ginter, J. C. Batchelder, C. R. Bingham, C. J. Gross,
   R. Grzywacz, J. H. Hamilton, Z. Janas, A. Piechaczek,
   A. V. Ramayya, K. Rykaczewski, W. B. Walters, and E. F. Zganjar
Search for Two-Proton Emitters at FRS-GSI 89
   M. Pfützner

Recent Studies of Proton Drip-Line Nuclei Using the Berkeley Gas-Filled Separator 95
    M. W. Rowe, J. C. Batchelder, V. Ninov, K. E. Gregorich,
    K. S. Toth, C. R. Bingham, A. Piechaczek, X. J. Xu, J. Powell,
    R. Joosten, and J. Cerny

Two-Proton Decay Experiments at MSU 105
    M. Thoennessen, M. J. Chromik, and P. G. Thirolf

In-Beam Studies of Proton Emitters Using the Recoil Decay Tagging Method 112
    D. Seweryniak, P. J. Woods, J. J. Ressler, C. N. Davids,
    A. Heinz, A. A. Sonzogni, J. Uusitalo, W. B. Walters,
    J. A. Caggiano, M. P. Carpenter, J. A. Cizewski, T. Davinson,
    K. Y. Ding, N. Fotiades, U. Garg, R. V. F. Janssens,
    T.-L. Khoo, F. G. Kondev, T. Lauritsen, C. J. Lister, P. Reiter,
    J. Shergur, and I. Wiedenhoever

SPECTROSCOPIC FACTORS AND ORBITAL MIXING

Spherical Proton Emitters and Spectroscopic Factors 125
    P. B. Semmes

Ground-State Properties of Deformed and Transitional Proton Emitters in the Relativistic Hartree-Bogoliubov Model 132
    G. A. Lalazissis, D. Vretenar, and P. Ring

Two-Proton Emission in the Hyperharmonics Approach 144
    I. G. Mukha

Exact Calculations for Deformed Proton Emitters 154
    E. Maglione and L. S. Ferreira

Theoretical Predictions for Beta-Delayed One and Two-Proton Emission 164
    T. Siiskonen and P. O. Lipas

Resonances in Deformed Nuclei: R-Matrix Theory and Oscillator Expansion 173
    A. T. Kruppa, W. Nazarewicz, and P. B. Semmes

Proton Emission from Gamow Resonance 184
    T. Vertse, A. T. Kruppa, B. Barmore, W. Nazarewicz,
    L. G. Ixaru, and M. Rizea

Dynamical Calculation of Proton Emission from a Deformed to a Spherical Nucleus 194
    P. Talou

Decay Rates for Spherical and Deformed Proton Emitters 200
    C. N. Davids and H. Esbensen

Theoretical Approaches and Experiments on Proton Decay 209
    S. G. Kadmensky

Asymptotic Behavior of the Wave Packet Propagation through a Barrier: The Green's Function Approach Revisited 217
    B. Mihaila, S. A. Gurvitz, D. J. Dean, and W. Nazarewicz

Proton-Emitting Nuclei in a Time Dependent Formalism 223
    N. Carjan, P. Talou, M. Rizea, and D. Strottman

BETA DELAYED PROTON EMISSION AND HIGH SPIN STATES

Beta-Delayed Proton Emission 229
    J. Hardy
The Effects of β-Delayed Proton Emission on the Path of the rp-Process 239
    R. N. Boyd
A Distribution of GT-Strength and βp-Emission Near ^{100}Sn 246
    M. Karny
Spectroscopy of β-Delayed Charged Particles at Projectile
Fragment Separators 255
    Z. Janas
Beta-Delayed Two-Particle Emission 264
    M. J. G. Borge
Isospin-Forbidden Beta-Delayed Proton Emission 275
    W. E. Ormand
Prompt Particle Decays from Deformed High-Spin States 285
    D. Rudolph

NOVEL EXPERIMENTAL TECHNIQUES AND NEW DIRECTIONS

Prospects for Future Proton Studies at HRIBF 297
    C. R. Bingham, J. C. Batchelder, T. N. Ginter, C. J. Gross,
    R. Grzywacz, Z. Janas, M. Karny, J. W. McConnell,
    K. Rykaczewski, K. S. Toth, and E. F. Zganjar
Applications of Real-Time Digital Pulse Processing in Nuclear Physics 307
    M. Momayezi, P. Grudberg, W. Skulski, and W. K. Warburton
The Statistical Properties of the Angular Distribution
of β-Delayed Protons from Oriented Nuclei 316
    J. Rikovska, N. J. Stone, and A. Wöhr
First Observation of Doubly-Magic ^{48}Ni 321
    J. Giovinazzo, B. Blank, C. Borcea, M. Chartier, S. Czajkowski,
    A. Fleury, G. de France, R. Grzywacz, Z. Janas, M. Lewitowicz,
    F. de Oliveira, M. Pfützner, M. S. Pravikoff, and J. C. Thomas
Proton-Radioactivity Studies at the FRS after the SIS Intensity Upgrade 326
    K. Schmidt

Attendee List 333
Conference Photos 343
Author Index 349

PREFACE

The past few years has seen an explosion of work both experimentally and theoretically on the topic of proton-emitting nuclei. The study of this exotic decay mode will continue to provide us with new nuclear structure information at the very limits of stability. Due to this, the International Symposium on Proton-Emitting Nuclei (PROCON99) was held to review the current situation and to discuss directions for the future. This conference was the first ever specialty conference devoted to this topic.

PROCON99 was held on October 7-9, 1999, at Pollard Auditorium in Oak Ridge, TN. It was attended by 87 physicists representing 38 universities or institutions from 14 countries. The conference featured both theoretical and experimental lectures by 33 speakers on the various subjects associated with proton emission from nuclear states.

The conference was made possible by the financial support of the Joint Institute for Heavy Ion Research (JIHIR), Oak Ridge Associated Universities (ORAU), the Physics Division of Oak Ridge National Laboratory (ORNL), and the UNIversity Radioactive Ion Beam consortium (UNIRIB). We also received financial support from the industries listed on the following pages.

I am indebted to the members of the organizing committees for their invaluable help and advice in the preparation of this conference. I would also like to thank the conference secretaries Carlene Stewart and Sherry Lamb for all their hard work. I would also like to thank our artist, Susan Jacques, for wonderful artwork produced for this conference. And lastly, thanks to all those who worked behind the scenes to make sure everything went smoothly throughout the conference.

Jon C. Batchelder

ORGANIZING COMMITTEES

Local Organizing Committee

J. C. Batchelder (chair) (UNIRIB/ORAU)

C. R. Bingham (U. Tennessee)

H. K. Carter (ORAU)

J. H. Hamilton (Vanderbilt)

W. Nazarewicz (U. Tennessee)

K. Rykaczewski (ORNL)

P. Semmes (Tennessee Tech)

K. S. Toth (ORNL)

E. F. Zganjar (Louisiana State U.)

International Advisory Committee

S. Åberg (Lund)

J. Äystö (Jyväskylä)

B. Blank (Bordeaux)

J. Cerny (LBNL)

C. N. Davids (Argonne)

S. Hofmann (GSI)

E. Maglione (Padova)

D. Strottman (Los Alamos)

J. Zylicz (Warsaw)

SPONSORS

Institutions

UNIversity Radioactive Ion Beam consortium (UNIRIB)

Oak Ridge Associated Universities (ORAU)

Joint Institute for Heavy Ion Research

Physics Division of Oak Ridge National Laboratory (ORNL)

Companies

LeCroy Corporation

Eurisys Mesures, Inc.

X-ray Instruments Associates

EG&G ORTEC

Micron Semiconductor

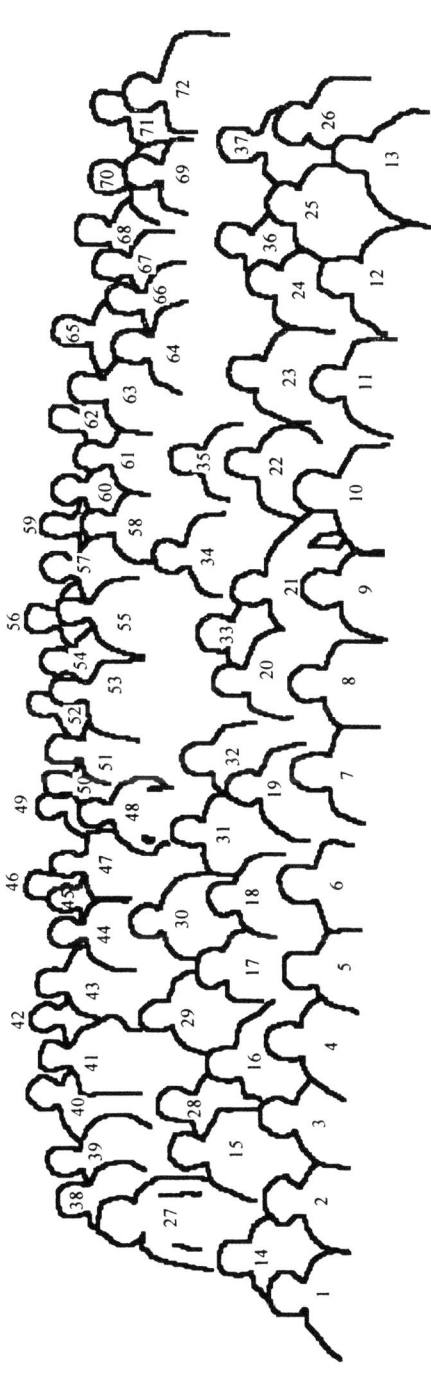

1. Wojtek Skulski   2. Yurdanur Akovali   3. Michael Momayezi   4. Marek Pfützner   5. Jan Kormicki   6. Jah Zylicz   7. Joseph Hamilton   8. Cary Davids   9. Jirina Rikovska   10. Jon Batchelder   11. Edward Zganjar   12. John Wood   13. Andreas Piechaczek   14. Thomas Ginter   15. Georgios Lalazissis   16. Stanislav Kadmensky   17. Bogdan Mihaila   18. Bryan Barmore   19. Ken Toth   20. Nicholas Stone   21. Krzysztof Rykaczewski   22. David Kulp   23. Nicolae Carjan   24. Thomas Faestermann   25. Andras Kruppa   26. Maria J. G. Borge   27. Lidia Ferreira   28. Andrei Andreyev   29. Omar Zeidan   30. Dimiter Balabanski   31. Volker Oberacker   32. Karsten Schmidt   33. Philip Woods   34. Carrol Bingham   35. Daryl Hartley   36. Patrick Talou   37. Carlene Stewart   38. Enrico Maglione   39. Matej Lipoglavsek   40. Alejandro Sonzogni   41. Jerome Giovinazzo   42. William Walters   43. Witold Nazarewicz   44. Marek Lewitowicz   45. Paul Mueller   46. Carl Gross   47. Robert Grzywacz   48. Francesca Soramel   49. Sigurd Hofmann   50. Eric Ormand   51. Alfredo Galindo-Uribarri   52. Michael Thoennessen   53. Ivan Mukha   54. David Radford   55. Paul Semmes   56. Thomas Davinson   57. Lee Reidinger   58. Ermias Gete   59. Teemu Siiskonen   60. Noah Johnson   61. Hubert Flocard   62. Jim Beene   63. John Hardy   64. Thomas Vertse   65. Joseph Cerny   66. Dirk Rudolph   67. Ari Jokinen   68. Marek Karny   69. Paola Spolaore   70. Zenon Janas   71. Cyrus Baktash   72. Ken Carter

Overview and Pioneering Experiments

Delayed Proton Emission from Nuclei: A Historical Perspective

Joseph Cerny

Department of Chemistry, University of California, Berkeley, and Nuclear Science Division, Ernest Orlando Lawrence Berkeley National Laboratory, Berkeley, California 94720

Abstract. Early experiments observing proton emission are reviewed, with an emphasis on the initial discovery. Beta-delayed proton emission (1963), direct proton radioactivity (1970), and beta-delayed two-proton emission (1983) have all been observed, while direct two-proton radioactivity is still being sought. This historical overview concludes in 1990.

INTRODUCTION

Though alpha-decay and its related beta-delayed alpha emission have been known for a relatively long time, it was not until the 1960s that decay processes resulting in the emission of a proton began to come under successful experimental observation.

In a series of seminal papers beginning in 1960, Goldanskii [1] described three phenomena involving proton emission: (a) beta-delayed proton emission, (b) proton radioactivity, and (c) his new concept of two-proton radioactivity. Then, in a further paper in 1980 [2], he predicted beta-delayed two-proton emission. What I will emphasize is the progress made in observing these decay modes through 1990, with a focus on the first experiment involving each decay mode. Since there are several other historical talks, I will only allude to the advances that other speakers will present in more detail. Interestingly, the first characterizations of nuclides whose decay process results in the emission of a proton (or protons) have all been observed in the lighter mass region (or are currently being sought there).

Although Karnaukhov and collaborators at the Dubna heavy ion cyclotron are credited with the discovery of beta-delayed proton emission in 1963 [3], from 130 MeV ^{20}Ne on nickel and tantalum targets, the first nuclide to be characterized as a beta-delayed proton emitter was ^{25}Si observed by Barton and collaborators at the McGill synchrocyclotron also in 1963 [4]. The influence of Goldanskii is clearly seen in the Dubna work and John Hardy told me that Professor Bell, leader of the McGill effort, had visited Russia about 1960 and had met Goldanskii. As a further historical note, Alvarez had reported in 1950 [5] attempts to find beta-delayed proton emission using the Berkeley Radiation Laboratory's 32 MeV proton linac, but instead discovered three new beta-delayed alpha-particle emitters, ^{8}B, ^{12}N, and ^{20}Na.

Figure 1 presents the lighter nuclei and the first nuclide to be characterized as decaying by a mode of proton emission. My discussion of beta-delayed proton emission will be relatively brief, since John Hardy will cover this topic later. Subsequently, I will discuss direct proton radioactivity, discovered in 1970 from an isomer in ^{53}Co, and the 1983 discovery of the first beta-delayed two-proton emitter ^{22}Al. Finally, direct two-proton radioactivity has yet to be observed, but I will conclude with the attempt by Détraz and collaborators at GANIL in 1990 to observe the decay of ^{39}Ti by this mode.

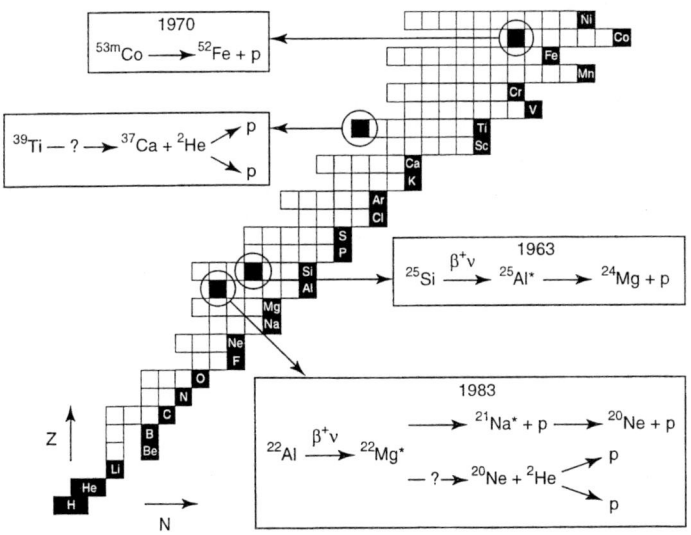

FIGURE 1. A partial chart of the nuclides showing the first nuclide characterized (or the best search) as decaying by a new mode of proton emission. See text.

DECAYS INVOLVING EMISSION OF ONE PROTON

Beta-Delayed Proton Emission

Figure 2 shows a decay scheme for ^{25}Si as a beta-delayed proton emitter. Of particular interest is the superallowed beta-decay to the isobaric analog state (IAS) in the daughter, which is followed by isospin-forbidden proton emission.

The nuclide ^{25}Si was produced at McGill with an external proton beam from their synchrocyclotron via the ^{27}Al(p,3n) reaction. Figure 3 presents the original ^{25}Si proton energy spectrum [4] measured with a single silicon detector which counted between beam bursts. The three observed proton groups were attributed to ^{25}Si by excitation function measurements and cross bombardments.

For comparison, Fig. 4 presents a recent beta-delayed proton spectrum from ^{25}Si produced in the ^{24}Mg(^{3}He,2n) reaction [6]. The region spanned by the McGill data is indicated, which includes one of the decay branches from the IAS. Figure 5 then

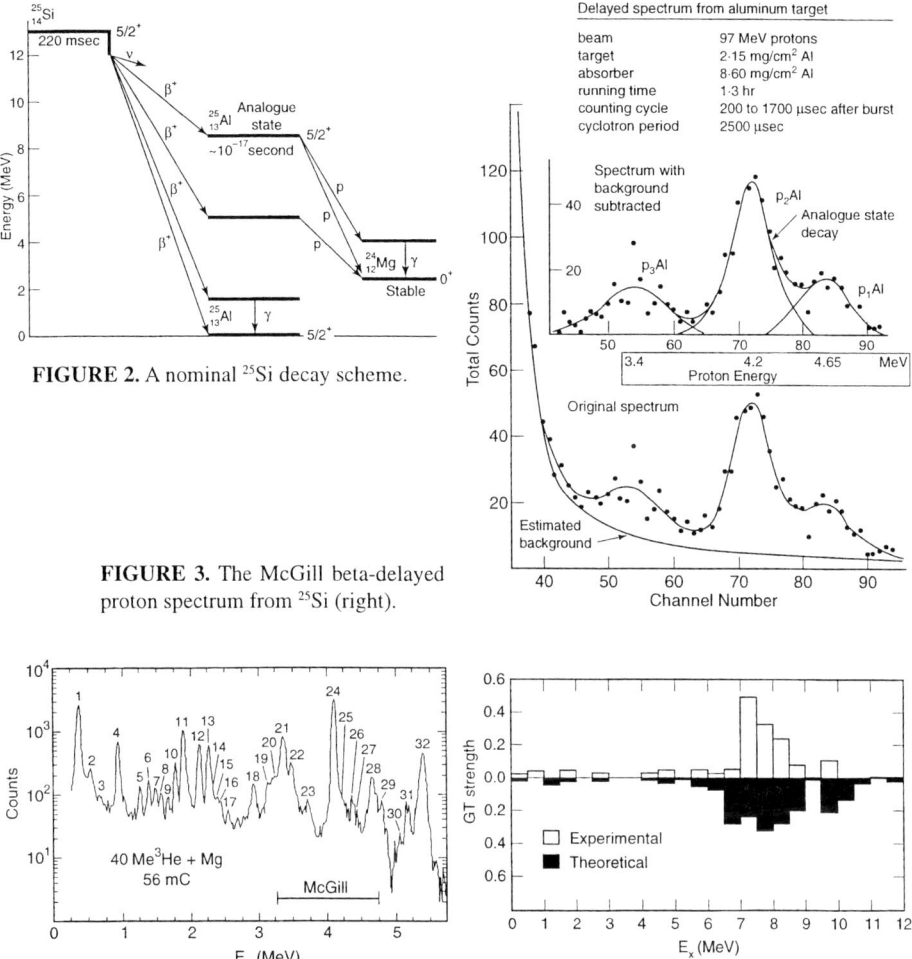

FIGURE 2. A nominal ^{25}Si decay scheme.

FIGURE 3. The McGill beta-delayed proton spectrum from ^{25}Si (right).

FIGURE 4. The beta-delayed proton spectrum of ^{25}Si obtained with helium-jet techniques and a gas ΔE-Si E detector telescope.

FIGURE 5. Gamow-Teller beta strengths as a function of excitation energy in ^{25}Al.

shows how such beta-delayed proton studies can be used to compare (successfully) experimental Gamow-Teller beta-decay strengths to predictions of large basis shell-model calculations [6].

Proton Radioactivity

I was on sabbatical at Oxford University during 1969–70 and was interested in utilizing both the tandem Van de Graaff at Oxford and the heavy ion cyclotron at Harwell to explore nuclei with $Z > N$ above the titanium isotopes, of which only a very few were known. In particular, the series of strong beta-delayed proton emitters with A = 4n+1 and $T_z = -3/2$ had been established through ^{41}Ti by (p,3n) reactions at McGill and (^{3}He,2n) reactions at Brookhaven. I wanted to try to use the (heavy ion, 3n) reaction at Harwell to extend this series.

The initial experiment at Harwell, the ^{40}Ca(^{12}C,3n) reaction with decay products observed in a ΔE-E silicon detector telescope, successfully observed ^{49}Fe [7] with a cross-section ~0.5 µb, a $\tau_{1/2}$ ~ 75 ms, and an $E_{c.m.}$ = 1.96 MeV β-delayed proton group. A subsequent experiment involved an attempt to produce ^{53}Ni via the ^{40}Ca(^{16}O,3n) reaction with an expected beta-delayed proton energy of 2.0 or 2.8 MeV. No proton groups near these energies were seen, but a proton group was observed near 1.5 MeV with a $\tau_{1/2}$ ~ 245 ms and about ten times the yield of the earlier reaction producing ^{49}Fe. Figure 6(a) shows the group in the detector telescope (and uncomfortably close to the telescope cut-off) while Fig. 6(b) again shows this proton group in a single silicon E detector. Measurement of the excitation function for the activity eliminated ^{53}Ni as a possible source. On analysis, the combination of proton energy, half-life and reaction threshold were completely inconsistent with the expected nuclear properties in this mass region.

Luckily, we had developed a setup at Oxford using βγ coincidences to search for other new proton-rich nuclei in the $f_{7/2}$ shell and this was utilized in the 49 MeV ^{16}O on ^{40}Ca reaction. A very high yield was observed for a recently-discovered isomer in ^{53}Fe—an isomer with J^π = 19/2$^-$, which was another example of a "spin-gap" isomer involving three nucleons just below or just beyond doubly closed shells. By applying mirror symmetry, with mass predictions and Coulomb displacement energy calculations, the prob-

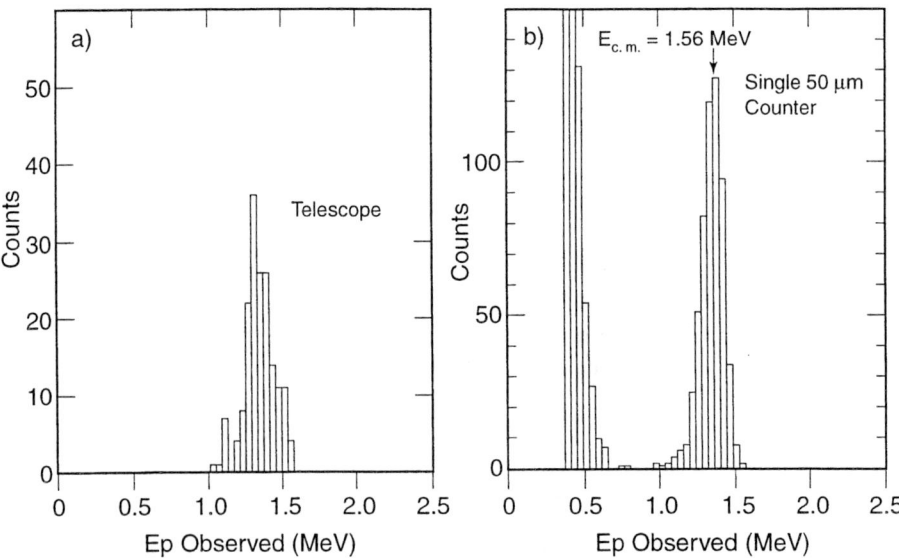

FIGURE 6. Proton energy spectra from the reaction of 80 MeV ^{16}O on ^{40}Ca: (a) using a detector telescope; (b) using a single E detector.

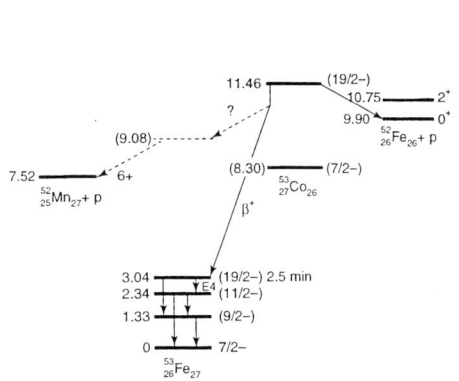 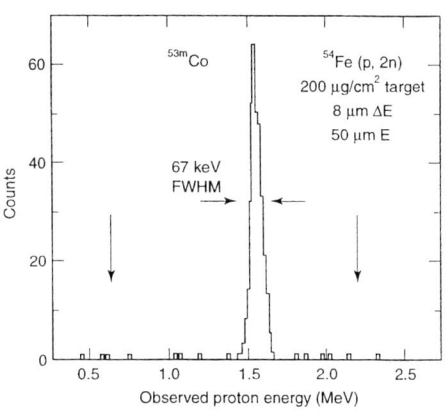

FIGURE 7. Probable decay scheme for ^{53m}Co, $\tau_{1/2} \sim 245$ ms.

FIGURE 8. The proton decay of ^{53m}Co produced in the ^{54}Fe(p,2n) reaction.

able decay of the mirror isomer ^{53m}Co [8] is shown in Fig. 7. The energy systematics shown in this figure indicate that ^{53m}Co would be proton radioactive, emitting an $E_{c.m.}$ = 1.56 MeV proton. Since the 245 ms half-life was only consistent with dominant β-decay, the direct proton radioactivity must be only a weak branch.

Since my sabbatical was over and the Oxford group had broken up, we continued these experiments at the Lawrence Berkeley Laboratory 88-in. Cyclotron. Figure 8 shows ^{53m}Co now produced via the ^{54}Fe(p,2n) reaction (as well as a search for a proton decay branch to the ^{52}Fe first excited state, which has never been observed). Experiments at Berkeley also eliminated the possibility that this proton group might arise by beta-delayed proton decay as indicated in Fig. 7; these results confirmed ^{53m}Co to be a direct proton emitter [8].

Figure 9 shows the ^{54}Fe(p,2n) ^{53m}Co and ^{54}Fe(p,pn) ^{53m}Fe excitation functions—the threshold for the former is in excellent agreement with that required by the proposed decay scheme. Also shown are fits to these excitation functions using the spin-dependent nuclear evaporation code GROGI-2, which permitted us to estimate the direct proton branch to be 1.5%.

These data then lead to the final decay scheme for ^{53m}Co shown in Fig. 10, with a partial half-life for direct proton radioactivity of 17s [9]. Decay of this $(f_{7/2})^{-3}$ configuration, coupled to 19/2⁻, by direct proton transmission through the Coulomb and $\ell = 9$ centrifugal barriers to the ^{52}Fe ground state leads to an expected half-life of ~60 ns, so the ~17s partial half-life implies a reduced width $\gamma_p^2 \sim 4 \times 10^{-9}$. Calculations by Bugrov and collaborators [10] using a multiparticle theory for the proton decay of spherical and deformed nuclei are in reasonable agreement with this long partial half-life.

Proton radioactivity from the ^{151}Lu ground state was discovered by Hofmann and collaborators in 1981 [11] in experiments at GSI. Additional discoveries of proton emitters at GSI and at Munich by 1990 are listed in Table 1. These experiments will be discussed by the next two speakers, Drs. Hofmann and Faestermann. A comprehensive review of proton radioactivity results through December 1, 1992 by Hofmann appears in Ref. 12.

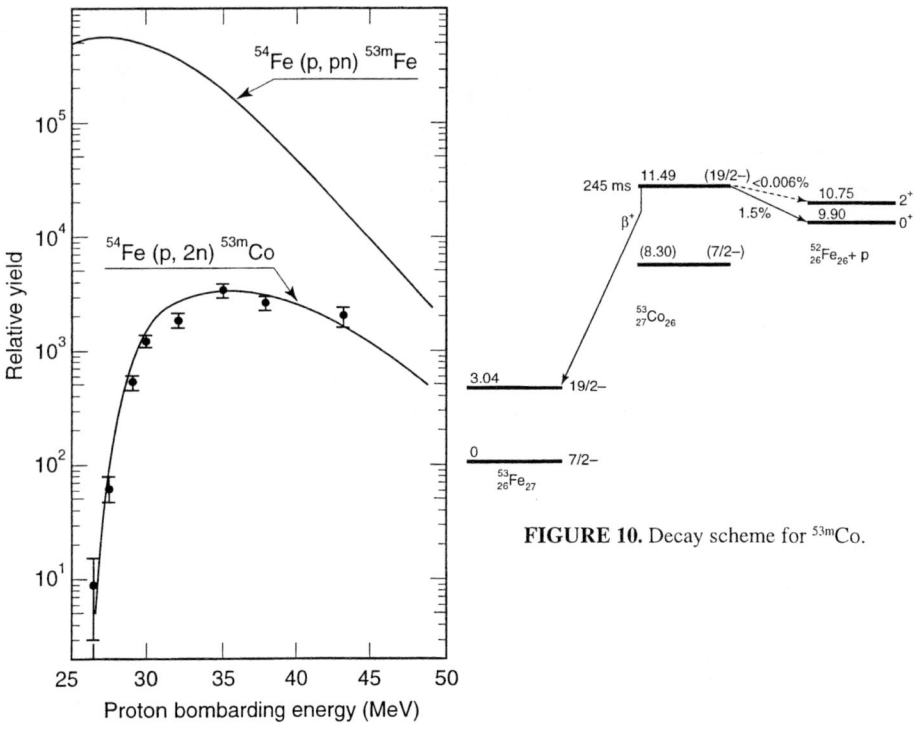

FIGURE 9. Excitation functions for p + ^{54}Fe. See text.

FIGURE 10. Decay scheme for ^{53m}Co.

TABLE 1. Discoveries of proton radioactivity by 1990.

Nuclide	Energy (keV)	Half-Life	Laboratory	Year
^{53m}Co	1590	245 ms	Oxford/LBL	1970
^{151}Lu	1231	85 ms	GSI	1981
^{147}Tm	1044	420 ms	GSI	1982
^{113}Cs	980	33 µs	Munich	1983
^{109}I	830	109 µs	Munich	1984
^{150}Lu	1261	>10 ms	GSI	1984
^{147m}Tm	1117	360 µs	GSI	1984

DECAYS INVOLVING EMISSION OF TWO PROTONS

Beta-Delayed Two-Proton Emission

By the early 1980s, our group had returned to an earlier interest—that of establishing the mass and/or the beta-delayed proton decay mode of the lightest bound member of each $T_z = -3/2$ or -2 mass series—and were characterizing ^{22}Al, $\tau_{1/2} \sim 70$ ms, as the first such member of the A = 4n + 2, $T_z = -2$ mass series [13]. With the timely prediction by Goldanskii [2] that nuclei such as ^{22}Al were also good candidates for a new decay process of beta-delayed two-proton emission, we decided to conduct a search for this decay mode.

The decay scheme for ^{22}Al shown in Fig. 11 indicates that the observed delayed proton decays to ^{21}Na from the ^{22}Mg isobaric analog state were of relatively high energy and that this IAS is unbound to two-proton emission to the ground state of ^{20}Ne by 6.1 MeV. Figure 12 shows the experimental setup used to observe this latter decay mode of ^{22}Al produced via the ^{24}Mg(^{3}He,p4n) reaction with 110 MeV ^{3}He beams from the 88-in. Cyclotron. A helium-jet system was used to transport the activity to a slowly rotating catcher wheel, so that ^{22}Al as a thin source could be observed by two high solid-angle detector telescopes. These three-element particle telescopes (with detectors denoted ΔE1, ΔE2, E) observed and identified decays involving two coincident protons and established ^{22}Al as the first known beta-delayed two-proton emitter [14].

Figure 13(a) shows the summed proton energy spectrum from the ^{22}Mg IAS to the ^{20}Ne ground state (g) and first excited state (x). Figure 13(c) shows the individual proton energy spectrum for protons forming group (g) in Fig. 13(a), while Fig. 13(b) shows the equivalent spectrum for protons forming group (x).

Clearly of much interest in the study of such two-proton emission processes is whether the decay is (a) sequential emission, (b) simultaneous but uncoupled emission, or (c) ^{2}He emission—decay through an L = 0 final-state interaction between the two protons [1,2]. Comparison of these small angle measurements (~45° separation of the two

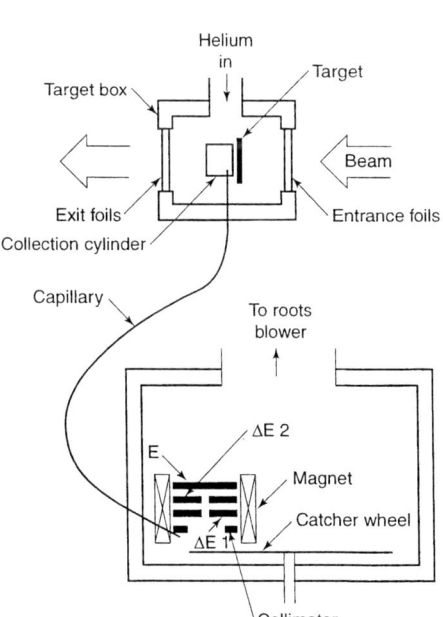

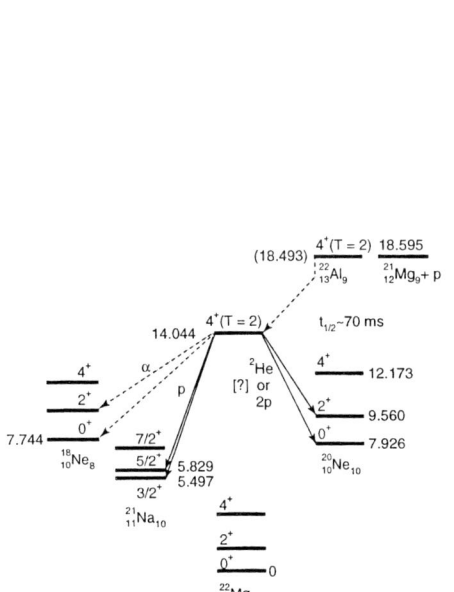

FIGURE 11. An initial decay scheme for ^{22}Al.

FIGURE 12. Schematic diagram of the experimental setup for observing two-proton decay at small relative angles.

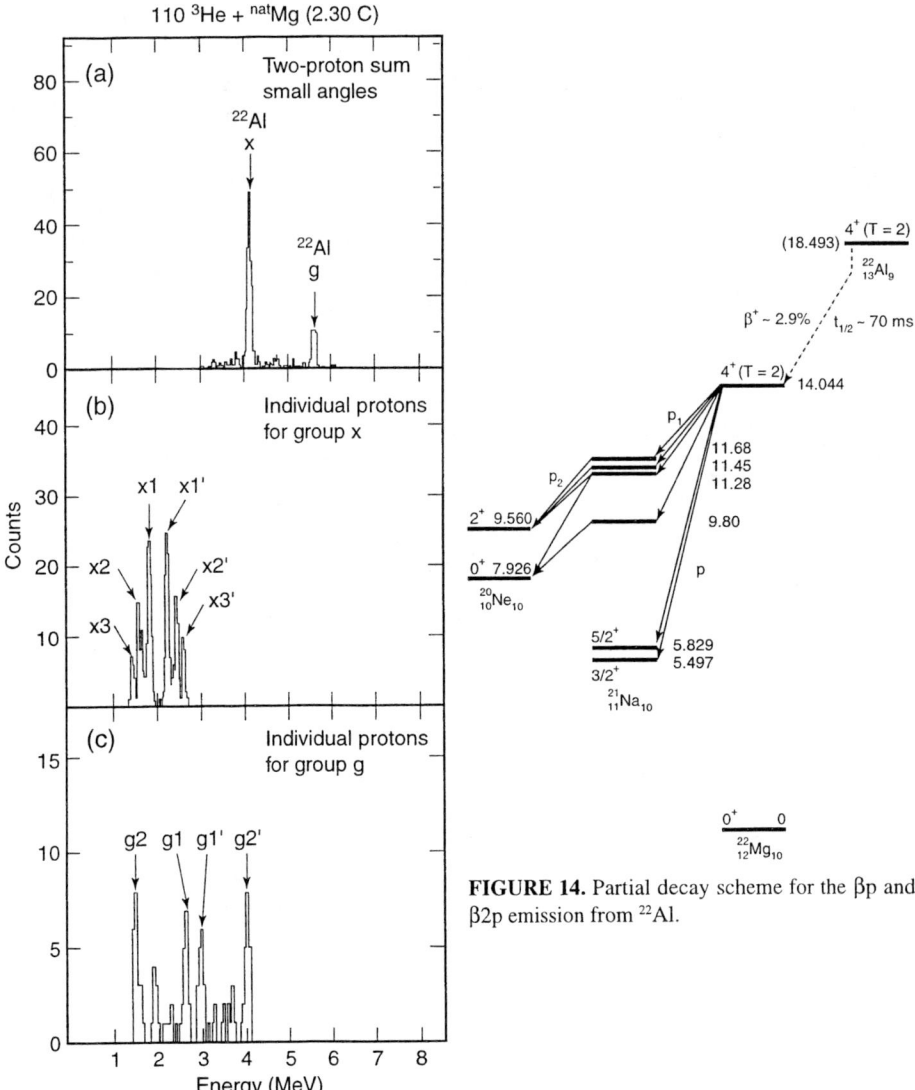

FIGURE 13. Proton-proton coincidence spectra obtained with the small angle detector system following the decay of ^{22}Al. See text.

FIGURE 14. Partial decay scheme for the βp and β2p emission from ^{22}Al.

protons) to large angle measurements (~120° separation) established the ^{22}Al decay mechanism to be sequential. Figure 14 presents the details of this sequential decay [15].

Table 2 lists the discoveries of the known beta-delayed two-proton emitters by 1990 (for references to all but ^{31}Ar, see [16]). The nuclide ^{31}Ar was discovered and shown to be a beta-delayed two-proton emitter in [17]. So far, only predominantly sequential proton decay of these nuclides has been observed.

TABLE 2. Discoveries of beta-delayed two-proton emission by 1990.

Nuclide	Observed IAS 2p Decay Energy (MeV)	Half-Life	Laboratory	Year
$T_z = -2$				
^{22}Al	5.6	~70 ms	LBL	1983
^{26}P	4.9	~20 ms	LBL	1983
$T_z = -5/2$				
^{35}Ca	4.1	~50 ms	LBL	1985
^{31}Ar	7.5	~15 ms	GANIL	1987
			LBL	1989

The Search for Two-Proton Radioactivity

By 1990 (and even by the time of this conference) there has been no experimental observation of this new decay mode of two-proton radioactivity as defined by Goldanskii [1]. Also see the review in Ref. [18]. The most promising search for this decay mode was by Détraz et al. [19 and Refs. therein] in observing the decay of ^{39}Ti with the LISE spectrometer at GANIL. The nuclide ^{39}Ti was expected to have a two-proton decay energy ~750 keV.

Figure 15 is from the GANIL experiment and shows the beta-delayed proton spectrum obtained from 75 decays of ^{39}Ti; it shows primarily one peak at ~3.6 MeV. Three independent results in the experiment ruled out ^{39}Ti as a significant direct 2p emitter from its ground state: (a) its observed half-life, ~26 ms, agreed with that expected from beta-decay theory; (b) the energy spectrum in Fig. 15 shows no evidence for beta-delayed protons from ^{37}Ca, which would be produced as the daughter nucleus in this decay mode; and (c) a search for correlated low-energy protons (several hundred keV each) in a digital-storage oscilloscope triggered by the implantation of ^{39}Ti nuclei showed no events.

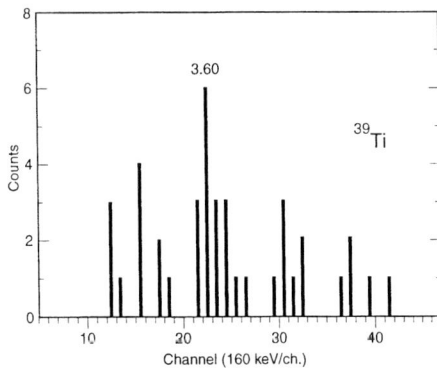

FIGURE 15. The energy spectrum of beta-delayed protons from ^{39}Ti.

When Détraz et al. [19] corrected for the Thomas-Ehrmann shift in this $T_z = -5/2$ nucleus, they found a significantly lowered two-proton decay energy of ~570 keV to be expected. A slightly later experiment [20] observed the beta-delayed two-proton decay of ^{39}Ti from the IAS and used Coulomb displacement energy calculations to estimate the mass of ^{39}Ti; results attributable to both experiments are shown in Fig. 16. The ground state of ^{39}Ti is found to be unbound to two-proton emission by 530 ± 65 keV (in agreement with the Thomas-Ehrmann shift calculations). Though a small direct two-proton decay branch from the ^{39}Ti ground state is not precluded by this lower mass, it is constrained to be less than 0.1%.

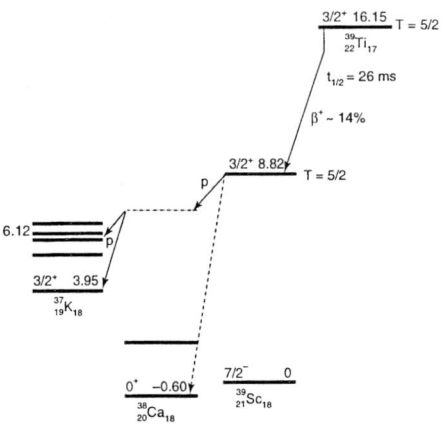

FIGURE 16. Proposed partial decay scheme for ^{39}Ti. The intermediate state in ^{38}Ca is not known.

CONCLUSIONS

This review has outlined the progress made in the observation of the emission of delayed protons from nuclei through 1990.

By the early 1990s, with the advent of double-sided silicon strip detectors, second generation recoil mass spectrometers, etc., the stage had been set for (a) a quintupling of known proton emitters; (b) a doubling of known beta-delayed two-proton emitters; and (c) a continuing search for direct ground-state two-proton radioactivity. Many of these newer results will be discussed throughout this conference.

This work was supported in part by the Director, Office of Energy Research, Division of Nuclear Physics of the Office of High Energy and Nuclear Physics of the U.S. Department of Energy under Contract No. DE-AC03-76SF00098.

REFERENCES

1. Goldanskii, V.I., Nucl. Phys. *19*, 482–495 (1960); Nucl. Phys. *27*, 648–664 (1961); Ann. Rev. Nucl. Sci. *16*, 1–30 (1966).
2. Goldanskii, V.I., Pis'ma Zh. Eksp. Teor. Fiz. *13*, 572–574 (1980). [Sov. Phys.–JETP Lett. *32*, 554–556 (1980).]
3. Karnaukhov, V.A., Ter-Akopian, G.M., and Subbotin, V.G., in *Proc. Conf. Reactions Between Complex Nuclei, 3rd*, Asilomar, CA, edited by A. Ghiorso, R.M. Diamond, and H.E. Conzett, Univ. of Calif. Press, 1963, pp. 434–437.
4. Barton, R., McPherson, R., Bell, R.E., Frisken, W.R., Link, W.T., and Moore, R.B., Can. J. Phys. *41*, 2007–2025 (1963).
5. Alvarez, L.W., Phys. Rev. *80*, 519–523 (1950).
6. Robertson, J.D., et al., Phys. Rev. C *47*, 1455–1461 (1993).
7. Cerny, J., Cardinal, C.U., Evans, H.C., Jackson, K.P., and Jelley, N.A., Phys. Rev. Lett. *24*, 1128–1130 (1970).
8. Jackson, K.P., Cardinal, C.U., Evans, H.C., Jelley, N.A., and Cerny, J., Phys. Lett. B *33*, 281–283 (1970). Cerny, J. Esterl, J.E., Gough, R.A., and Sextro, R.G. Phys. Lett. B *33*, 284–286 (1970).
9. Cerny, J., Gough, R.A., Sextro, R.G., and Esterl, J.E., Nucl. Phys. A *188*, 666–672 (1972).
10. Bugrov, V.P., Bunakov, V.E., Kadmenskii, S.G., and Furman, V.I., Yad. Fiz. *42*, 57–66 (1985). [Sov. J. Nucl. Phys. *42*, 34–39 (1985).]
11. Hofmann, S., et al., in Proc. 4th Intl. Conf. on Nuclei Far from Stability, CERN 81-09, Geneva, 1981, pp. 190–201.
12. Hofmann, S., in *Nuclear Decay Modes*, edited by D.N. Poenaru, Bristol, UK: Inst. Physics, 1996, pp. 143–203.
13. Cable, M.D., et al., Phys. Rev. C *26*, 1778–1780 (1982).
14. Cable, M.D., Honkanen, J., Parry, R.F., Zhou, Z.H., Zhou, Z.Y., and Cerny, J., Phys. Rev. Lett. *50*, 404–406 (1983).
15. Cable, M.D., et al., Phys. Rev. C *30*, 1276–1285 (1984).
16. Äystö, J., and Cerny, J., in *Treatise on Heavy-Ion Science, Vol. 8*, edited by D.A. Bromley, Plenum Publishing Corp., 1989, pp. 207–258.
17. Borrel, V., et al., Nucl. Phys. A *473*, 331–341 (1987). Reiff, J.E., et al., Nucl. Instr. Methods A *276*, 228–232 (1989).
18. Détraz, C., and Vieira, D.J., Ann. Rev. Nucl. Part. Sci. *39*, 407–465 (1989).
19. Détraz, C., et al., Nucl. Phys. A *519*, 529–547 (1990).
20. Moltz, D.M., et al., Zeit. Phys. A *342*, 273–276 (1992).

Proton Emission Studies at GSI in the 1980s

Sigurd Hofmann

Gesellschaft für Schwerionenforschung (GSI),
Planckstrasse 1, D-64220 Darmstadt, Germany

Abstract. This article describes the experiments that were performed during the first decade of the operation of UNILAC, GSI-Darmstadt, at the recoil separator SHIP and the on-line mass separator. The measurements resulted in the discovery of the first radioactive ground state proton emitters, ^{151}Lu and ^{147}Tm.

HISTORICAL BACKGROUND

It was supposed that proton radioactivity would be the fourth decay mode which transmutes nuclei besides α decay, β decay, and fission. It was considered that proton emission would limit the existence of nuclear matter on the neutron-deficient side of the chart of nuclides. Basic theoretical studies had been performed in the 1960s by Goldansky [1]. He showed that, whenever the proton drip-line was crossed, the lifetime of the nuclei decreased rapidly down to picoseconds within a few number of isotopes. The decrease was faster for the light elements due to the lower Coulomb barrier. Goldansky also pointed out the importance of a centrifugal barrier for a reduction of the proton emission probability. Due to the low mass of the proton, the centrifugal barrier is four times higher than in the case of α decay. Although the theoretical work was reliably based on the simplicity of the decay process and the possible comparison with the already well investigated α decay, the ambiguity of the nuclear mass models caused uncertainty in predicting which specific nuclei would be proton emitters. This resulted in differences of up to 1 MeV for the proton binding energy of neutron-deficient nuclei in the medium mass region, which led to an uncertainty of 3 - 4 isotopes for the location of the proton drip-line.

Another uncertainty was related to the production mechanism. Cross-section measurements for spallation reactions at ISOLDE, CERN, had shown that the yield for neutron-deficient cesium isotopes dropped rapidly with decreasing proton binding energies. Similar results were obtained from fusion reactions with light projectiles at high excitation energies. The details of excitation functions were fairly unknown in the case of heavy ion reactions that produced compound nuclei located in an extremely neutron-deficient region. It was speculated that proton emitters from the ground state could not possibly be produced because the proton to neutron ratio of the evaporation process would become too small for residues located beyond the drip-line.

Many interesting physical questions were waiting to be answered. However, when the UNILAC started to operate in 1976, we had an excellent heavy ion beam, but neither the experience nor the equipment to start answering fundamental questions immediately. 'We' is the group of postdocs and diploma students employed by P. Armbruster to perform experiments at SHIP, the Separator for Heavy Ion reaction Products. The people who were mainly working at SHIP were W. Faust, K. Güttner, G. Münzenberg, W. Reisdorf, K.H. Schmidt and myself. The first 'fundamental' question we had to answer, was: "How high is the background behind SHIP?" For experimentalists, this is certainly a fundamental question, because the background will determine which experiments can be performed, which detectors can be used, and which lowest cross-section level can be reached. We could not profit from the experience gained at other laboratories, because SHIP was the first instrument of this type constructed for the separation of 'heavy' ions.

The situation was different at the GSI 'on-line mass separator'. Similar instruments were in operation elsewhere, e.g., ISOLDE at CERN, and the GSI mass separator group had gained experimental experience searching for proton radioactivity there, already before the GSI experiments started. One of the early experiments proposed by R. Kirchner, O. Klepper, E. Roeckl, D. Schardt (the mass-separator staff) and co-workers was the search for proton radioactivity of light cesium isotopes. This was a continuation of experiments performed at ISOLDE using spallation reactions, and at GSI, where, however, heavy ion fusion-evaporation reactions were used.

It was already concluded from the ISOLDE experiments that ^{113}Cs should be with high probability a radioactive proton emitter with a short half-life, because it was not observed in the experiment, although the cross-section should have been high enough for its detection [2]. The same result was obtained at the GSI on-line mass separator [3]. Although no ground state proton emitters were observed, a number of new neutron-deficient isotopes could be identified by spectroscopy of α particles and β-delayed protons.

Hints on the radioactive proton emission of ^{121}Pr had been reported in the early 1970s from experiments in Dubna by Karnaukhov et al. [4] and Bogdanov et al. [5]. However, these early experiments in Dubna were not continued, and their results still remain to be confirmed.

At the time, proton emission from an isomeric state in ^{53}Co was definitely known, having been discovered in 1970 [6, 7] (see also contribution to this Symposium by J. Cerny). So far, proton radioactivity of ^{53}Co^m is the only known case of proton emission from a nucleus that is proton stable in its ground state. Also known was β-delayed proton emission discovered in the early 1960s by Karnaukhov, TerAkopian and co-workers [8] (see also contribution to this Symposium by J. Hardy). Beta-delayed proton emission is a clear indication that the proton drip-line is approached, although its exact location could not be deduced.

In the following section, I will report on the first experiments at SHIP, which resulted in the development of detectors and the identification of a number of new isotopes close to the drip-line. A continuation of these experiments led in 1981 to the detection of the first case of proton emission from the ground state of a nucleus. The activity was assigned to ^{151}Lu. Follow-up experiments at the GSI on-line mass separator and at SHIP resulted in the detection of three more proton transitions assigned to ^{147}Tm, ^{147m}Tm, and ^{150}Lu. I will also present experiments performed to search for proton emission in the range of elements from rubidium to bismuth and an experiment to confirm the results obtained in Munich by Faestermann, Gillitzer et al. [9, 10] on proton emission from ^{113}Cs and ^{109}I (see also contribution to this Symposium by T. Faestermann). Finally, some reflections are added about what we learned, why the experimental investigation of proton emitters was stopped at GSI and whether a revival may become possible in the future.

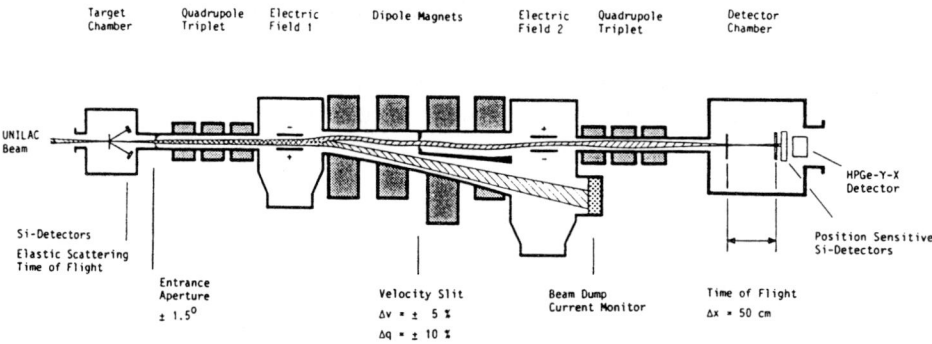

FIGURE 1. Separation of the evaporation residues from the projectiles by the velocity filter SHIP and their implantation into silicon detectors.

DETECTOR DEVELOPMENT AND FIRST RESULTS

The strategy adopted at SHIP for the exploration of proton radioactivity was to approach the drip-line stepwise. Already in one of the early irradiations, ^{40}Ar+^{144}Sm, we had learned that the background suppression is high enough to mount silicon detectors directly in the focal plane of SHIP. Fig. 1 shows the experimental setup. The method had the advantage that very clean α-decay spectra of high resolution were obtained. Because α particles could be detected with high sensitivity, we selected the region of nuclei on the right of the closed shell N=82 and above element gadolinium for further exploration. There, α emission is the dominant decay mode of nuclei, and half-lives are short ($\approx$1-100 ms) and therefore well suited to the short separation time of SHIP ($\approx$1 μs). After having gained enough information on the

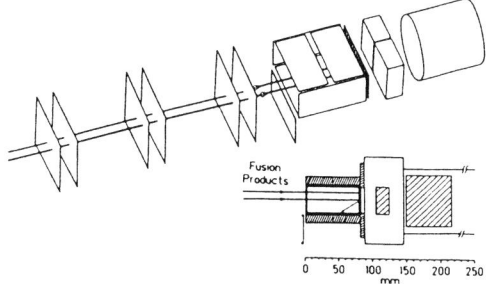

FIGURE 2. Assembly of detectors in the focal plane of SHIP composed of large area secondary electron time-of-flight detectors, position-sensitive silicon strip detectors and germanium detectors.

production probability of nuclei close to the drip-line, we wanted to switch over to the region on the left side of N=82, where the α-decay probability is strongly reduced due to low Q_α values, and β decay was expected to have a long enough half-life that proton emission may compete with it.

In order to safely identify the produced nuclei, we used position sensitive silicon detectors. These had the advantage that we could control the ion optical properties of SHIP, like the position and width of the reaction products in the focal plane. Using the relative measured position of the implanted nuclei and the subsequent

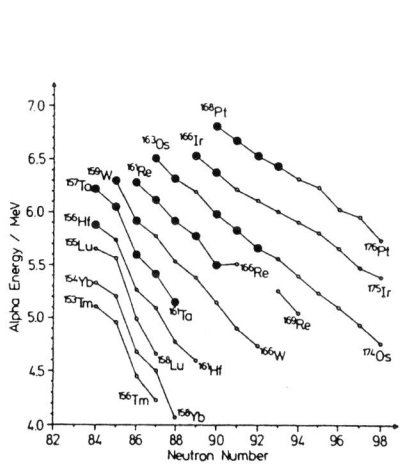

FIGURE 3. Alpha systematics of neutron deficient isotopes of elements from Tm to Pt. The new isotopes were marked as bigger dots. Their data fit well into the trend of the previously known isotopes (from [11]).

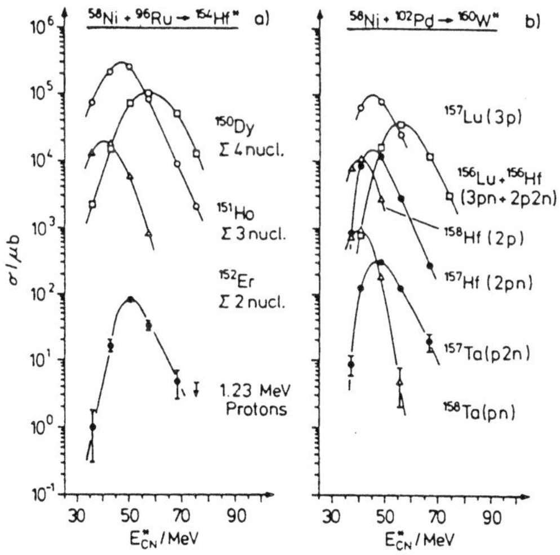

FIGURE 4. Comparison of measured excitation functions for the 2-, 3-, and 4-nucleon deexcitation channels in the reactions ^{58}Ni + ^{96}Ru (a) and ^{58}Ni + ^{102}Pd (b). The excitation function for the proton activity is shown in the lower part of the left hand figure (from [12]).

α decays, unknown species could safely be identified by generic correlation. The latest version of our detector arrangement is shown in Fig. 2. The silicon-detector array is complemented by three time-of-flight detectors, which are also used as anti-coincidence detectors, and germanium detectors for the measurement of coincident γ rays or X-rays.

A number of newly identified isotopes and the systematic trend of α energies near the N=82 shell are shown in Fig. 3. In Fig. 4b, excitation functions are displayed for the production of some of these nuclei in heavy ion fusion-evaporation reactions. It shows an obvious, rapid decrease of the production yield of the neutron evaporation channels at the expense of increasing proton evaporation.

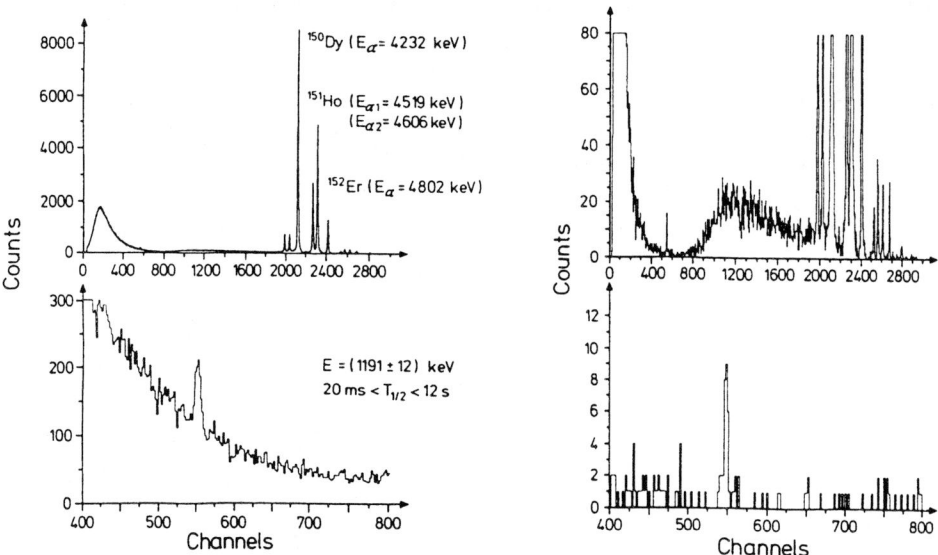

FIGURE 5. Energy spectra taken from the decay of evaporation residues that were implanted into detectors after separation by SHIP. The detector thickness was 300 μm (left side) and 140 μm (right side). Reaction: ^{58}Ni + ^{96}Ru → ^{154}Hf* at E* = 47 and 54 MeV.

PROTON RADIOACTIVITY OF ^{151}Lu

Using the cross-sections shown in Fig. 4b, we estimated that the reaction ^{58}Ni + ^{96}Ru → ^{154}Hf* is probably the best case in which to search for proton radioactivity. The compound nucleus has 82 neutrons, and the p2n evaporation channel was expected to have a cross-section only slightly smaller than 1 mb. The reason it was expected to be smaller is that the Γ_p/Γ_n ratio should be smaller for the more neutron-deficient ^{154}Hf* than for ^{160}W*.

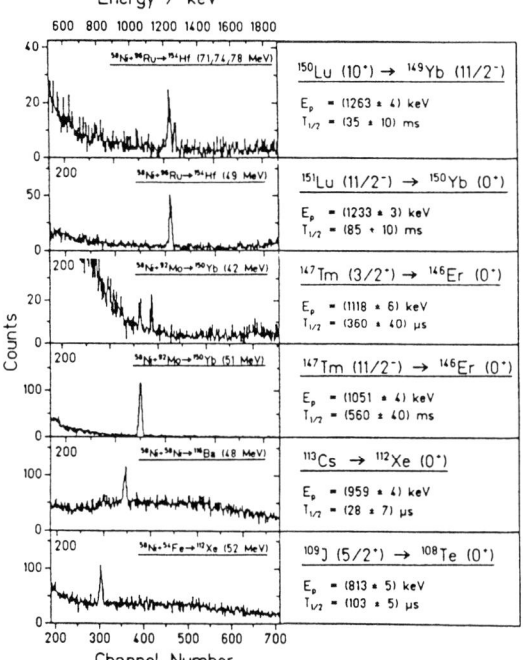

FIGURE 6. Energy spectra measured in experiments at SHIP from decays of nuclei that were implanted into silicon detectors. The background in spectra 1 to 4 from the top is due to electrons and tails from α lines, in spectra 5 and 6 due to the implantation signals of energy-degraded evaporation residues. These were not completely suppressed by the anti-coincidence system. The short half-life of ^{113}Cs and ^{109}I (the values are taken from Ref. [10]) made a reduction of the implantation energy necessary in order to avoid a pile-up of the signals from the proton decay on the tails of the signals from implantation.

The irradiation was performed in April 1981. We chose a beam energy of 276.7 MeV, which resulted in a mean excitation energy of 54.3 MeV, close to the expected maximum of the p2n channel. Already after one hour of irradiation, we observed a small peak in the low energy part of the spectrum arising from the background of β particles and electrons. The detector was 300 μm thick, which resulted in a relatively high background, but it was relatively cheap (at the time, our resources were very limited). The better suited, but more expensive 140 μm thin detector was irradiated instead of the thick detector at a yield of 10 %. We were afraid that the detector would not stand the full 2 days of irradiation because of radiation damage. The off-line analysis showed that this detector had the optimum thickness. The new line was located in a region of minimum background between the rising flanks from electrons on the low energy side and escaping α particles on the higher energy side. The two spectra from this very first experiment are shown in Fig. 5. The result was presented in June at the Helsingør conference on "Nuclei Far from Stability" [11].

In a subsequent experiment in July, 1981, we completed the excitation function (see Fig. 4a). The assignment of the new line to the p2n channel was based on that measurement. In the same experiment, the new line was identified as a proton line by ΔE-E measurement, the half-life, $T_{1/2} = (85 \pm 10)$ ms, was determined using a rotating slit cylinder (at the time the beam could not be chopped fast enough), and the energy $E_p = (1232.8 \pm 2.8)$ keV was determined using conversion electrons from ^{137}Cs and ^{207}Bi sources. The maximum cross-section was 70 μb at an excitation energy of 50 MeV.

Already in the first experiment we observed another weak line in the reaction ^{58}Ni + ^{92}Mo. However, the analysis was abandoned in favor of the stronger ^{151}Lu line.

SEARCH FOR FURTHER PROTON EMITTERS

In 1981, the irradiation of ^{96}Ru and ^{92}Mo targets with ^{58}Ni was repeated at the GSI on-line mass separator. The isotope ^{151}Lu could not be observed because of a short half-life and the low efficiency of the ion source. However, in the second reaction, a proton line was measured at mass A = 147. The line was assigned to the ground state proton decay of ^{147}Tm, and the half-life was $T_{1/2} = (560 \pm 40)$ ms [13].

Subsequent to the observation of proton emission from ^{151}Lu and ^{147}Tm, a search for further proton emitters started at SHIP and at the on-line mass separator. Beams of ^{40}Ca, ^{58}Ni, and ^{92}Mo were used. The produced compound nuclei were distributed across a range of nuclei from ^{80}Zr to neutron-deficient polonium isotopes. The beam energy was varied to search for the p2n and p3n evaporation channels. No further ground state proton emitters were observed at the mass separator, but new data on β-delayed proton emitters were measured [14]. In the SHIP experiments, two new transitions were detected and assigned to the decay of the ground state of ^{150}Lu and the decay of a low spin isomer in ^{147}Tm. The results obtained in Munich on the proton decay of ^{113}Cs and ^{109}I [9, 10] could be confirmed. A summary of the measured data is shown in Fig. 6.

INTERPRETATION

The observation of proton emission from ^{147}Tm and ^{151}Lu fixed the proton drip-line in that region of elements. The measured proton energy, however, did not allow to judge of the quality of mass models, but rather only of the proton pairing energy used in the model. Keeping this restriction in mind, we found that there was good agreement with the predictions for the proton energy given in Ref. [15, 16]. The physical reasons that determine the proton drip-line are visualized in Fig. 7.

The half-life for the proton decay of ^{151}Lu was calculated using the semi-classical WKB method. The proton-nucleus potential used for the calculation of the Gamov factor is shown in Fig. 8. The strong influence of the centrifugal potential, due to the small value of the reduced mass in the denominator, is obvious. In the case of ^{151}Lu, the experimental half-life could only be reproduced by assuming a change of the angular momentum by 5 units. This change is in agreement with a proton transition from a $h_{11/2}$ ground state, as suggested by the shell model, to the 0^+ ground state in ^{150}Yb. Similar good agreement is obtained for the other proton transitions of the Lu and Tm isotopes using the spin assignments as given in Fig. 6.

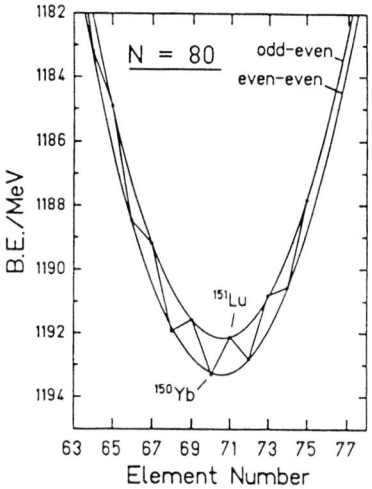

FIGURE 7. Calculated binding energies of neutron-deficient N = 80 isotones [16]. Isotonic binding-energy parabolas, not to be mixed with isobaric mass parabolas, connect the data points of odd-even and even-even isotones separated by the proton-pairing energy Δp. Position and slope of the parabolas are mainly determined by the Coulomb and asymmetry terms, both quadratic in Z, of the macroscopic liquid-drop part of the mass formula. The proton-decay energy of ^{151}Lu near the vertex is determined by the pairing energy.

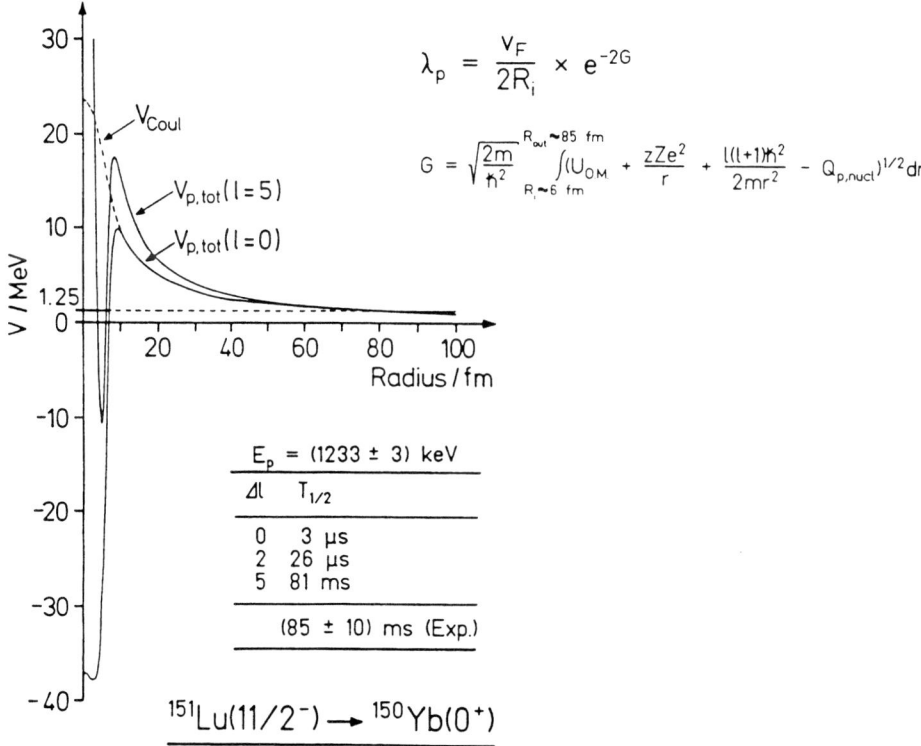

FIGURE 8. Proton-nucleus potential for the semi-classical calculation of the ^{151}Lu partial proton half-life.

REFLECTIONS

"The means is the end." This is usually the answer of mountaineers when they are asked what is so fascinating in their enterprise to climb up to the top of a mountain. In the exploration of the unknown, certainly, the way is part of the aim, and having reached the aim is like viewing the landscape from the top of a mountain. The efforts are made on the track, the result is the reward.

When I was asked, after our observation of proton radioactivity, how important this result would be, I answered that it would depend on how many of these species would be found in the future. During the course of this symposium, we could see that proton radioactivity is not an isolated phenomenon, but one that has been observed by now in about 30 cases along the drip-line for elements from iodine to bismuth. Proton emitters were found near closed proton and neutron shells as well as in the region of strongly deformed neutron-deficient nuclei of rare-earth elements. The data provide useful, in some cases unique, information on nuclear matter far from stability.

Protons can be detected with high efficiency. Therefore proton emitters are used for the identification of nuclei in recoil-decay tagging experiments. Results from various laboratories were presented at this symposium.

At GSI, the investigation of proton radioactivity was stopped in 1987. At 'heavy' ion accelerators, and not only at GSI, the study of neutron-deficient nuclei was always in strong competition with the search for superheavy elements.

Often, the decision was made to the advantage of the superheavies, and that was the case in 1987. The reason was not only a matter of preference. The promised 'island of superheavies' could be reached only using 'heavy' ion beams at intense beam currents, whereas neutron-deficient nuclei were already studied using beams of lighter nuclei and at a lower intensity. At GSI, the schedule simply did not allow for the investigation of both, especially in the case of the low cross-sections and long beam times that were needed for the investigation of p3n and p4n evaporation channels.

Meanwhile, alternative methods came into the field, fragmentation reactions and ion trapping. The unique identification methods that could be applied for high energetic fragments resulted in an exploration of the proton drip-line in the region of light elements. Presently, experiments to extend the study into the region of heavier elements at the GSI-fragment separator are under preparation. SHIPtrap, which is under construction, will allow for the separation of isobars. An isobaric separation of nuclei close to the drip-line should already be possible at a moderate mass resolution, because the binding energy differences are high. If the lifetimes are long enough, the investigation of nuclei with only a small proton branching ratio will become feasible. An outstanding example could be the search for the proton radioactivity of ^{104}Sb and ^{105}Sb, which is still uncertain.

From the theoretical point of view, the simplicity of the decay process provides valuable insight into nuclear structure effects like angular momentum and configuration mixing. One can hope that even more information will be obtained in the future with the application of additional experimental methods, e.g, angular correlation and orientation experiments at low temperature.

REFERENCES

[1] Goldansky, V.I., *Ann. Rev. Nucl. Sci.* **16**, 1 (1966).
[2] D'Auria, J.M., et al., *Nucl. Phys.* **A301**, 397 (1978).
[3] Schardt, D., et al., *Nucl. Phys.* **A326**, 65 (1979).
[4] Karnaukhov, V.A., et al., Proc. Int. Conf. on the Properties of Nuclei far from the Region of Beta-Stability, Leysin, Switzerland, 1970, p. 457.
[5] Bogdanov, D.D., et al., *Yad. Fiz.* **16**, 890 (1972); *Sov. J. Nucl. Phys.* **16**, 491 (1973).
[6] Jackson, et al., *Phys. Lett.* **33B**, 281 (1970).
[7] Cerny, J., et al., *Phys. Lett.* **33B**, 284 (1970).
[8] Karnaukhov, V.A. and Ter-Akopyan, G.M., *Phys. Lett.* **12**, 339 (1964).
[9] Faestermann, T., et al., *Phys. Lett.* **137B**, 23 (1984).
[10] Gillitzer, A., et al., *Z. Phys.* **A326**, 107 (1987).
[11] Hofmann, S., et al., Proc. 4th Int. Conf. on Nuclei Far From Stability, Helsingør, Danmark, June 1981, CERN report 81-09, Geneva, 1981, p. 190.
[12] Hofmann, S., et al., *Z. Phys.* **A305**, 111 (1982).
[13] Klepper, O., et al., *Z. Phys.* **A305**, 125 (1982).
[14] Larsson, P.O., et al., *Z. Phys.* **A314**, 9 (1983).
[15] Liran, S. and Zeldes, N., *Atomic Data and Nucl. Data Tables* **17**, 431 (1976).
[16] Möller, P. and Nix, J.R., *Atomic Data and Nucl. Data Tables* **26**, 165 (1981).

Munich Efforts To Search For Proton Radioactivity In Beam

Thomas Faestermann

*Fakultät für Physik E12, Technische Universität München,
James-Franck-Strasse, D85748 Garching, Germany*

Abstract. This report will cover four topics:
1) The development of catcher techniques to detect shortlived particle radioactivities. With this setup proton emission from ^{113}Cs and ^{109}I has been observed for the first time (1983). Their slow decay rate has been realized and for the first time expressed in terms of a spectroscopic factor.
2) A discussion of various types of recoil separation schemes and experiments using an electrostatic deflector, which yielded improved decay data for ^{113}Cs and ^{109}I.
3) The production of proton rich nuclei by fragmentation of relativistic ^{124}Xe and ^{112}Sn ions at GSI and the determination of their decay properties. The beta-decay halflives of ^{105}Sb and ^{77}Y as well as of many other nuclei along the path of rp-process nucleosynthesis have been measured. In $T_Z = -1/2$ nuclei proton decay has to compete with superallowed Fermi-decay, which is also observed for the $T_Z = 0$ odd-odd nuclei between ^{78}Y and ^{94}Ag.
4) A possible solution to the problem how to detect the proton decay of ^{39}Sc ($S_p = -0.602$ MeV, $T_{1/2} \sim 0.3$ ps) and to measure ist half-life.

PROLOGUE

Since this is going to be a historical account, I shall start with my own first contacts with proton radioactivity. The subject of my Ph.D. thesis was g-factor measurements of nuclear isomers whose halflives were in the region below 1 µs. As a postdoc I spent the years 1975-1977 in Chalk River working mostly in the group of John Hardy on beta-delayed proton emitters. There the half-lives were measured in seconds. But in my second year, while John Hardy had left for a sabbatical at CERN, Peter Jackson - first author of the first paper (1) reporting the proton radioactivity of the high-spin isomer in ^{53}Co - who was an outpost of the University of Toronto in Chalk River, proposed to search for self-delayed proton decay of ^{69}Br. As I recall it, our setup was simple: a pulsed ^{32}S beam bombarded a ^{40}Ca target and the evaporation residues were caught on a thin foil which was inclined some 45° with respect to the beam. A Si detector besides the target but well shielded against radiation from it could detect protons emitted by nuclei on the catcher in backward direction. We did not find any off-beam protons and concluded (2), that *"the partial lifetime for proton decay of ^{69}Br is probably not in the range from 0.1µs to 1s."* This statement is still valid today and there is strong evidence that the lifetime is below 0.1µs (3).

CATCHER TECHNIQUES (1980 - 1985)

After being back in Munich I worked mainly on γ-spectroscopy and ns-isomers. At the end of 1977 Paul Kienle, who had been my teacher, came back from one of his frequent visits to GSI and told me that the ISOL group there had searched for proton radioactivity of ^{113}Cs in the ^{58}Ni(^{58}Ni,p2n) reaction, but without success. After a short discussion we were convinced, that ISOL facilities had little chance to detect proton radioactivity, because their separation times were of the same order as the β-decay halflives with which the proton-radioactivity would have to compete. Much shorter halflives would be inaccessible. Thus we decided, that we could have a better chance with a method adopted to ns-isomers.

Our MP-Tandem accelerator in Munich was at that time operating up to 12 MV, not sufficient to bring ^{58}Ni ions up to about 230 MeV, necessary for the ^{58}Ni(^{58}Ni,p2n)^{113}Cs reaction. But H. Morinaga and E. Nolte had built a linear postaccelerator with a very efficient RF structure (4), which is now used in many labs including GSI and CERN. Eckehart Nolte, at that time still active in nuclear spectroscopy, shared our interest. With his LINAC we could add some 5 MV of accelerating voltage and with a second stripping after the Tandem from the 11^+ (or 12^+) to the 22^+ charge state one could reach ^{58}Ni energies up to 250 MeV, although with a drastic loss in intensity due to stripping and pulsing the beam.

We found an undergraduate student, Albrecht Gillitzer, to work on the detection system. The first schemes we tried were along the same line as I described before: After the target a catcher and - in backward direction - a detector, well shielded against radiation from the target. But we used also gas detectors like parallel plate avalanche counters (PPAC) which are insensitive for β-particles (but sometimes even for protons). These tries didnt bring a real breakthrough.

When Albrecht Gillitzer started his PhD-thesis we adapted an annular detector system, which had been developed by another PhD-student, Kurt Hartel, to search for high energy α-particles from superheavies. The final setup was the following: A pulsed (about 2 ns pulse width) ^{58}Ni beam impinged on a ^{58}Ni target, which was mounted on the backside of an annular detector system. The target was backed with more material, such that evaporation residues were slowed down to small energies and stopped not very deep in the catcher. This cather foil made of polyimide was provided with a small hole to let the primary beam pass on to the Faraday cup. The decay radiation was detected in backward direction with two gas detectors both operated with the fast CF_4 gas. A two-stage PPAC gave a fast timing signal. The following ionization chamber was of the Bragg-curve-spectroscopy (BCS) type with longitudinal electric field (5). It yielded a total energy signal, a highly Z-dependent and from the timing between PPAC and anode a range information.

Our first observation of self-delayed protons from the reaction ^{58}Ni + ^{58}Ni was presented (6) with a poster at the Florence Conference on Nuclear Physics 1984. This first spectrum (Fig. 1) of proton events between beam pulses had still poor statistics. The time distribution of the events suggested a decay with a half-life of about 1 μs, but the data were with 5% probability still compatible with a much longer half-life.

Later experiments then yielded a clean proton line for ^{113}Cs as well as for ^{109}I produced with the ^{54}Fe(^{58}Ni,p2n) reaction (7). The determination of the halflives was rather time consuming, since the period of the beam pulsing had to be adjusted until it was a few times longer than the half-life, in order to fix the constant background due to the long lived β-delayed proton emitters. And since we were biased by our first observation of a 1μs half-life, which also fitted to the expected value for the observed ^{113}Cs proton energy of 0.98

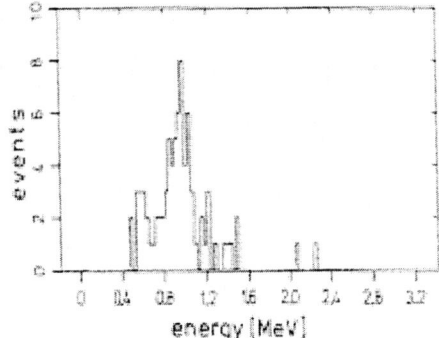

FIGURE 1. First evidence for a proton line at 0.98 MeV in ^{113}Cs.

MeV, we had to double the pulsing period many times to finally arrive at our best value for the half-life as 33±7 μs. For the 0.83 MeV line in ^{109}I we measured 109±17 μs (8). Both proton emitters were reproduced shortly after by S. Hofmann et al. at GSI (9).

Although there was only little theoretical help at the time, these halflives were obviously too long, if compared with simple WKB calculations for the emission of a $d_{5/2}$ proton from a spherical nucleus. Therefore we introduced for the first time (8) a spectroscopic factor in analogy to one-proton pickup reactions. Using the shell model value or the BCS approximation we could understand the small hindrance in the decay of ^{151}Lu and ^{147}Tm, but for ^{113}Cs and ^{109}I the additional hindrance of about a factor of 10 could only in recent years be quantitatively understood.

Using the same setup we searched also for proton radioactivity of ^{105}Sb. Not observing any protons we could exclude proton emission except for a window of decay energies between 0.31 MeV and 0.56 MeV. For the potential proton emitter ^{77}Y we could exclude decay energies between 0.45 MeV and 0.90 MeV (10).

RECOIL SEPARATORS (1986 - 1992)

The cross sections for the (p2n) evaporation channels leading to ^{113}Cs and ^{109}I were about 30 μb. In order to look for the (p3n) channels with about 1 μb cross section the selectivity of our method seemed not good enough. A real separation of evaporation residues from the projectiles seemed necessary.

The Mass Separator LARA

Wolfgang Wilhelm, who had gained experience in ion optics during his Ph. D. work, designed a Large Acceptance Recoil mass Analyzer (LARA) mainly influenced by the Legnaro system (11). It consisted actually of two mass separators which were set up back to back (12). The acceptance of LARA was really large (10msr), the mass resolution after the first half of the separator was already m/Δm = 340 and improved in the second half to m/Δm=450 (FWHM) and the vertical width of the residues was about 20mm in the middle but only 3mm at the end of the separator. This design was considered superior to the others at the time by J.D. Larson (13). But these numbers were all based on beam transport calculations and unfortunately we could not get the funding to build the instrument.

A Gas-Filled Spectrograph

A less expensive option for a separator appeared to be a gas-filled magnet. This principle had already proven to be very powerful for the separation of fission fragments (14). The magnetic rigidity is $B\rho = mv/<q>$ with an average charge state $<q>$ in the gas due to frequent charge changing collisions. The average charge state can be parametrized (15) as $<q> \propto v ?Z^\alpha$ with $0.4<\alpha<0.6$. With this approximation we get $B\rho \propto A/Z^\alpha$ independent of the velocity v. If one plugs in numbers, it turns out, that isotones will have about the same rigidity. This would be advantageous to reduce background from the more abundantly produced evaporation residues. In fusion-evaporation reactions the most prolific products are usually the isobars to the exotic nuclide one is interested in and a mass separator cannot reduce those.

Influenced by work at Argonne, where the gas filled split pole spectrograph was used as a separator to distinguish isobaric ions in Accelerator Mass Spectrometry (16), we tested the performance of our Q3D spectrograph as a gas filled magnet. With optimized gas pressure 120 MeV ^{127}I ions as well as 140 MeV ^{58}Ni ions arrived at the focal plane in a distribution with a FWHM corresponding to $\Delta B\rho/B\rho = 1.2\%$. Even with a 10 MeV wide energy distribution the width increased only to 1.3% (17). These results were very encouraging. But since the postaccelerated beam of the tandem could not be transported to the Q3D spectrograph, we used the last 90° deflection magnet after the LINAC with gas. Its resolution was much worse and unfortunately the magnetic rigidity of the residues from the ^{58}Ni+^{58}Ni reaction was too close to that of the beam. Thus we might have had an isotone separation but, at least in this symmetric reaction, we did not have a recoil separation. (Nevertheless we use the gas filled isobar separation with great success in accelerator mass spectrometry for radioisotopes with A≅60.)

The Electrostatic Separator ESEL

In order to get at least a recoil separation we then built an electrostatic deflection system. Most of the work was done by Frank Heine for his diploma and Ph. D. thesis and is described in more detail in Refs. [18,19]. Two pairs of deflector plates, each 25cm long and 40mm apart, deflected the evaporation residues by 8°. For products from the ^{58}Ni + ^{58}Ni reaction we applied +70kV/-50kV to the plates. The beam was only deflected by 3.4° and left the setup through a slit in the second anode to be caught in a Faraday-Cup. The residues traversed a PPAC to produce a timing signal for discrimination against scattered projectiles and were stopped in an array of 10x10 Si PIN diodes each with an active area of 10x10mm². The following p- or α-decays were then observed in the same detector.

With this setup we obtained improved half-life and Q-value data on α-emitters above ^{100}Sn and also on the proton emitters ^{109}I and ^{113}Cs, the latter nowadays being used as a calibration source in p-radioactivity experiments. We could for the first time show, that the 0.81 MeV proton line from the ^{58}Ni + ^{54}Fe reaction is followed by the 3.31 MeV alphas from ^{108}Te and thus prove that ^{109}I is the proton emitter. A search for p-emission from ^{112}Cs, in which we required a ^{111}Xe alpha to follow, only resulted in an energy dependent upper limit of the cross section in the ^{58}Ni(^{58}Ni,p3n) reaction. At the decay energy of 0.81 MeV, where Page et al. (20) later found the line with a production cross section of 0.5μb using the Daresbury recoil mass separator, we had reached only a limit of 1μb. So in this case, nature (and fortune) was not on our side.

EXPERIMENTS USING THE FRAGMENT SEPARATOR FRS (SINCE 1992)

An alternative method to produce exotic nuclei has been developed at GANIL and at GSI: the projectile fragmentation. The high energy of the produced radioactivities allows one to identify not only their mass number A, but also their nuclear charge Z. At GSI this is done with the fragment separator FRS. In collaboration with GSI we have done two experiments to identify proton rich nuclei in the neighbourhood of ^{100}Sn and to measure their decay properties, one in 1994 using fragmentation of a ^{124}Xe beam and one at the end of 1998 with a ^{112}Sn beam. The main credit in the preparation and the evaluation of these experiments goes to Robert Schneider, Andreas Stolz and Elmar Wefers.

The FRS consists of a series of 4 mass separators with a focal plane after each. In the middle focal plane F2, where already a small band in mass/charge (m/q) values is selected, a passive absorber (wedge shaped) introduces a Z-dependent energy loss and the modified magnetic rigidity is analyzed in the second half of the FRS. We measure the position of every ion at F2 and at the final focal plane F4. Together with the measured angle at F4 we get the exact rigidity Bρ=mv/q. The velocity v we measure with scintillators at F2 and F4 and correct for the actual path length. For most of the ions at energies of about 400 A MeV the ionic charge q equals the nuclear charge Z. Interference of H-like ions would mean that Z=q+1 and, if the separator is set for p-rich nuclei, that the "contaminant" is even more p-rich. But we do not rely on this assumption that the nuclei are completely stripped, we measure the nuclear charge Z twice, at F2 with a fast ionization chamber developed at Munich and at F4 with the standard MUSIC ionization chamber of the GSI. Both measurements have similar resolution of $\Delta Z \cong 0.25$ (FWHM). Thus the fragments can be clearly identified.

The fragments were stopped in a stack of Si detectors. For a given nuclide we could adjust the degrader thickness in front of the implantation detectors such that some 80% stopped in the middle 2 of 4 position sensitive (x and y) 0.25mm thick Si detectors. Thus the stopping position of the ions was measured in 3 dimensions and we could correlate deay events in space and time. To determine the β^+-energy we had 10mm of Si detectors in front and 10mm behind the implantation detectors and thus 100% detection efficiency for betas up to about 6 MeV. To detect γ-rays these Si detectors were surrounded in the first experiment by a BGO detector covering 90% of the solid angle and for the second experiment by a huge Ge clover detector in forward plus a NaI shell in sideward and backward directions covering 3/4 of the solid angle.

In the ^{124}Xe fragmentation experiment we succeeded not only to identify for the first time the doubly magic nucleus ^{100}Sn, but also to measure from 7 implanted nuclei its half-life and rough numbers for the β-decay and total γ-decay energy (21). We spent a few hours to search for ^{105}Sb and ^{104}Sb and we could measure their halflives as $1.12 \pm 0.16\,s$ and $0.44^{+0.14}_{-0.11}\,s$ respectively (22,23). For ^{105}Sb we found 1 out of 99 decay events which was consistent with a proton (with 85% probability, based on the requirements that only 1 Si detector had responded and no coincident γ or 511 keV quantum was detected by the BGO). For this event we measured an energy of $550 \pm 30\,keV$ which only marginally is consistent with the Berkeley value (24) of $478 \pm 15\,keV$. But if it is a true proton the branching ratio is 1% with the uncertainty corresponding to this single event. For ^{104}Sb all 16 events observed were betas.

In the recent ^{112}Sn fragmentation experiment we tried to improve the experimental data to describe the rp-process. This process of nucleosynthesis is hoped to explain the rather high abundances of the proton rich isotopes of elements between Ge and Ru. It is assumed that in a very hot and dense (explosive) environment a series of successive proton capture reactions could lead up to the proton drip line, where the next proton would be barely bound or unbound. As in the r-process the capture reactions cannot continue until these socalled waiting point nuclei have decayed by β^+ emission. Therefore the halflives of these nuclei will determine the total time needed to synthesize proton rich nuclei up to a mass number of about 100. In Fig. 2 we show a recent calculation of the rp-process path by Schatz et al. (25) who have even considered the possibility to cross the proton drip line by capturing an additional proton on a nucleus which is proton unstable and therefore lives only for a very short time. This path essentially goes along the N=Z line up to ^{100}Sn. But Schatz et al. had to use theoretical estimates for the proton binding energies and also for most of the halflives of these nuclei. To improve the experimental situation we have measured halflives for nearly all relevant nuclei along the rp-process path.

Important for the rp-process path is also the location of the proton dripline. Therefore we investigated the possibly proton-unstable nuclei ^{77}Y and ^{81}Nb. ^{77}Y ions could be identified and we observed their decay with a short half-life, consistent with a superallowed Fermi β-decay. From the non-observation of ^{81}Nb we deduce a half-life shorter than 200 ns, considering the flight time through the fragment separator. Thus ^{81}Nb most probably decays by proton emission.

Our preliminary results for the halflives are displayed in Fig. 3 as a function of the mass number A. The halflives of waiting point nuclei (even proton number Z) are on the average a factor of 2 to 3 longer than the values (+ sign) used in the calculation by Schatz et al (25). Therefore the total time to synthesize the heaviest rp-nuclei may be

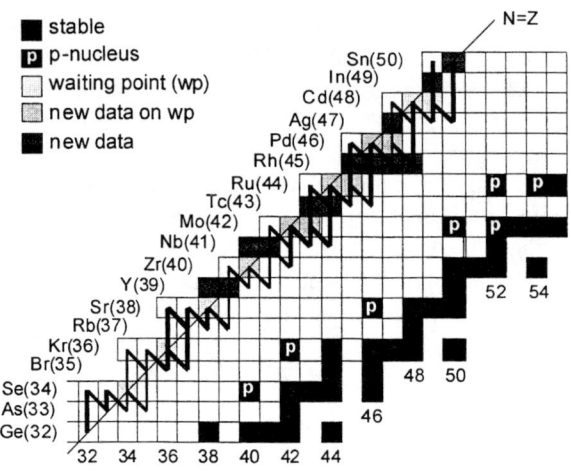

FIGURE 2. Path of the rp-process (according to Ref. 25)

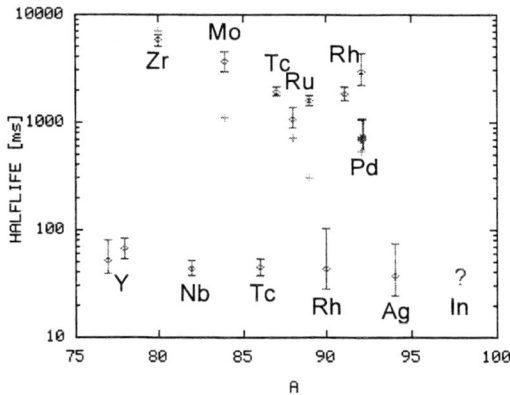

FIGURE 3. Our preliminary results for the halflives. The values from Ref. [25] are shown as +.

longer. We also investigated the six cases of odd-odd N=Z nuclei between ^{78}Y and ^{98}In, which are the heaviest nuclei where one can hope to study pure Fermi β-transitions. In a very preliminary analysis of our data we are able to show that even ^{90}Rh and ^{94}Ag

decay by superallowed Fermi β-decay. For the three lighter members of this series, ^{78}Y to ^{86}Tc, a low lying T=1, 0$^+$ state decaying with a superallowed Fermi transition was recently detected at GANIL (26). Our new data on these nuclei are in good agreement. The half-life of ^{77}Y is similarly short because a superallowed β-decay to its mirror nucleus ^{77}Sr is possible.

^{39}Sc - A PROTON EMITTER WITH $T_{1/2} = 0.3$ ps (since 1993)

A real challenge is the nucleus ^{39}Sc. Its ground state mass has been measured with transfer reactions, namely ^{40}Ca(^{14}N,^{15}C)^{39}Sc (27) and ^{40}Ca(^{7}Li,^{8}He)^{39}Sc (28). Thus the proton separation energy is determined as -602 ± 24 keV. With our WKB-calculation (8) this results in a half-life against proton emission of $0.25^{+0.17}_{-0.10}$ ps, if we assume the emission from the f$_{7/2}$ shell. So this could be the best case of a single-particle proton emitter (maybe in competition with ^{103}Sb) with just 2 neutron holes.

The problems are: how to produce and detect ^{39}Sc and how to measure its half-life. Separation is impossible, since the nucleus travels only about 10μm in a half-life. So one has to measure the protons directly and the only way to reduce the abundant proton background from the reaction is to deflect the protons with a magnetic field to get rid of those with smaller and larger energies and of heavier ions (with $1/A \leq 1/2$). For the production of ^{39}Sc fusion evaporation reactions seem to be ruled out, because they do not populate the ground state directly, and all excited states will decay by proton emisson. The first excited state, probably a 3/2$^-$ state, is at 950 keV (27) and lives only 10^{-19}s and even the next yrast state, an 11/2$^-$ state according to the mirror nucleus ^{39}Ar, has a proton decay half-life of 10^{-15}s. Therefore only a transfer reaction seems suitable. But how to get a sharp proton energy, if the emitter has an angular distribution and a corresponding energy distribution?

The trick is to use the (^{14}N,^{15}C) reaction. The CM scattering angle is restricted to angles smaller than the grazing angle of 16°. With an ^{14}N energy of 103 MeV, the ^{39}Sc has in the CM system nearly the same velocity as the protons from its decay. If we look for the protons emitted in backwards direction, they are nearly at rest in the CM system. But protons emitted from the compound system have much higher energies. Therefore we would expect "monoenergetic" protons with about the CM velocity.

We have tried to use our Q3D magnetic spectrograph to identify these protons. A student, Christian Albrecht, built a focal plane detector with 128 Si PIN diodes for an energy and a TOF measurement to avoid background from scattered projectiles (29). It turned out, that hydrogen in the Ca target was the major background due to double scattering events. With the skill of our target makers (30) we could reduce the H-content in the target from 100% to 1%, but the background was still 2 orders of magnitude above the expected effect. This might be discriminated with a coincident detection of the recoil nucleus after the decay. But Christian Albrecht then went into medical physics and the project was not continued.

The "ultimate" experiment to measure the half-life - sketched in Fig. 4 - would use inverse kinematics to get higher recoil velocity, a retarding foil on a plunger with variable distance to the target, a magnetic spectrometer to distinguish, if the proton was emitted before or after the plunger, and an annular detector (PPAC) for the coincident detection of the daughter nucleus ^{38}Ca. Anybody planning such an experiment is invited to ask me for details (31).

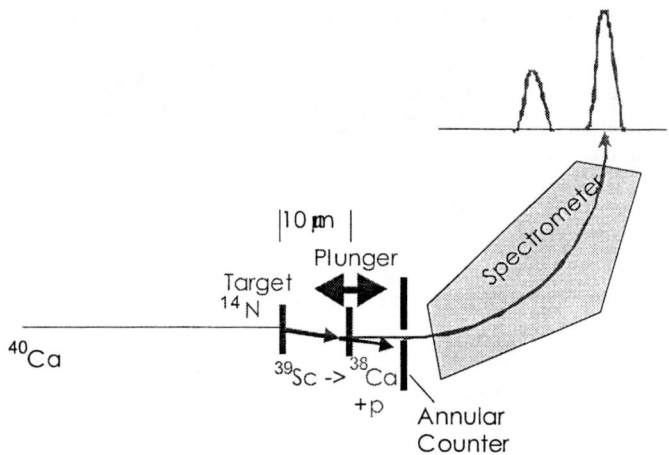

FIGURE 4. Possible setup to measure the half-life of ^{39}Sc.

ACKNOWLEDGMENTS

I would like to dedicate this report to my teacher Paul Kienle who retired from his duties as a professor of physics a week before this talk (Oct. 1., 1999). He has contributed significantly to many of the research activities described in this report and I hope he will stay active and influential.

This work was supported by the *Bundesministerium für Forschung und Technologie*, the *Bundesministerium für Bildung und Wissenschaft* and the *Deutsche Forschungsgemeinschaft* (SFB 375).

REFERENCES

1. Jackson, K.P., Cardinal, C.U., Evans, H.C., Jelley, N.A., Cerny, J., *Phys. Lett.* **33B**, 281-283 (1970)
2. Jackson, K.P., Clifford, E.T.H., Azuma, R.E., Faestermann, T., and Schmeing, H., *Progress Report Physics Division, Chalk River Nuclear Laboratories* **PR-P-112**, 24-25 (1976)
3. Blank, B. et al., *Phys. Rev. Lett.* **74**, 4611-4614 (1995)
4. Nolte, E. et al., *Nucl. Instr. Meth.* **158**, 311-324 (1979)
5. Schießl, Ch., Wagner, W., Hartel, K., Körner, H.J., Mayer, Wa., Rehm, E., *Nucl. Instr. Meth.* **192**, 291-294 (1982)
6. Faestermann, T., Gillitzer, A., Hartel, K., Kienle, P., Nolte, E., *Proc. Int. Conf. on Nucl. Phys.*, Florence, 1983, B263
7. Faestermann, T., Gillitzer, A., Hartel, K., Kienle, P., Nolte, E., *Phys. Lett.* **137B**, 23-26 (1984)
8. Gillitzer, A., Faestermann, T., Hartel, K., Kienle, P., Nolte, E., *Z. Phys.* **A326**, 107-119 (1987)
9. Hofmann, S. et al, *Proc. AMCO-7*, 184-195 (1984)
10. Faestermann, T., Gillitzer, A., Hartel, K., Kienle, P., Nolte, E., *Proc. 5th Int. Conf. on Nuclei far from Stability*, Rosseau Lake, Canada, 1987, pp. 739-748
11. Spolaore P., *Lecture Notes in Physics* **317**, 305-312 (1988)
12. Wilhelm, W., Faestermann, T., Körner, H.-J., *Lecture Notes in Physics* **317**, 320-327 (1988)
13. Larson, J.D., *Lecture Notes in Physics* **317**, 282-288 (1988)
14. e.g. Armbruster, P., Eidens, J., Grüter, J.W., Lawin, H., Roeckl, E., Sistemich, K., *Nucl. Instr. Meth.* **91**, 499-507 (1971)
15. Betz, H.-D., *Rev. Mod. Phys.* **44**, 465-537 (1972)
16. Paul, M. et al., *Nucl. Instr. Meth.* **A277**, 418-430, (1989)
17. Faestermann, T., Gillitzer, A., Kutschera, W., Nolte, E., Wilhelm, W., *Progress Report of the Accelerator Lab.* Munich, 129-130 (1986)
18. Heine, F., Faestermann, T., Gillitzer, A., Homolka, J., Köpf, M., Wagner, W., *Z. Phys.* **A340**, 225-226 (1991)
19. Heine, F., Faestermann, T., Gillitzer, A., Körner, H.-J., *Proc. 6th Int. Conf. on Nuclei far from Stability & AMCO-9*, Bernkastel-Kues, (1992) pp. 331-336
20. Page, R.D., Woods, P.J., Cunningham, R.A., Davinson, T., Davis, N.J., James, A.N., Livingston, K., Sellin, P.J., Shotter,A.C., *Phys. Rev. Lett.* **72**, 1798-1801 (1994)
21. Schneider, R., Friese, J., Reinhold, J., Zeitelhack, K., Faestermann, T., Gernhäuser, R., Gilg, H., Heine, F., Homolka, J., Kienle, P., Körner, H.-J., Geissel, H., Münzenberg, G., Sümmerer, K., *Z. Phys.* **A348**, 241-242 (1994)
22. Heine, F., Schneider, R., Faestermann, T., Friese, J., Homolka, J., Kienle, P.,Körner, H.-J., Reinhold, J., Zeitelhack, K., Geissel, H., Münzenberg, G., Sümmerer, K., *Proc. Int. Conf. on Exotic Nuclei and Atomic Masses*, Arles, France, 1995, pp. 565-570
23. Schneider, R. et al., to be published
24. Tighe, R.J., Moltz, D.M., Batchelder, J.C. Ognibene, Rowe, M.W., Cerny, J., *Phys. Rev.* **C49**, R2871-R2874 (1994)
25. Schatz, H. et al., *Phys. Rep.* **294**, 167-263 (1998)
26. Longour, C., et al., *Phys. Rev. Lett.* **81**, 3337-3340 (1998)
27. Woods, C.L., Catford, W.N., Fifield, L.K., Orr, N.A., *Nucl.Phys.* **A484**, 145-154 (1988)
28. Mohar, M.F., Adamides, E., Benenson, W., Bloch, C., Brown, B.A., Clayton, J., Kashy, E., Lowe, M., Nolan, J.A., Ormand, W.E., van der Plicht, J., Sherrill, B., Stevenson, J., Winfield, J.S., *Phys. Rev.* **C38**, 737-740 (1988)
29. Albrecht, C., Faestermann, T., Gillitzer, A., Heine, F., Schneider, R., *Progress Report of the Accelerator Lab.* Munich, 188-190 (1993)
30. Dollinger, G., Faestermann, T., Frey, C.M., Maier-Komor, P., *Nucl. Instr. Meth.* **A362**,60-62 (1995)
31. e-mail: thomas.faestermann@physik.tu-muenchen.de

Proton Decay Rates and Nuclear Structure

Philip J. Woods

Department of Physics and Astronomy, Edinburgh University, EH9 3JZ UK

Abstract. This paper outlines the major advances in the study of proton radioactive nuclei in the era of the 1990s achieved using the Daresbury Recoil Mass Separator and the Argonne Fragment Mass Analyzer.

INTRODUCTION

This paper aims to broadly outline the developments in studies of proton radioactive nuclei that have transformed the field in the the 1990s. As will be discussed in earlier papers in these proceedings, the initial experimental breakthrough was the serendipitous discovery of proton radioactivity from a multi-particle isomer in ^{53}Co at the Harwell Variable Energy Cyclotron in the UK [1]. This discovery was expected to presage many further examples, but these were not subsequently found, and ^{53}Co remains a unique isotope to this day in having a proton bound ground-state but exhibiting proton radioactivity. It was not until the advent of studies on the in-flight velocity filter SHIP at GSI that ground-state proton radioactivity was discovered in the early 1980s for the isotope ^{151}Lu [2], which was rapidly followed by ^{147}Tm discovered at GSI using the on-line mass separator [3]. The sensitivity of the proton partial half-life to the orbital angular momentum was used to assign the transitions to an $h_{11/2}$ proton shell by a WKB analysis. Measurements of proton decays from ^{109}I and ^{113}Cs were made using the fast catcher foil technique developed by the Munich group [4] and their half-lives subsequently measured [5]. These decay rates could not be reproduced using a simple WKB model. Bugrov and Kadmensky [6] demonstrated that these half-lives could be reproduced using a so called multi-particle model approach which required modest quadrupole deformations of $\beta \sim 0.10$-0.15 consistent with their location in a transitional region. This theme of relating decay rates to nuclear structure was to resonate throughout the 1990s.

DARESBURY EXPERIMENTS

At the end of the 1980s the author initiated a programme of research into proton radioactive nuclei taking advantage of the newly commissioned Daresbury Recoil Mass Separator (RMS) [7]. The RMS offered the dual benefits of fast in-flight separation with mass analysis. This meant that the recoils could be restricted to the mass of interest thereby increasing the luminosity of experimental searches. Furthermore the ions could be mass focused onto a Residue Implantation Detection System (RIDS) and the position of the radioactivity used to assign the parent mass. RIDS consisted of a transmission channel plate detector and a 2-dimensionally position sensitive silicon PSD into which ions were implanted [8]. This system was used to identify the decay of ^{108}I. Surprisingly this nucleus alpha-decayed indicating that the proton in ^{108}I was less unbound than that in ^{109}I [9]. The great majority of mass models failed to predict this behaviour. This experiment also demonstrated that the alpha-decay daughter product ^{104}Sb decayed predominantly by β emission.

The 2D-PSD suffered from certain defects. It had a large area and high leakage current with poor energy resolution, and high sensitivity to the prolifically produced β^+ background. Furthermore, the resistive division of the signals produced a variable recovery response under overload conditions. When the author moved to Edinburgh a new implantation detector - a Double-sided Silicon Strip Detector (DSSD) - was developed [10] utilising the low cost high performance electronics already being developed within the group in collaboration with the Rutherford Appleton Laboratory [11]. This represented the first use of a DSSD in a nuclear spectroscopy application - such devices now being commonplace. The new DSSD system was initially used to confirm the identication of a short-lived proton decaying isomer in ^{147}Tm tentatively assigned to this nucleus in earlier experiments at SHIP [12,13]. Similarly the proton decay of ^{150}Lu was confirmed and the half-life measured for the first time showing that the proton was emitted from an $h_{11/2}$ orbital. This represented a significant breakthrough since it opened up searches on 1p3n fusion evaporation channels with $\sigma \sim \mu$bs compared with 10's μb for 1p2n channels. Subsequently a sequence of odd-odd proton emitters from Z = 69-75 were identified establishing proton emission in Ta and Re isotopes in the process [14-17]. The decay rates and shell structures could be understood well within a spherical shell model framework. The 1p3n evaporation channel was also used to study the decay of ^{112}Cs [18]. This study showed directly that the proton decay energy of ^{112}Cs was less than that of ^{113}Cs exhibiting the same odd-even staggering effect for Q_p as implied for the neighbouring I isotopes. The decay rate of ^{112}Cs was also inconsistent with WKB calculations lending further weight to the evidence for deformation effects in this region.

Anomalous decay rates of themselves do not constitute conclusive evidence for the influence of deformation. Deformation can be independently inferred from inbeam gamma ray spectroscopy studies. However at the time such studies were not possible on proton emitters owing to the insensitivity of existing techniques. Typically these would involve either particle detectors around the target position or using an RMS to mass gate gamma-rays. In the latter case the ions would be stopped in a gas ionization counter. While this method could provide adequate Z selection up to Z~40 [19] it yielded no additional information beyond total energy for Z ~50, the domain of ground state proton radioactivity. In a parasitic experiment using the RMS, the Eurogam (phase 1) Ge array and the DSSD system the Recoil Decay Tagging technique (RDT) [20] was successfully pioneered (it later emerged that this method had already been applied to heavy ion radiative capture studies [21] at GSI although the sensitivity was limited by the use of a NaI array). In this experiment the characteristic charge particle decays were used to retrospectively genetically fingerprint gamma-rays detected at the the target position (see Figure 1). This technique was applied to alpha-decaying isotopes (^{108,109}Te), beta-delayed proton emitting isotopes (^{113}Xe and ^{109}Te) and the ground-state proton emitter ^{109}I. In the latter instance poor statistics were obtained, and no information could be deduced on the structure of the ground-state. As anticipated in [20] the use of the RDT technique has expanded to revolutionise in-beam gamma-ray studies of neutron deficient high Z nuclei [22]. However, this was the 1st and last application of the RDT technique at Daresbury since it was to close 2 months later in the spring of 1993.

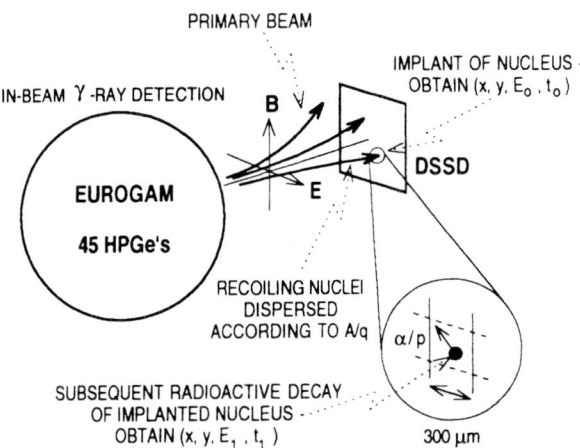

Figure 1. Schematic diagram of the Recoil Decay Tagging (RDT) technique applied using the Daresbury RMS to identify in-beam γ-rays from the ground-state proton emitter ^{109}I .

EXPERIMENTS ON THE ARGONNE FMA

Introduction

Prior to the closure of Daresbury the author and Cary Davids met and established a future collaboration for proton radioactivity measurements which became based on the use of the Argonne Fragment Mass Analyzer (FMA) [23] in combination with the DSSD system [10]. The FMA offered the advantages of a much greater energy acceptance than the Daresbury RMS and the acceptance of more than one charge state, leading to an increase in efficiency of around a factor of 5, along with a greater mass resolving power. The mass focus of the FMA was on the Parallel Grid Avalanche Counter (PGAC) and the transmitted recoils were dispersed across the surface of DSSD situated several cms downstream. This had the advantage of reducing the recoil implantation rate per quasi-pixel on the DSSD. This both improved the correlation performance and reduced the rate of radiation damage suffered by the DSSD. A further major benefit to these measurements was the use of the versatile ATLAS accelerator complex at Argonne. This system unlike the Daresbury Tandem could accelerate Noble gas species, most notably ^{78}Kr, which was to prove vital in opening up new regions of the proton drip-line. The ATLAS system could accelerate these beams to high enough energies to study higher multiplicity evaporation channels such as the 1p4n which allowed meassurements to be extended further beyond the proton drip-line than previously. The high beam energies also allowed the use of near-symmetric reactions to populate high Z proton drip-line nuclei and inverse reactions to improve the recoil transmission efficiency of the FMA. This latter feature was to prove invaluable for the exploitation of the RDT technique.

The programme of measurements of proton radioactive nuclei at the Argonne FMA was to result in the most significant extension of such studies. These data provided detailed and systematic information on proton radioactive nuclei which has transformed our knowledge of the proton drip-line, enabling us to explore the rich phenomena associated with different regions of nuclear structure [24]. The major advances will be outlined below.

Proton emitters from Z = 69 - 83

^{78}Kr beams were used to bombard targets of ^{92}Mo and ^{96}Ru in order to produce the new proton emitters ^{165,166,167}Ir and ^{171}Au [25]. This represented the first instance of three proton emitting isotopes being discovered from a single element and in the case of ^{165}Ir it represented the first use of a 1p4n evaporation channel to discover a proton emitter. It was found that the isotopic cross section decreased by around a factor of 20 for each additional neutron evaporated. Low and high spin proton emitting states were discovered for ^{166,7}Ir each also having an alpha decay branch. Alpha decay cascades were observed connecting the low(high) spin

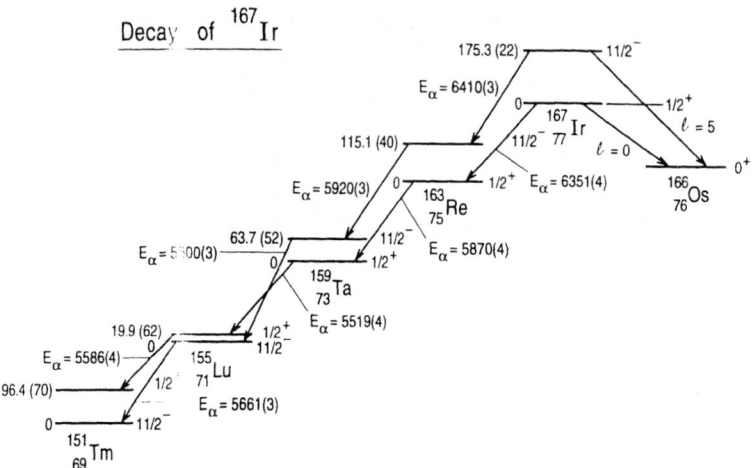

Fig 2. Ground and isomeric decay chains from ^{167}Ir

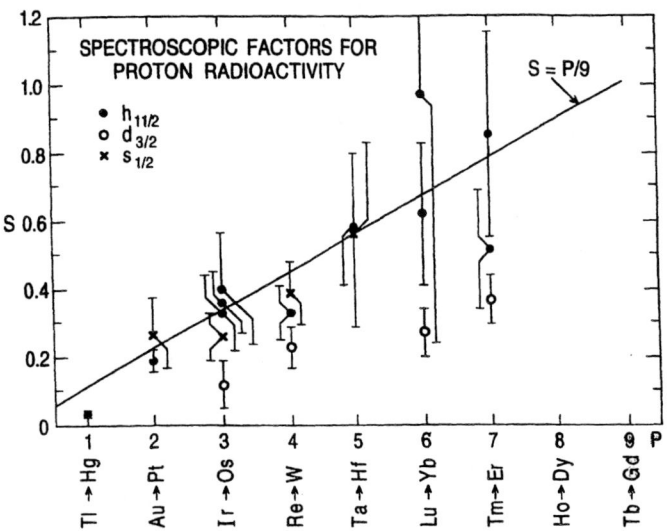

Figure 3. Experimental proton decay spectroscopic factors compared with the low seniority shell model calculation prediction of P/9 for the region $64<Z<82$, where P represents the number of proton hole pairs wrt the $Z = 82$ shell closure in the daughter nucleus.

states in the daughter nuclei (see Figure 2). These data were used to identify shell model states and their excitation energies for the daughter nuclei, the original parent structure being determined by the proton decay branch. Energy conservation arguments were used to infer proton separation energies even where the decay itself was unobserved. In the case of ^{167}Ir the chain terminated on a known mass isotope thereby enabling the masses of all nuclei connected by alpha or proton decay from this nucleus to be determined. These data demonstrated the rich nuclear structure information that could be unlocked by proton decay measurements in regions of alpha decay. However, the greatest significance of this work was the first systematic survey of experimental proton decay spectroscopic factors. With the availability of these new data, spectroscopic factors from $s_{1/2}$, $d_{3/2}$ and $h_{11/2}$ proton emitting states could be traced as a function of Z from Z = 69-79. In [25] a low seniority shell model approach assuming degenerate $s_{1/2}$, $d_{3/2}$ and $h_{11/2}$ states using a constant pairing force was introduced which predicted a spectroscopic factor of P/9, where P represented the number of pairs of proton holes in the daughter nuclei with respect to the Z = 82 shell closure for the Z = 64-82 shell model space covered by these states. Figure 3 shows how this approach nicely reproduces the trends of spectroscopic factors. Such detailed agreement gave great confidence in our understanding of these nuclei and of the proton decay mechanism from spherical nuclei. Subsequently a more rigorous theoretical approach using a BCS model of pairing also reproduced these trends well [26]. In both instances it was notable that the calculations systematically underestimate spectroscopic factors from the $d_{3/2}$ level which may be due to admixing of this state with other configurations. Such considerations are not likely to be significant for $h_{11/2}$ transitions since these should be relatively pure configurations having high spin and opposite parity compared to neighbouring levels. This was clearly demonstrated when proton decay was identified from an $h_{11/2}$ isomer in the N=82 closed neutron shell nucleus ^{155}Ta. Here the measured proton decay spectroscopic factor of 0.58 is in excellent agreement with the value of 0.56 predicted by the low seniority shell model approach [27]. Proton decay measurements on ^{157}Ta and ^{161}Re [28] demonstrated that the $h_{11/2}$ state crosses above and increases in excitation energy with respect to the $s_{1/2}$ ground-state with increasing Z. This trend continues until at Z = 81, corresponding to the proton emitter ^{177}Tl, the energy difference reaches a value of 807 keV. Here the measured spectroscpic factor for the isomeric state is very small, 0.034, compared to the prediction of 0.11 from the low seniority shell model calculation [29]. This is not surprising since this calculation assumes degenerate shell model states which is no longer a reasonable approximation for this nucleus, and transitions to the predominantly $(s_{1/2})^{-2}$ ^{176}Hg ground-state configuration should be hindered. For neighbouring odd-A Tl isotopes lying closer to stability the isomeric state consists of an $h_{9/2}$ intruder configuration which reaches its nadir around the neutron mid-shell closure N $\sim$ 104. However, this state is ruled out from the low spectroscopic factor obtained, since no hindrance should occur for this transition and a spectroscopic factor around unity would be anticipated. This is consistent with the expected parabolic rise in excitation energy of the intruder state configuration about the neutron mid-shell closure. For neutron deficient odd A Bi isotopes lying across the Z = 82 major shell closure the $s_{1/2}$ state becomes a low-lying excited intruder state configuration associated with a mildly oblate shape. Proton radioactivity was identified from such a configuration in the heaviest known proton emitter yet discovered ^{185}Bi [30]. The low spectroscopic factor of 0.05 was attributed to a small admixture of a 0^+ oblate excited state configuration in the spherical ground state configuration of the daughter nucleus ^{184}Pb.

Discovery of highly deformed proton emitters

As discussed earlier, modest deformations ($\beta\sim$0.10-0.15) had first been proposed to account for the anomalous decay rates of ^{109}I and ^{113}Cs. Searches in the rare earth region using the FMA led to the discovery of proton decay from the ground-states of ^{131}Eu and ^{141}Ho [31], predicted to have quadrupole deformations of 0.29 and 0.33, respectively [32]. The application of the low seniority shell model calculation to the proton decay of the Z=67 nucleus ^{141}Ho produced a sharp disagreement with experiment whereas the same calculation works very well for the neighbouring proton emitter ^{147}Tm. A similar calculation for ^{131}Eu also produced disagreement. Deformed DWBA calculations based on the approach of Bugrov and Kadmensky [6] were applied to these proton decays and excellent agreement was obtained for high quadrupole deformations ($\beta\sim$0.3), see Figure 4 , [31]. The ground state of ^{141}Ho was assigned to a $7/2^-$[523] Nilsson configuration in agreement with the theoretical prediction of Moller et al. [33]. However, in the case of ^{131}Eu both $3/2^+$[411] and $5/2^+$[413] assignments were allowed.

Discovery of Proton Decay Fine Structure

Following the identification of proton radioactivity from the highly deformed nucleus ^{131}Eu it was decided to revisit this isotope in an experiment at Argonne in order to search for the previously unobserved phenomenon of proton decay fine structure on the basis that the first excited 2^+ level in the daughter nucleus ^{130}Sm should be low enough for a significant decay branch [34]. Figure 5 shows the energy spectrum for decays occurring within 100ms of an A = 131 ion implanting into the same quasi-pixel of a Double-sided Silicon Strip Detector situated behind the focal plane of the Argonne FMA. The more intensely produced peak at higher energy corresponds to the previously identified ground-state proton transition from ^{131}Eu. The second peak produced with approximately one tenth of the intensity is at an energy 120 keV lower. The peak has the same half-life within errors as the previously identified transition and is assigned to the proton decay fine structure of ^{131}Eu. Using the Grodzin's formula [35,36] this implies a value of $\beta\sim$0.34 for the daughter nucleus ^{130}Sm in excellent agreement with the value of 0.33 predicted by Moller et al. [32]. This also provides a consistency check on the high deformation necessary to reproduce the partial half-life of the main ground-state proton

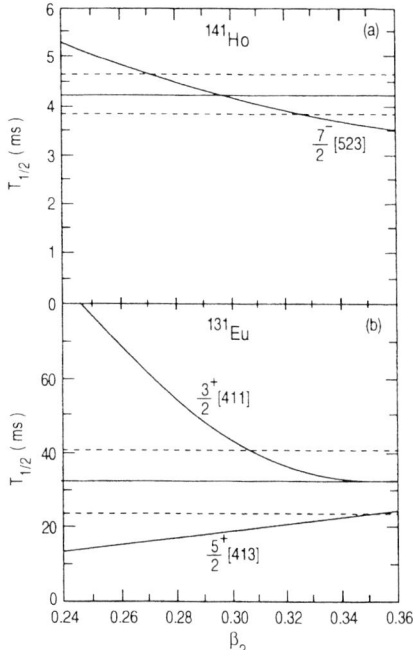

Fig 4. Deformed DWBA calculations for Nilsson states of ^{141}Ho and ^{131}Eu compared to measured proton decay half-lives.

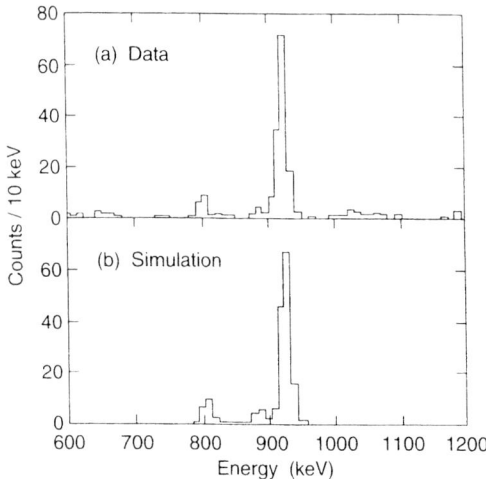

Fig 5. Proton decay energy spectrum showing fine structure in ^{131}Eu compared with a Monte Carlo simulation.

transition from ^{131}Eu using the deformed DWBA calculational approach [31]. The proton branching ratio is well reproduced by the calculation for a 3/2$^+$[411] configuration (see Figure 6) whereas the 5/2$^+$ configuration can now be ruled out [34].

GAMMA-RAY SPECTROSCOPY OF PROTON EMITTERS USING RECOIL DECAY TAGGING

Although theories of spherical and deformed proton emitters are now being tested over a wide range of nuclei, including the new phenomenon of proton decay fine structure, it is desirable to have independent information on the structure of these nuclei. In particular high resolution in-beam gamma-ray studies can provide insights into the nuclear deformation. The technique of Recoil Decay Tagging (RDT) [20] is an ideal tool that was developed with this particular goal in mind. The RDT technique was successfully applied to ground and isomeric proton emission from ^{147}Tm in an experiment on the Argonne FMA [37] using a $\sim$ 1 per cent efficiency Ge array. The gamma-ray band built on the ground-state is consistent with β=0.13. Interestingly, Kadmensky and Bugrov have applied their model for deformed proton emission to this case [38] and the results do not disagree significantly with spherical calculations, unlike the effect of such a deformation on proton decay rates in the region above Z = 50. It appears that Tm and Ho proton emitters lie right at the interface of the region of rapid shape change to high prolate deformations. Clearly it is desirable to identify in-beam gamma-rays from the highly deformed proton emitters. In a recent RDT experiment using the Argonne FMA coupled to the Gammasphere array ($\sim$10 per cent efficiency), gamma-rays were successfully identified from the ground and isomeric states in ^{141}Ho, the latter having a cross-section$\sim$50nb. These data shown in Figure 7 clearly demonstrate the existence of rotational bands built on the ground and isomeric states providing further evidence for the rapid increase in deformation below Z = 69. A preliminary analysis of the ground-state band yielded a value of β=0.28 in excellent agreement with both theoretical predictions of the ground-state shape and the deformation range required to reproduce the anomalous proton decay rate. Even more recently rotational bands have been identified in ^{131}Eu using the RDT technique. In this experiment it was shown that the fine structure peak and ground-ground proton peak were associated with the same gamma ray transitions thereby providing independent confirmation of the fine structure decay mechanism. This required a sensitivity $\sim$10 nb which can be compared with a value $\sim$10μb which represented the limit of sensitivity of in-beam gamma-ray experiments prior to the introduction of the RDT technique [19].

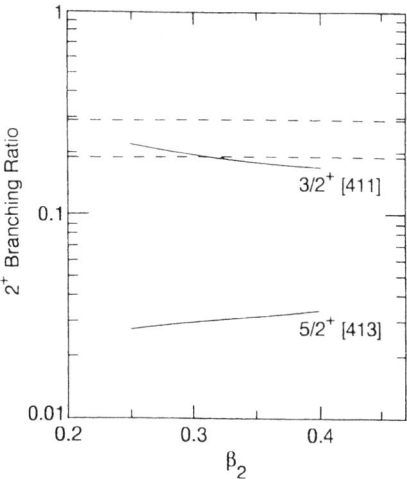

Fig 6. A comparison of the proton branching ratio with deformed DWBA calculations for possible Nilsson configurations in ^{131}Eu.

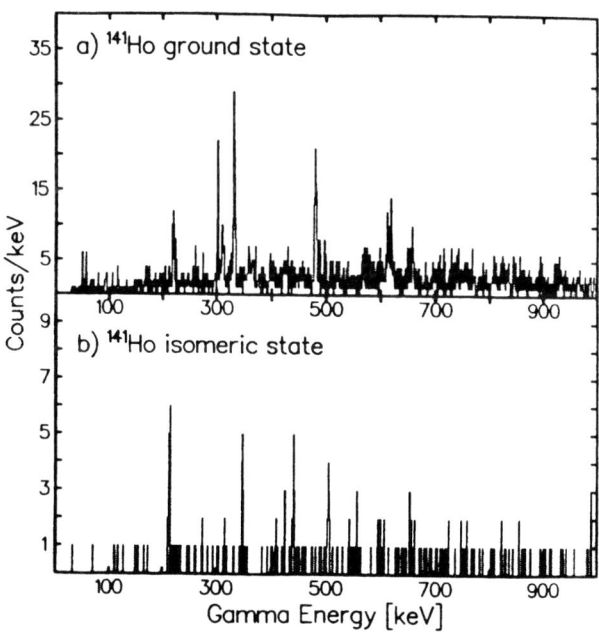

Fig 7 Gamma ray spectra obtained using the RDT technique. Rotational bands are observed populating the ground and isomeric proton decaying states in ^{141}Ho.

CONCLUDING REMARKS

The 1990s have seen an explosion of information on proton radioactive nuclei. Theoretical models are able to reproduce in detail the systematic variation of proton decay spectroscopic factors for a wide range of spherical nuclei thereby sensitively testing the nuclear shell model at the extreme edge of stability. The first examples of proton emission from highly deformed nuclei have been discovered. These transitions have provided a direct insight into the fragmentation of single particle strength within highly deformed nuclei. Decay rates from these nuclei are found to agree well with theoretical calculations assuming Nilsson states. Proton decay fine structure has been observed for the first time providing independent confirmation of the high deformations involved. In-beam studies of the gamma-rays using the RDT technique are providing complementary nuclear structure information on proton-radioactive nuclei that is assisting in constraining theoretical calculations of proton decay rates. Furthermore, such studies will provide new insights into the behaviour of proton unbound nuclei at high spin and excitation energy. In summary there has been a great advance in our detailed understanding of the phenomenon of proton radioactivity. This has derived from the large increase in known transitions and the varied nuclear landscape in which this process is now found.

ACKNOWLEDGEMENTS

I would like to thank all my colleagues involved in experiments reviewed here. In particular I would like to acknowledge Cary Davids, Darek Seweryniak and Bill Walters for the work at Argonne, and Rob Page, Paul Sellin, and Arthur James for the work at Daresbury, Tom Davinson being omnipresent.

REFERENCES

1. K.P. Jackson et al., *Phys. Lett.* **B33**, 281 (1970).
2. S. Hofmann, W. Reisdorf, G. Munzenberg, F.P. Hessberger, J.R.H. Schneider, P. Armbruster, *Z. Phys.* **A305**, 111 (1982).
3. O. Klepper et al., *Z. Phys.* **A305**, 125 (1982).
4. T. Faestermann et al., *Phys. Lett.* **B137**, 23 (1984).
5. A. Gillitzer, T. Faestermann, K. Hartel, P. Kienle and E. Nolte, *Z. Phys.* **A326**, 107 (1987).
6. V.P. Bugrov and S.G. Kadmensky, *Sov. J. Nucl. Phys.* **49**, 967 (1989).
7. A.N. James et al., *Nucl. Inst. Meth.* **A267**, 144 (1988).
8. P.J. Woods et al., *Nucl. Inst. Meth.* **A276**, 195 (1989).
9. R.D. Page, P.J. Woods, S.J. Bennett, M. Freer, B.R. Fulton, M.A.C. Hotchkis, A.N. James and R.A. Cunningham, *Z. Phys.* **A338**, 291 (1991).

10. P.J. Sellin, P.J. Woods, T. Davinson, N.J. Davis, K. Livingston, A.C. Shotter, A.N. James, S. Hofmann and R. Hunt, *Nucl. Inst. Meth.* **A311**, 217 (1992).
11. S.L. Thomas et al., *Nucl. Inst. Meth.* **A288**, 212 (1988).
12. S. Hofmann et al., Proceedings of the 7th International Conference on Atomic Masses and Fundamental Constants, Darmstadt, 184 (1984).
13. P.J. Sellin, P.J. Woods, T. Davinson, N.J. Davies, A.N. James, K. Livingston, R.D. Page, A.C. Shotter, *Phys. Rev.* **C47**, 1933 (1993).
14. P.J. Woods, T. Davinson, N.J. Davis, K. Livingston, A.N. James, R.D. Page, P.J. Sellin, and S. Hofmann, *Nucl. Phys.* **A553**, 485 (1993).
15. K. Livingston, P.J. Woods, T. Davinson, N.J. Davis, S. Hofmann, A.N. James, R.D. Page, P.J. Sellin and A.C. Shotter, *Phys. Lett.* **B312**, 46 (1993).
16. K. Livingston, P.J. Woods, T. Davinson, N.J. Davis, S. Hofmann, A.N. James, R.D. Page, P.J. Sellin and A.C. Shotter, *Phys. Rev.* **C48**, 3113 (1993).
17. R.D. Page, P.J. Woods, T. Davinson, N.J. Davis, K. Livingston, A.C. Shotter, A.N. James and S. Hofmann, *Phys. Rev. Lett.* **68**, 1287 (1992).
18. R.D. Page, P.J. Woods, R.A. Cunningham, T. Davinson, N.J. Davis, A.N. James, K. Livingston, P.J. Sellin and A.C. Shotter, *Phys. Rev. Lett.* **72**, 1798 (1994).
19. C.J. Lister et al., *Phys. Rev. Lett.* **55**, 810 (1985).
20. E.S. Paul, P. J. Woods, T. Davinson, R. D. Page, P. J. Sellin, C. W. Beausang, R. M. Clark, R. A. Cunningham, S. A. Forbes, D. B. Fossan, A. Gizon, J. Gizon, K. Hauschild, I. M. Hibbert, A. N. James, D. R. LaFosse, I. Lazarus, H. Schnare, J. Simpson, R. Wadsorth and M. D. Waring, *Phys. Rev.* **C51**, 78 (1995).
21. R. S.Simon, K-H. Schmidt, F. P. Hessberger, S. Hlavae, M. Honusek, G. Munzenberg, H. G. Clerc, U. Gollerthan, W. Schwab, *Z. Phys.* **A325**, 197 (1986).
22. W. Gelletly and P.J. Woods, *Phil. Trans. R. Soc. Lond.* **A356**, 2033 (1998).
23. C. N. Davids et al., *Nucl. Instrum. Methods* **B70**, 358 (1992).
24. P.J. Woods and C.N. Davids, *Ann. Rev. Nucl. Part. Sci.* **47**, 541 (1997).
25. C. N. Davids, P. J. Woods, J. C. Batchelder, C. R. Bingham, D. J. Blumenthal, L. T. Brown, B. C. Busse, L. F. Conticchio, T. Davinson, S. J. Freeman, D. J. Henderson, R. J. Irvine, R. D. Page, H. T. Pentilla, D. Seweryniak, K. S. Toth, W. B. Walters and B. E. Zimmerman, *Phys. Rev.* **C55**, 2255 (1997).
26. S. Aberg, P.B. Semmes, W. Nazarewicz, *Phys. Rev.* **C56**, 1762 (1997).
27. R.J. Irvine et al., *Phys. Rev.* **C55**, 1621 (1997).
28. J. Uusitalo et al., *Phys. Rev.* **C59**, 2975 (1999).
29. G. Poli et al., *Phys. Rev.* **C59**, 2979 (1999).
30. C.N. Davids et al., *Phys. Rev. Lett.* **76**, 592 (1996).
31. C.N. Davids, P.J. Woods, D. Seweryniak, A.A. Sonzogni, J.C. Batchelder, C.R. Bingham, T. Davinson, D.J. Henderson, R.J. Irvine, G.L. Poli, J.Uusitalo, W.B. Walters, *Phys. Rev. Lett.* **80**, 1849 (1998).
32. P. Moller et al., *At. Data Nucl. Data Tab.* **59**, 185 (1995).
33. P. Moller, J.R. Nix, K.-L. Kratz, *At. Data Nucl. Data Tables* **66**, 131 (1997).
34. A. Sonzogni, C.N. Davids, P.J. Woods, D. Seweryniak, M. Carpenter, J. Ressler, J. Uusitalo and W.B. Walters, *Phys. Rev. Lett.* **83**, 1116 (1999).
35. L. Grodzins, *Phys. Lett.* **2**, 88 (1962).
36. F. Stephens et al., *Phys. Rev. Lett.* **29**, 438 (1972).

37. D. Seweryniak, C. N.Davids, W. B. Walters, P. J. Woods, I. Ahmad, H. Amro, D. J. Blumental, L.T. Brown, M. P. Carpenter, T. Davinson, S.M. Fischer, D.J. Henderson, R.V.F. Janssens, T. L. Khoo, I. Hibbert, R.J. Irvine, R. J. Irvine, C.J. Lister, J. A. Mckenzie, D. Nisius, C. Parry, and R. Wadsworth *Phys. Rev.* **C55**, R2137 (1997).
38. S.G. Kadmensky and V.P. Bugrov, *Phys. At. Nucl.* **59**, 399 (1996).

Experimental Direct Proton Radioactivity Measurements

Proton Drip-Line Studies at HRIBF

K. P. Rykaczewski,[1,2] J. C. Batchelder,[3] C. R. Bingham,[1,4]
R. E. Bryan,[5] T. Davinson,[6] T. N. Ginter,[7] C. J. Gross,[1,3]
R. Grzywacz,[2,4] J. H. Hamilton,[7] Z. Janas,[2,8] M. Karny,[2,4,8]
B. D. Macdonald,[9] J. W. McConnell,[1] A. Piechaczek,[10] J. Szerypo,[8]
K. S. Toth,[1] W. B. Walters,[11] P. J. Woods,[6] and E. F. Zganjar[10]

[1] *Physics Division, ORNL, Oak Ridge, TN 37830, USA*
[2] *Warsaw University, IEP, PL-00681 Warsaw, POLAND*
[3] *Oak Ridge Institute for Science and Education, Oak Ridge, TN 37831, USA*
[4] *University of Tennessee, Knoxville, TN 37996, USA*
[5] *Webb School, Knoxville, TN 37923, USA*
[6] *University of Edinburgh, Edinburgh, EH9 3JZ, UK*
[7] *Vanderbilt University, Nashville, TN 37235, USA*
[8] *Joint Institute for Heavy Ion Research, Oak Ridge, TN 37831, USA*
[9] *Georgia Institute of Technology, Atlanta, GA 30332, USA*
[10] *Louisiana State University, Baton Rouge, LA 70803, USA*
[11] *University of Maryland, College Park, MD 20742, USA*

Abstract. Proton radioactivity studies performed at the Holifield Radioactive Ion Beam Facility (HRIBF) within the last few years are reviewed. The discovery of five new proton radioactivities ^{140}Ho, ^{141m}Ho, ^{145}Tm, ^{150m}Lu and ^{151m}Lu is presented together with a recent observation of fine structure in proton emission from ^{146gs,m}Tm. These proton emitters were produced by means of fusion-evaporation reactions and studied with the HRIBF Recoil Mass Separator and detection system based on a Double-sided Silicon Strip Detector. For ^{113}Cs and ^{151}Lu, the studies of level structure were extended beyond the proton-emitting states via the measurements with a clover array CLARION using Recoil Decay Tagging.

I INTRODUCTION

Studies at the limits of nuclear stability represent a hot topic in today's nuclear physics. Recent developments of production and detection methods have resulted in a great extension of experimental information on the structure of nuclei at and beyond the proton drip-line (see the proceedings of this meeting).

The structure of loosely bound nuclei and their decay modes, the nuclear energy surface, and cross sections for capture processes related to nucleosynthesis within the rp-process are among the key themes of nuclear physics for the beginning of the next millenium [1,2]. These problems can be addressed via the studies in the proton drip-line region. With the intense postaccelerated radioactive beams like e.g. ^{56}Ni ions, and new counting techniques [3] one should be able to reach even more exotic short-lived nuclear states.

Observation of proton radioactive nuclei greatly helps to understand the limits of the nuclear landscape. Usually, a precision at the level of 10 keV is achieved in such measurements of proton separation energies. The observed decay rates allow us to develop and test our understanding of quantum tunneling through the Coulomb and centrifugal barriers for spherical and deformed shapes. Measured decay properties can be used to establish proton single-particle energies and a structure of proton emitting orbitals, both far beyond the proton drip-line. Recent observation of the proton multi-line structure in a decay of an odd-odd nucleus ^{146}Tm [4,5] opens up a field of studies on neutron single particle states via proton radioactivity measurements.

II HRIBF EXPERIMENTS ON PROTON EMITTERS

The proton radioactivities of deformed ^{140}Ho and ^{141m}Ho [6], and spherical ^{145}Tm [7], ^{150m}Lu [8] and ^{151m}Lu [9], were observed for the first time at the Holifield Radioactive Ion Beam Facility (HRIBF) in Oak Ridge. Also, more precise data on proton emission from ^{113}Cs, ^{146gs,m}Tm, ^{150gs}Lu and ^{151gs}Lu were obtained during HRIBF experiments. These results make an important contribution to the total number of 36 proton emitting ground and isomeric states in nuclei reported till now [10–12,14–16].

Fine structure in proton emission has only been observed for the radioactivities of deformed ^{131}Eu [17] and spherical ^{146}Tm [4,5], at Argonne and Oak Ridge National Laboratories, respectively.

Modern experiments on proton emitters at the Fragment Mass Analyzer (FMA) at Argonne [13,14] and at the Recoil Mass Separator (RMS) at HRIBF Oak Ridge [18] are utilizing the technique pioneered at the velocity filter SHIP (GSI Darmstadt) [19] and developed further at the Daresbury Recoil Separator [20]. The products of fusion-evaporation reactions are selected according to their mass–to–charge ratio (A/Q). Separation is followed by the recoil detection using a position-sensitive gas detector (e.g. an Avalanche Counter) and by implantation into a Double-sided Silicon Strip Detector (DSSD) [20,21]. The detection technique of proton radioactivity used at HRIBF is described in [6–9] as well as the most recent experimental review [18]. It is important to note the high selectivity, i.e. good mass resolution and primary beam rejection of the RMS set-up at HRIBF. The experiments with the ^{92}Mo targets were performed at relatively high intensity of ^{50}Cr,^{54}Fe and ^{58}Ni beams (typically of 15 pnA), but without an overloading rate of scattered primary beam particles or recoiling reaction products at the PSAC or DSSD detectors. High selectivity together with an overall transmission of a few percent allowed to reach the experimental observation limits of about 13 nanobarns for ^{140}Ho [6]. This activity was produced in a *p5n* reaction channel used for the first time for the discovery of a proton emitter. The standard DSSD signal processing technique [20,21] was tuned to its limits, with proton decay events measured as early as 4 μs after the recoil implantation. However, such events were detected with a large energy shift, of about 500 keV above the 960 keV proton line from the ^{113}Cs decay. These signals were not counted with full efficiency, due to the amplifier overload by recoiling ions and the hardware coincidence condition between the front and back strips of the DSSD required for decay event processing. A new method, based on the digital preamplifier signal processing applied at HRIBF to overcome these limitations, was presented at this meeting [3,22,23].

Recently, two important changes have been made at the RMS, both improving the yields of exotic nuclei [18]. A converging solution for the ion optics was implemented. A focusing of the recoils transmitted in the two neighbouring charge states increases the yield at the final focus by a factor $\approx$ 1.6. The latter number was obtained from counting proton events following the decay of known emitter ^{113}Cs. The counting rate achieved at RMS with 6 pnA of 235 MeV ^{58}Ni beam on a 500 μg/cm^2 ^{58}Ni target, for proton emission from ^{113}Cs produced in a (p,2n) reaction channel, was about 500 events per hour.

Also, a thin charge-reset carbon foil was added 10 cm downstream the target. This foil restores the charge state equilibrium for the recoils produced in a short-lived isomeric state and decaying via an electron-converted transition between the target and the foil [4,5,18]. For the studies of the proton-decaying isomer in an odd-odd nucleus ^{150}Lu [5,8], the foil alone resulted in a ^{150m}Lu proton rate increase by a factor of 3, see Figure 1. This RMS-transmission increase for the ^{150m}Lu ions indicates the presence of a short-lived isomeric state, produced in the reaction and decaying to the proton-emitting metastable state [5].

III SPHERICAL DESCRIPTION OF PROTON EMISSION RATES

For most reported proton activities, the proton emission rates can be interpreted within a spherical description of the nuclei involved in the decay process [11,24]. The spectroscopic factor corresponding to the ratio of calculated to observed partial proton half-lives represents the vacancy of the respective proton orbital in a daughter nucleus [24]. This means that the study of proton emitting states gives direct information on the composition of the wave function of the unbound proton orbital. However, for several emitters in the heavy rare-earth region, it was impossible till now to obtain precise experimental values for the proton partial half-lives. This was due to the unknown probability of beta decay contributing to the decay process. The proton emitting orbitals were identified; however, the error bars on the spectroscopic factors were too large to make a meaningful conclusion on the structure of the wave function involved. Therefore, the study of very short-lived

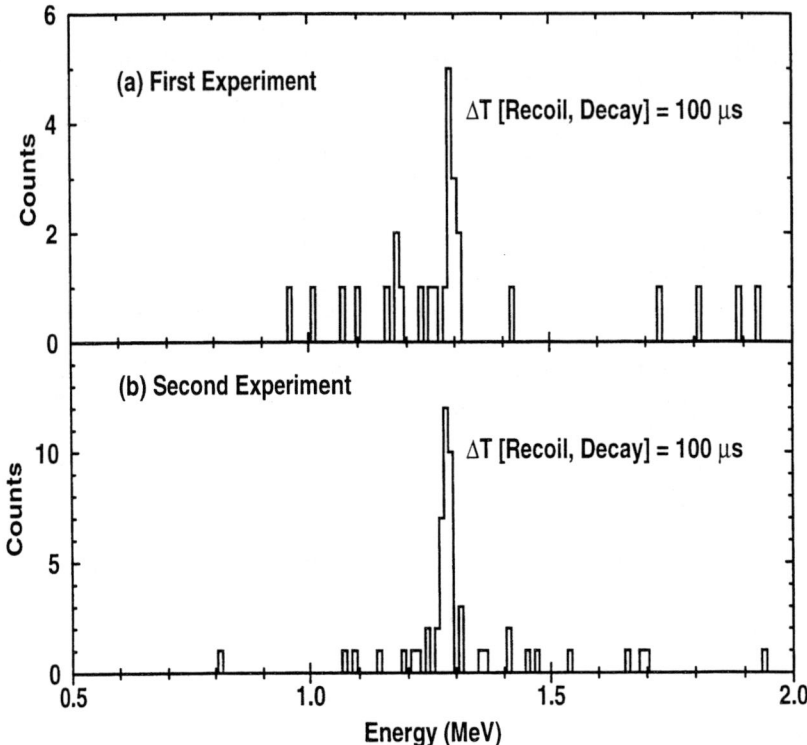

FIGURE 1. The low-energy part of particle spectra recorded during two experiments, without (a) and with (b) charge reset foil at RMS (from [5]). The 5 pnA ^{58}Ni beams at 292 MeV on a ^{96}Ru target of 0.54 mg/cm^2, were used for (a) and (b) runs. The 100 μs gating conditions for the recoil-decay correlation times were applied to enhance the short-lived new activity of ^{150m}Lu.

proton radioactivities, in which proton emission dominates the total decay width, has been initiated at HRIBF. New short-lived proton emitters in the region of spherical nuclei discovered at HRIBF : ^{145}Tm ($T_{1/2}$=3.5 μs [7]), ^{151m}Lu ($T_{1/2}$=16 μs [9]) and ^{150m}Lu ($T_{1/2} \approx 32\mu$s [5,8]) have contributed to the understanding of the structure of the $\pi h_{11/2}$ and $\pi d_{3/2}$ proton emitting states.

In Fig. 2, the experimental spectroscopic factors for odd–Z, even–N $\pi h_{11/2}$ proton emitters are compared to the vacancies u^2 calculated (see e.g. Refs. [25,26]) for the daughter even-even spherical nuclei. These spectroscopic factors were obtained from measured partial proton half-lives and theoretical proton rates obtained within the Two-Potential-Approach (TPA) of Ref. [24]. The radioactivities, where proton emission dominates the decay process, or the competing alpha branching was measured, are selected for this comparison. For Z=70, the spectroscopic factor includes new more precise halflife measurements of ^{151gs}Lu ($T_{1/2}$=80±2 ms [9]) and ^{150gs}Lu ($T_{1/2}$=49±5 ms [8]). However, the estimated beta branching [11,27] makes

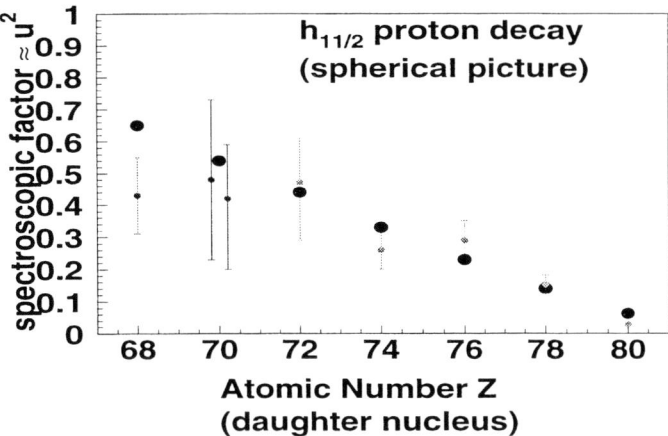

FIGURE 2. Comparison of the experimental spectroscopic factors for the proton emission from the $\pi h_{11/2}$ orbital for $Z \geq 68$ nuclei and the vacancies $u^2(\pi h_{11/2})$ calculated for the daughter nuclei (bold black dots). The HRIBF data : ^{145}Tm [7], ^{150}Lu [5,8] and ^{151}Lu [9].

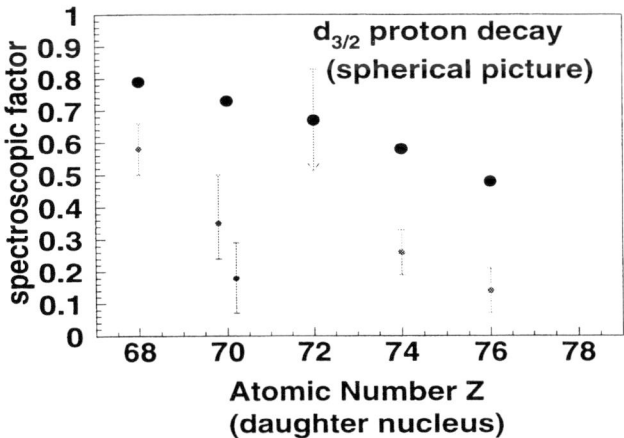

FIGURE 3. Experimental spectroscopic factors for the proton emission from the $\pi d_{3/2}$ orbital for $Z \geq 68$ nuclei and the vacancies $u^2(\pi d_{3/2})$ calculated for pure $\pi d_{3/2}$ configuration in the daughter nuclei. The HRIBF results : ^{150m}Lu [5,8] and ^{151m}Lu [9].

the error bars large. The u^2 values, calculated within a Two-Potential-Approach [24], fit very well the experimental systematics. Such agreement indicates that the spherical $\pi h_{11/2}$ orbital is relatively pure and dominates the proton-emitting wave function for these very exotic nuclei. The observed proton rates directly reflect the occupation of $\pi h_{11/2}$ along the proton drip line.

In contrast, the difference is quite striking when the spectroscopic factors for $\pi d_{3/2}$ emitters are compared to the calculated u^2 values, see Fig. 3. The observed $\pi d_{3/2}$ proton rates are below the values expected from the model which reproduced well the $\pi h_{11/2}$ related emission. This discrepancy indicates the level of mixing of the $I^\pi=3/2^+$ state. The wave function component, $\pi d_{3/2}$, is responsible for the observed proton transition rate. The presence of a $\pi s_{1/2} \otimes 2^+$ component in this $3/2^+$ state reduces the decay width to the ground-state of the daughter nucleus. For the first quantitative estimation of this effect, see the contribution by P. Semmes to this meeting [28].

The study resulting in the first observation of fine structure in proton emission from a spherical nucleus ^{146}Tm, performed recently at Oak Ridge, is described in a separate contribution to this meeting [4]. The collected data are presently under evaluation. However, the presence of the new proton lines is evident, see spectra in Ref. [4]. The increase of statistics in the HRIBF experiment was about a factor 20 in comparison to the previous study [29]. The halflives of new transitions at about 0.88 MeV, 0.93 MeV and 1.03 MeV seem to be similar to the known 1.19 MeV and 1.12 MeV lines, respectively. This suggests a fine structure for both, isomeric and ground-state emission. The branchings to neutron levels, with a spacing of about 300 keV between the $\nu s_{1/2}$ and $\nu h_{11/2}$, might be responsible for such decay pattern.

IV PROTON EMITTERS IN A REGION OF DEFORMED NUCLEI

The observed proton decay probabilities of ^{109}I [30–32] and ^{113}Cs [33,34,31,35] were found to be retarded in comparison to the standard spherical approach estimates. Both emitters are in the transitional region between the strong spherical shell $Z=50$ and the deformed (β_2 of about 0.3) rare-earth nuclei midway between $N=50$ and $N=82$ neutron magic numbers. The structure of proton emitting states for ^{109}I and ^{113}Cs nuclei is still a subject of experimental and theoretical investigations. The deformed $1/2^+[420]$ Nilsson orbital originating from the spherical $\pi d_{5/2}$ level was proposed for the ground-state of ^{109}I [36,37]. Such assignment was a result of an extrapolation of proton ground-state configurations for heavier odd-mass iodine isotopes [38,39] supported by a theoretical estimate for the observed proton rate [36,37].

In calculations [36,37], the best fit to experimental halflife was obtained with a quadrupole deformation parameter β_2 of about 0.1. The Recoil Decay Tagging experiments on ^{109}I [39,40], however, do not confirm the ground-state configuration suggested in Refs. [36,37]. In particular, the recently deduced rotational states in the $\pi h_{11/2}$ band of ^{109}I support reduced deformation for the more proton–rich iodine isotopes [40], an effect probably related to the proximity of doubly-magic nucleus ^{100}Sn. The configuration of the proton-emitting state in the ^{113}Cs is also unclear. Note the first half-life measurement for ^{113}Cs, ($T_{1/2}$=33 ±7 µs [35]) differs from more recent measurements, with the HRIBF result of 18.3 ±0.3 µs [34] being the most precise. The calculations of Ref. [37] suggest the 3/2$^+$[421] orbital originates from the $\pi d_{5/2}$ spherical state. More recent work [41] suggests the observed ^{113}Cs half-life [34] can also be explained by a proton emission from an I^π=1/2$^+$ state coming from a deformed $\pi g_{7/2}$ orbital. The deformation parameters are $\beta_2 \approx 0.2$ for the discussed configurations. Recent RDT studies [34,42] suggest a link between the dominantly populated $\pi h_{11/2}$ band and the lower energy state of positive parity and presumably spin I^π=3/2.

The proton radioactivity from the highly deformed nuclei ^{131}Eu and ^{141}Ho was observed for the first time in the FMA-based experiment in Argonne [14]. The proton emission half-lives were calculated within the DWBA approach of Refs. [36,37], as a function of deformation for the $\pi 7/2^-$[523] state in ^{141}Ho, and for $\pi 3/2^+$[411] and $\pi 5/2^+$[413] states for ^{131}Eu. The observed decay rates could be reproduced with $\beta_2 \geq 0.3$. A number of theoretical approaches to the observed decay rates have been presented at the present meeting.

Two new proton emitting deformed states, ^{140}Ho and ^{141m}Ho, were identified by their direct proton radioactivity at HRIBF [6]. The single–quasiproton band heads for the region holmium isotopes have been analyzed following the description given in Ref. [26]. In Ref. [26], the shell correction method with an average Woods-Saxon potential and a monopole pairing residual interaction was applied. The total energy of each nucleus was minimized in the [β_2,β_4] deformation lattice. For Z=67 holmium isotopes, proton orbitals were studied between mass numbers A=153 and A=171. Here, for lighter holmium isotopes down to ^{140}Ho, such calculations were performed in an extended [β_2,β_4,β_6] deformation space.

The proximity of spherical shell at N=82 promotes the spherical minimum for the $\pi h_{11/2}$ ground states of odd-mass ^{151}Ho, ^{149}Ho and ^{147}Ho. The positive parity orbitals, $\pi d_{3/2}$ and $\pi s_{1/2}$, appear in the calculations somewhat high, about 350 to 400 keV above the ground state. A "transitional" nucleus ^{145}Ho has $\pi d_{3/2}$ orbital splitted down to the region of the $\pi h_{11/2}$-originating states. The more proton rich ^{143}Ho and ^{141}Ho isotopes are well deformed, with the values of $\beta_2 \approx 0.27$, $\beta_4 \approx -0.07$ (and small $\beta_6 \approx 0.01$). Practically identical β-values were obtained for the three lowest proton orbitals in ^{141}Ho, the $\pi 1/2^+$[411] originating from $\pi d_{3/2}$ orbital, and the $\pi 7/2^-$[523] and $\pi 5/2^-$[532] from splitted $\pi h_{11/2}$ orbital. The computed states, $\pi 1/2^+$[411] and $\pi 7/2^-$[523], have the same energy within a couple of keV, and the $\pi 5/2^-$[532] Nilsson orbital is about 250 keV higher. This theoretical picture of Nilsson orbitals in the ^{141}Ho alone suggests that in addition to the l=3 proton emission with $T_{1/2} \sim 4$ ms from the $\pi 7/2^-$[523] ground–state, we observed a much faster l=0 decay from the isomeric 1/2$^+$[411] level in ^{141}Ho.

The width of proton resonances observed in ^{141}Ho was interpreted within the recently developed theoretical formalism based on the coupled channel Schrödinger equation with outgoing boundary conditions (see Ref. [6] for the description and references). More recent versions of the model were presented at this meeting, and are described in forthcoming papers [43,44]. The main conclusions given in [6] are still valid. The decay process of ^{141m}Ho is primarily governed by a small admixture of the $\pi s_{1/2}$ wave function in the $1/2^+[411]$ isomeric state (note $s_{1/2}$, not $d_{3/2}$). As given in [6], the corresponding spherical amplitude $(c_{lj})^2$ was calculated to be about 0.18, with the main wave function components arising from $\pi d_{3/2}$, $\pi d_{5/2}$ and $\pi g_{7/2}$ orbitals. For the latter components, the respective $(c_{lj})^2$ values were calculated as 0.24, 0.18 and 0.34. However, the $\pi g_{7/2}$ being the main wave function component, has a negligible contribution to the decay width of proton emission to the ^{140}Dy ground state. Similarly, for ^{141gs}Ho, the $\pi 7/2^-[523]$ Nilsson orbital originates mainly from $\pi h_{11/2}$ having an amplitude $(c_{lj})^2=0.86$, but the proton decay width is governed again by a small admixture, $(c_{lj})^2=0.08$, of the lowest-l orbital, a $\pi f_{7/2}$ spherical state.

It is interesting to note that the energy of the proton line from odd-odd ^{140}Ho decay is lower by about 100 keV than the one from the neighbouring, less exotic odd-even ^{141}Ho. Such a pattern was measured before [33] for nuclei in the transitional region above $Z=50$. It might be related to the presence of proton-decaying isomeric states in odd-odd nuclei resulting from the odd proton - odd neutron coupling, see the contribution of W. B. Walters to this meeting [45].

Such proton energy pattern for the deformed region between $Z=55$ and $Z=67$ does not show up to such extent in the calculations of ground-state mass differences [46,47,27]. However, it may explain the non-observation of proton decay from ^{136}Tb and ^{137}Tb. At HRIBF, a reaction of a 15 pnA ^{50}Cr beam on a 0.91 mg/cm^2 ^{92}Mo target, was studied at the RMS (tuned to the diverging mode) for over 35 hours. The beam energy, 290 MeV, was optimized [48] for ^{136}Tb production, but a part of $A=137$ recoils were also implanted into the DSSD similar to the $A=140$ vs $A=141$ mass distribution, see Figure 2 in [6]. No convincing evidence for proton lines was found. Such non-observation might be a result of dominant beta–decay channel, which implies a relatively low energy of proton emission. For the ^{137}Tb, various mass predictions indicate the proton line energy below 900 keV, which, together with a retardation caused by the deformation makes the above interpretation probable. However, the more proton rich ^{136}Tb was calculated to be more proton–unstable, at least for the respective ground-state mass differences. The results for ^{140}Ho - ^{141}Ho emitters might indicate an opposite pattern of proton separation energy.

V SUMMARY

The studies of proton radioactivity have became an important part of the scientific program at Oak Ridge National Laboratory. The experiments performed at HRIBF have added meaningful data to experimental systematics. Advanced theoretical interpretion of spherical [24] and deformed emitters [6,43,44] has been carried out by the UT/ORNL Theory Group in collaboration with physicists from Hungary, Romania, Sweden and US universities, see the proceedings of this meeting for recent developments. The future studies at HRIBF based on a radioactive beams like ^{56}Ni (under development) and/or new method of digital signal processing [3,22,23] should allow us to reach even more exotic nuclei near and beyond the proton drip-line.

ACKNOWLEDGMENTS

Oak Ridge National Laboratory is managed by Lockheed Martin Energy Research Corporation under contract DE-AC05-96OR22464 with the U. S. Department of Energy. Nuclear physics research is supported by the U. S. Department of Energy through Contracts Nos. DE-FG02-96ER40963 and DE-FG02-96ER40983 (University of Tennessee), DE-FG05-88ER40407 (Vanderbilt University), DE-FG02-92ER40694 (Tennessee Technological University), DE-FG05-88ER40330 (Georgia Institute of Technology), DE-FG02-96ER40978 (Louisiana State University) and DE-AC05-76OR00033 (ORISE), respectively. The Joint Institute for Heavy Ion Research has as member institutions the University of Tennessee, Vanderbilt University, and Oak Ridge National Laboratory; it is supported by the three members and the U. S. Department of Energy. ZJ and MK are partially supported by the Polish Committee for Scientific Research KBN, contract KBN 2 PO3B 086 17.

REFERENCES

1. Scientific Opportunities With an Advanced ISOL Facility, Report, November 1997, see *www.er.doe.gov/production/henp/isolpaper.pdf*
2. Opportunities in Nuclear Astrophysics, Report, September 1999, see *www.nscl.msu.edu/ austin/nuclear-astrophysics.pdf*
3. C. R. Bingham *et al.*, "Prospects for Future Proton Studies at HRIBF", These Proceedings.
4. T. Ginter *et al*, "A Search for Neutron Single-Particle States Populated Via Proton Emission from ^{146}Tm", These Proceedings.
5. T. Ginter, "The New HRIBF Recoil Mass Spectrometer - Performance And First Results", dissertation, Vanderbilt University, December 1999, unpublished.
6. K. Rykaczewski *et al.*, Phys. Rev. **C60**, 011301 (1999).
7. J. C. Batchelder *et al.*, Phys. Rev. **C57**, R1042 (1998).
8. T. Ginter *et al.*,"Study of Proton Emission from ^{150}Lu", in press, Phys. Rev. C.
9. C. R. Bingham *et al.*, Phys. Rev. **C59**, R2984 (1999).
10. K. P. Jackson *et al.*, Phys. Lett. **33B**, 281 (1970).
11. S. Hofmann, Radiochimica Acta **70/71**, 93 (1995).
12. P. J. Woods and C. N. Davids, Annu. Rev. Nucl. Part. Sci. **47**, 541 (1997).
13. C. N. Davids *et al.*, Nucl.Instr. Meth. B**70**, 358 (1992)
14. C. N. Davids *et al.*, Phys. Rev. Lett. **80**, 1849 (1998).

15. J. Uusitalo et al., Phys. Rev. C**59**, R2975 (1999).
16. F. Soramel et al., "First Observation of Proton Emission From ^{117}La", These Proceedings.
17. A. A. Sonzogni et al., Phys. Rev. Lett. **83**, 1116 (1999).
18. C. J. Gross et al., "Performance of the Recoil Mass Spectrometer and its Detector Systems at the Holifield Radioactive Ion Beam Facility", in press, Nucl. Instr. Methods Phys. Res. A.
19. S.Hofmann et al., Z. Phys. A**305**, 111 (1982).
20. P. J. Sellin et al, Nucl. Instrum. Methods Phys. Res. A**311**, 217 (1992).
21. S. L. Thomas et al., Nucl. Instrum. Methods A**288**, 212 (1990)
22. B. Hubbard-Nelson, M. Momayezi and W.K. Warburton, Nucl. Instr. Meth. **A422** (1999) 411, see *www.xia.com*
23. M.Momayezi et al, "Overcoming Limitations on Electronics", These Proceedings.
24. S. Åberg, P. B. Semmes, and W. Nazarewicz, Phys. Rev. C **56**, 1762 (1997);Phys. Rev. C **58**, 3011 (1998).
25. S. Ćwiok, J. Dudek, W. Nazarewicz, J. Skalski, and T. Werner, Comp. Phys. Comm. **46**, 379 (1987).
26. W. Nazarewicz, M. A. Riley and J. D. Garrett, Nucl. Phys. A**512**, 61 (1990)
27. P. Möller, J.R. Nix and K.-L. Kratz, At. Data Nucl. Data Tables, **66** 131 (1997)
28. P. B. Semmes, "Spherical Proton Emitters", *ibid.*
29. K. Livingstone et al., Phys. Lett. B**312** 46 (1993).
30. P. J. Sellin et al, Phys. Rev. C**47**, 1933 (1993).
31. T. Faestermann et al., Phys. Lett. B**137**, 23 (1984).
32. F. Heine et al., Z. Phys. A**340**, 107 (1987).
33. R. D. Page et al., Phys. Rev. Lett. **72**, 1798 (1994).
34. C. J. Gross et al, in Proc. of Int. Conf. ENAM 98, Bellaire, Michigan, June 1998, AIP Proc 455, Woodbury, New York,1998, p.444
35. A. Gillitzer et al., Z. Phys. A**326**, 107 (1987).
36. V. P. Bugrov and S. G. Kadmensky, Sov. J. Nucl. Phys. **49**, 967 (1989)
37. S. G. Kadmensky and V. P. Bugrov, Phys. of Atomic Nucl. **59**, 399 (1996).
38. M. Karny et al., Z. Phys. A **350**, 179 (1994).
39. E. S. Paul et al., Phys. Rev. C**51**, 78 (1995).
40. C. H. Yu et al., Phys. Rev. C**59** R1834 (1999).
41. E. Maglione, L.S. Ferreira and R.J. Liotta, Phys. Rev. Lett. **81**, 538 (1998).
42. C. J. Gross, private communication 1999.
43. T. Vertse et al, "Proton emission from Gamow resonance", These Proceedings.
44. A. T. Kruppa, B.Barmore, W. Nazarewicz and T.Vertse, "Fine Structure in the Decay of Deformed Proton Emitters: Non-adiabatic Approach", submitted to Phys. Rev. Lett.
45. W. B. Walters, "Nuclear Structure Effects on Proton Emission from Odd-Odd Nuclides", *ibid.*
46. P. E. Haustein (ed.), At. Data Nucl. Data Tables **39**, 185 (1988)
47. P. Möller, J.R. Nix, W.D. Myers and W.J. Swiatecki, At. Data Nucl. Data Tables **59**, 185 (1995)
48. W. Reisdorf, Z. Phys. A**300**, 227 (1981).

Fine Structure in Deformed Proton Emitters

A.A. Sonzogni[1], C.N. Davids[1], P.J. Woods[2],
D. Seweryniak[1], M.P. Carpenter[1], J.J. Ressler[3],
J. Schwartz[1], J. Uusitalo[1], and W.B. Walters[3]

[1] *Argonne National Laboratory, Argonne IL 60439, USA*
[2] *University of Edinburgh, Edinburgh, EH9 3JZ, UK*
[3] *University of Maryland, College Park, MD 20742, USA*

Abstract. In a recent experiment to study the proton radioactivity of the highly deformed ^{131}Eu nucleus, two proton lines were detected. The higher energy one was assigned to the ground-state to ground-state decay, while the lower energy, to the ground-state to the 2^+ state decay. This constitutes the first observation of fine structure in proton radioactivity. With these four measured quantities, proton energies, half-life and branching ratio, it is possible to determine the Nilsson configuration of the ground state of the proton emitting nucleus as well as the 2^+ energy and nuclear deformation of the daughter nucleus. These results will be presented and discussed.

I INTRODUCTION

The proton drip-line meets a region of high quadrupole deformation around the light rare earth nuclei [1]. Calculations by Möller et al. [2] and Aboussir et al. [3] indicate deformations of $\sim$ 0.3-0.35 for proton emitting nuclei ranging from La to Ho. There is sufficient experimental information confirming these ideas. In fig. 1 the energy of the first 2^+ state [4] is plotted as a function of neutron number for a number of different nuclei. Although the information for light rare earths is not particularly abundant, a trend is easily visible and for these nuclei, energies of around 100-150 keV can be expected. The discovery of highly deformed proton emitters [5] in this mass region prompted us to speculate about the existence of fine structure in proton radioactivity, i.e. the population of excited states in the daughter nuclei, since as was just mentioned, the energy of the 2^+ state would be quite low.

Because proton decay rates are very sensitive to the energy of the emitted proton, one may expect that chances of observing fine structure are quite small. For instance, if in the daughter nucleus there is a level at energy E, which we will assume can be populated with the same proton angular momentum as the ground state, the expression:

$$B = T_{1/2}(Q_{gs})/(T_{1/2}(Q_{gs}) + T_{1/2}(Q_{gs} - E)) \tag{1}$$

can be used as a first approximation to estimate the branching ratio of this particular state, where Q_{gs} is the ground-state proton Q-value. Using Distorted Wave calculations - assuming spherical shapes - we conclude that if this hypothetical excited state lies at 100-150 keV, the estimated branching ratio appears to be smaller than 5 % for rare earth nuclei with $Q_{gs} \sim 0.95$ MeV. Deformed nuclei can effectively provide that low value of excitation energy that is needed, and despite the approximative nature of Eqn. (1), some insight can be obtained from it. If one aims at finding fine structure, one should produce the proton emitter with the highest deformation - and therefore lowest 2^+ energy - and measure a proton energy spectrum with high statistics.

With these ideas in mind, an experiment was performed at Argonne National Laboratory in August 1998. The results of this experiment have been published and can be found in ref. [6], a few extra details and some ideas that have been developed since then will be discussed here.

II EXPERIMENT

^{131}Eu nuclei were produced through the reaction ^{58}Ni(^{78}Kr,p4n) at E_{lab}=402 MeV. The Fragment Mass Analyzer [7] was used to separate the recoiling nuclei of interest. The detection system consisted of a parallel grid avalanche counter - giving position and timing information - and a double sided Silicon detector (DSSD), where implant and decay events were recorded. The DSSD was surrounded by extra Si detectors to veto out escape events. The calibration of the DSSD was performed with standard α sources and with protons from ^{167m}Ir. An energy spectrum of decays events which occurred within 100 ms of an implant event is shown in fig. 2 (a). Two proton peaks can be seen, the higher energy one corresponds to the previously identified ^{131}Eu ground-state to ground-state proton transition ($E_p =$ 932(7) keV, $T_{1/2}$=17.8(19) ms), and the lower one having E_p=811(7) keV and $T_{1/2} = 23^{+10}_{-6} ms$.

The fact that the two proton peaks, with an energy difference of 121(3) keV, have nearly identical values of half-life, suggests that we are observing the decay from the ground state of ^{131}Eu through two branches. The higher energy line feeds the ground state of ^{130}Sm while the lower energy one, an excited state of the same nucleus. With such an excitation energy value, the only option for this state is to be the 2^+ of the ground state rotational band. which agrees well with one may expect on the basis of previous measurements, as shown in fig. 1. Additionally, if the Grodzins' equation [9] is used to relate this value of 2^+ energy with β_2, we obtain a β_2 value of 0.33, which agrees quite well with the predictions of Möller et al (β_2=0.331) [2] and Aboussir et al (β_2=0.370) [3].

From the peak areas, the measured branching ratio to the 2^+ state is obtained as:

$$B_{2^+}^m = \frac{Area(2^+)}{Area(2^+) + Area(0^+)}, \qquad (2)$$

and one may expect to deduce additional information about the decay process from the knowledge of this experimental quantity. However, before we can proceed, we need to remember that the decay of the 2^+ state to the ground state in ^{130}Sm can occur by two competing ways, an E2 photon or electron conversion. If a gamma ray is emitted, the probability that it will interact with the DSSD is negligible. However, if an electron is emitted, it may leave some of its kinetic energy in the DSSD, which will sum up with that left by the proton. For instance, if both electron and proton are fully stopped within the DSSD, the energy of the event will be equal to the energy of a ground-state to ground-state event minus the atomic binding energy of the electron. As a consequence, the numbers of events in the lower energy peak is smaller that the total number of 2^+ decays, and $B_{2^+}^m$, has to be considered a lower limit of the intrinsic value of branching ratio B_{2^+}. To obtain the intrinsic value, a correction was applied to the measured one, by means of a MonteCarlo simulation of the decay process and its detection by the DSSD. The input parameters of the simulation were the implantation depth of the ^{131}Eu nuclei, the energy of the 2^+ and the conversion coefficient of the 2^+ to 0^+ transition [8]. The lesson learned from this procedure is that a deep implantation will produce a large correction to $B_{2^+}^m$, since the electrons are more likely to leave some energy in the DSSD. This is certainly a negative aspect of deep implantation, which is counteracted by a large efficiency to detect protons, an important issue when one considers the very low production cross section for these nuclei. A shallow implantation, on the other hand, will minimize the correction factor, but decrease the collection efficiency of the DSSD.

As it is mentioned in ref. [6], most of these data were taken with a deep implantation, with the resulting branching ratio of $24 \pm 5\%$. A simulated proton spectrum can be seen in Fig. 1 (b), which agrees well with the experimental result. On the basis of what was discussed in the Introduction, the B_{2^+} value obtained seems to be particularly high, understanding it reveals a new aspect proton emission from deformed nuclei.

III FORMALISM

The proton decay width, from the initial state $J_i K_i$ feeding the ground state, is given by:

$$\Gamma_{0^+} = \Gamma_{J_f=0^+ j_p \ell_p}^{J_i K_i}, \qquad (3)$$

where j_p and ℓ_p are the total and orbital momentum quantum numbers of the emitted proton and

$$\Gamma_{J_f=0^+j_p\ell_p}^{J_iK_i} = 2\pi |B_{J_f=0^+j_p\ell_p}^{J_iK_i}|^2, \tag{4}$$

with the transition amplitude $B_{J_f=0^+j_p\ell_p}^{J_iK_i}$ given in refs. [5], [10]. For protons feeding the 2^+ state,

$$\Gamma_{2^+} = \sum_{j_p\ell_p} \Gamma_{J_f=2^+j_p\ell_p}^{J_iK_i}. \tag{5}$$

The total decay width is:

$$\Gamma = \Gamma_{0^+} + \Gamma_{2^+}, \tag{6}$$

and the branching ratios are given by:

$$B_{0^+} = \frac{\Gamma_{0^+}}{\Gamma_{0^+} + \Gamma_{2^+}}. \tag{7}$$

$$B_{2^+} = \frac{\Gamma_{2^+}}{\Gamma_{0^+} + \Gamma_{2^+}}. \tag{8}$$

For a proton in a prolate/oblate nucleus, the only good quantum numbers are the projection of its total angular momentum on the symmetry axis, Ω, and its parity, π. The wave function for this proton can be written as:

$$\Phi^{\Omega^\pi} = \sum_{\ell j} C_{\ell j}^{\Omega^\pi} \phi_{\ell j}^{\Omega^\pi}, \tag{9}$$

where ℓ and j are the orbital and total angular momentum of the quasibound proton and the expansion coefficients $C_{\ell j}^{\Omega^\pi}$ satisfy $\sum_{\ell j} |C_{\ell j}^{\Omega^\pi}|^2 = 1$. For instance, a $3/2^+$ wave function will look like:

$$\Phi^{3/2^+} = C_{2,3/2}^{3/2^+}\phi_{2,3/2}^{3/2^+} + C_{2,5/2}^{3/2^+}\phi_{2,5/2}^{3/2^+} + C_{4,7/2}^{3/2^+}\phi_{4,7/2}^{3/2^+} + C_{4,9/2}^{3/2^+}\phi_{4,9/2}^{3/2^+} + \ldots \tag{10}$$

Due to angular momentum conservation, the decay to the 0^+ will require that j_p be equal to 3/2, but if it feeds the 2^+ state, j_p can be equal to 3/2, 5/2 and 7/2. We will examine the contribution of each term to the B_{2^+} value. As was mentioned in the Introduction, the term with $j_p=3/2$ will produce a small contribution (less than 10 %) due to the lower proton energy. The term with $j_p=7/2$ is of not particular significance, since its decay is further hindered by a large centrifugal barrier. The decay with $j_p=5/2$ on the other hand, has the potential of becoming important if $|C_{2,3/2}| \leq |C_{2,5/2}|$. In other words, the hindrance caused by the reduction in proton energy can be overcome with a favorable proton wave function structure.

Similarly, a $5/2^+$ wave function will look like:

$$\Phi^{5/2^+} = C_{2,5/2}^{3/2^+}\phi_{2,5/2}^{3/2^+} + C_{4,7/2}^{3/2^+}\phi_{4,7/2}^{3/2^+} + C_{4,9/2}^{3/2^+}\phi_{4,9/2}^{3/2^+} + \ldots \tag{11}$$

A decay to the 2^+ will require $j_p=5/2$, $7/2$ and $9/2$. This time, however, the condition that favor an enhancement in B_{2^+} is not present.

One then concludes that Nilsson orbitals $[Nn_3\Lambda]\Omega^\pi$, for which one can find an integer number k so that $\Omega = N - 2k - 1/2$, are likely to produce a higher B_{2^+} value than orbitals with $\Omega = N - 2k + 1/2$. For instance the orbitals $[411]3/2^+$ and $[532]5/2^-$ belong to the first group, while $[413]5/2^+$ and $[523]7/2^-$ belong to the second group.

When ^{131}Eu was first observed, it was not possible to discern between the $[411]3/2^+$ and $[413]5/2^+$ orbitals on the basis of proton energy and half-life alone. The results of our calculations are shown in fig. 3. Based on what was discussed on the previous paragraphs, it shouldn't come as a surprise that only the $[411]3/2^+$ orbital can produce a B_{2^+} value as large as the measured one.

IV ENERGY DEPENDENCE OF BRANCHING RATIO

Fine structure with a measurable value of B_{2^+} has been so far found in ^{131}Eu. Other cases could be the lighter Eu isotopes, for which the proton Q-value is also larger. One may wonder then what B_{2^+} values will be found for these nuclei. Fig. 4 shows the proton $T_{1/2}$ and B_{2^+} as a function of the proton energy for the $[411]3/2^+$ orbital. A value of E_{2^+} of 120 keV was assumed for these calculations. These results indicate a rapid increase of the branching ratio with increasing proton energy. In particular, it may eventually reach the point where both ground state and 2^+ are populated with the same intensity.

The experimental observation of this feature will be very interesting. Unfortunately, production cross sections are quite small. Depending on the input parameters one uses in a Statistical Model calculation, we obtain values close to 1 nb or smaller, which will require considerable amounts of beam time. In addition, the proton Q-values could be too large and a development of an experimental technique to study very fast proton emitters would have to be done.

This dependence of B_{2^+} with Q-value is also seen for α particles. Fig. 5 shows experimental values of B_{2^+} for even-even, actinide alpha emitters for different values of Q_α [4]. The points are restricted to relatively constant values of 2^+ energies ($\sim$ 45 keV) to isolate the branching ratio dependence on Q-value. The increasing trend with increasing Q_α is easily seen, but with a more modest magnitude than for protons.

The cause of this difference between alpha and proton emitters can be found in the dependence of the respective decay rates with the Q-values, with protons being far more sensitive. This last statement can be more clearly visualized with the help of the WKB method [11] to calculate the penetrabilities:

$$P = exp(-2G), \tag{12}$$

where G is the so-called Gamow factor. For even-even α-emitters, B_{2^+} is basically given by:

$$R_\alpha = P(\ell_\alpha = 2, Q - E_{2^+})/(P(\ell_\alpha = 0, Q) + P(\ell_\alpha = 2, Q - E_{2^+})). \tag{13}$$

As we have seen, the expression for proton emitters is more complicated, but a rough estimate can be obtained with:

$$R_p = P(\ell_p, Q - E_{2^+})/(P(\ell_p, Q) + P(\ell_p, Q - E_{2^+})). \tag{14}$$

Plots of R_α (left panel) and R_p (right panel) - where for simplicity we took $\ell_p = 2$ - as a function of Q can be seen in fig. 6. As can be appreciated, R_α changes little - from 17% to 30 % - over a large range of Q, R_p on the other hand, experiences a dramatic increase with increasing Q.

In summary, the results from the first observation of proton radioactivity fine structure have been presented and the interpretation discussed. Additionally, the dependence of fine structure probability with proton Q-value was investigated.

ACKNOWLEDGMENTS

This work was supported by the U.S. Department of Energy, Nuclear Physics Division, under Contract W-31-109-ENG-38.

REFERENCES

1. P.J. Woods and C.N. Davids, Annu. Rev. Nucl. Part. Sci. **47**, 541 (1997).
2. P. Möller, J.R. Nix, W.D. Myers, W.J. Swiatecki, At. Data Nucl. Data Tables **59**, 185 (1995).
3. Y. Aboussir et al., At. Data Nucl. Data Tables **61**, 127 (1995).
4. Table of Isotopes (8th edition), R. Firestone and V.S. Shirley, Editors. John Wiley and Sons (1996).
5. C.N. Davids et al., Phys. Rev. Lett. **80**, 1849 (1998).
6. A.A. Sonzogni et al., Phys. Rev. Lett. **83**, 1116 (1999).
7. C.N. Davids et al, Nucl. Instrum. Meth. **B70**, 358 (1992).
8. F. Rösel, H.M. Fries, K. Alder and H.C. Pauli, At. Data Nucl. Data Tables, **21**, 91 (1978).
9. L. Grodzins, Phys. Lett. **2**, 88 (1962).
10. V.P. Bugrov and S.G. Kadmensky, Sov. J. Nucl. Phys. **49**, 967 (1989).
11. S. Aberg, P.B. Semmes and W. Nazarewicz, Phys. Rev. **C56**, 1762 (1997); Phys. Rev. **C58**, 3011 (1998).

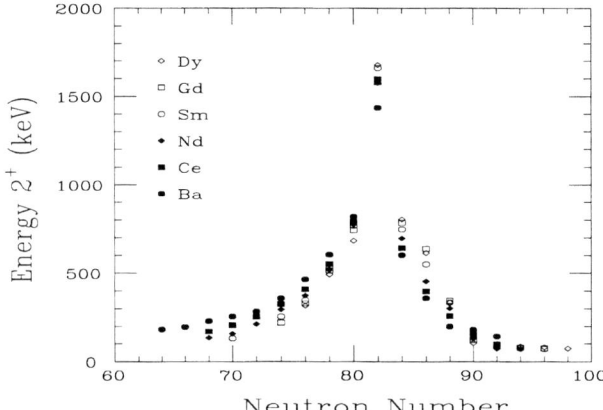

FIGURE 1. Energy of the lowest 2^+ state as a function of neutron number for different nuclei.

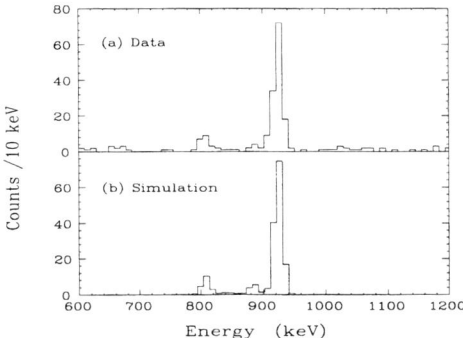

FIGURE 2. Top panel: Measured proton energy spectrum from the decay of ^{131}Eu. Bottom panel: Simulated proton energy spectrum.

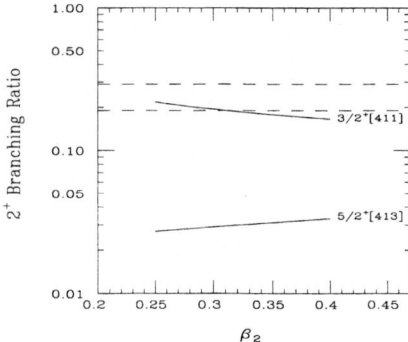

FIGURE 3. Calculated 2^+ decay branching ratios for ^{131}Eu as a function of the quadrupole deformation parameter β_2, based on the [411]3/2$^+$ and [413]5/2$^+$ Nilsson orbitals. The observed branching ratio lies in the band between the dashed lines.

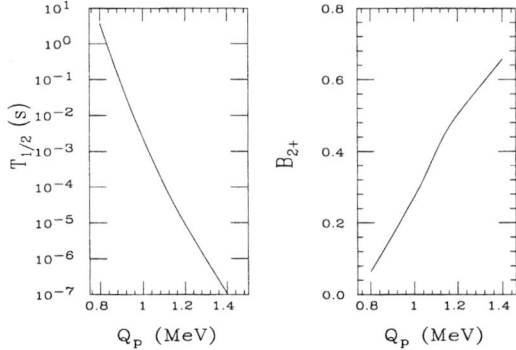

FIGURE 4. Calculated $T_{1/2}$ and B_{2+} values as a function of the proton Q-value.

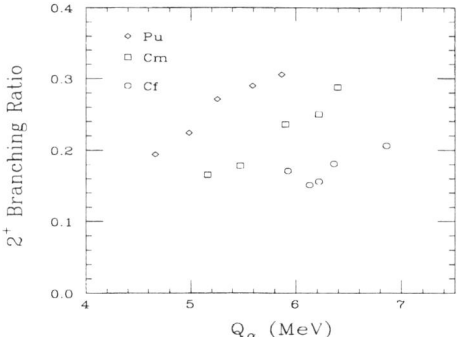

FIGURE 5. Experimental B_{2+} values for different even-even α emitters as a function of Q_α.

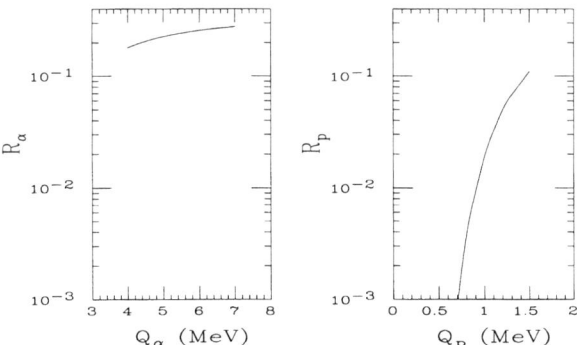

FIGURE 6. Ratios of penetrabilities calculated using the WKB method for α emitters (left) and for proton emitters (right).

First observation of proton emission from ^{117}La

F. Soramel*, A. Guglielmetti†, L. Stroe°,+, L. Müller°, R. Bonetti†,
F. Malerba†, G.L. Poli†, C. Boiano†, A. Andrighetto°, Z.C. Li°,•,
F. Scarlassara°, C. Signorini°, A. Dal Bello°, R. Isocrate°,
Z.H. Liu°,•, M. Ruan°,•, M.Ivascu+, P.Bednarczyk‡, C. Broude^d

*Physics Department and INFN University of Udine, Udine, Italy
†General Physics Insitute and INFN University of Milan, Milan, Italy
°Physics Department and INFN University of Padua, Padua, Italy
°INFN - Laboratori Nazionali di Legnaro, Legnaro, Italy
+NIPNE, Bucharest, Romania
‡IFJ Kraków, Kraków, Poland
•CIAE Beijing, People Republic of China
^dWeizmann Institute of Science, Rehovot, Israel

Abstract. We report the first measurament, at the XTU Tandem + LINAC accelerator of the Laboratori Nazionali di Legnaro, of the decay of the very neutron deficient nucleus ^{117}La using a 310 MeV ^{58}Ni beam on a ^{64}Zn target; the ^{117}La nucleus was populated via the (p,4n) evaporation channel. The Recoil Mass Spectrometer (RMS) was used to select M/q = 117/30 recoils that were implanted in a (40×40) mm^2 Double Sided Silicon Strip Detector (DSSD) detector. The analysis has revealed that ^{117}La decays to ^{116}Ba via proton emission with $E_p = (783 \pm 6)$keV and $T_{1/2} = (20 \pm 5)$ ms. From this result deformation parameters of $\beta_2 = 0.3$ and $\beta_4 = 0.1$ have been deduced for the ^{117}La ground state which was assigned to $J^\pi = 3/2^+$.

INTRODUCTION

In recent years great progress has been achieved in mapping the p-dripline, also thanks to the identification of p-emitting nuclei [1]: The region between the two closed shells Z = 50 and Z = 82 has been mapped and for several elements more than one p-emitting isotope has been identified. For Z ≥ 69 the p-emitting nuclei are spherical and their halflives and decay probabilities are well reproduced by various theoretical approaches; on the other hand, below Z = 68 p-emitters are strongly deformed in their ground state and the interpretation of the experimental results is not straightforward. Identification of many such cases, beside being important in obtaining information on nuclear binding energies of proton and daughter nuclei and in studying nuclear structure effects, is crucial for testing theoretical models.

A couple of these cases have been studied in earlier days [2] and three more, with higher deformation, have been identified recently [3] [4]. Several important odd-Z species ar still lacking to completely map the region 50 < Z < 82. Among these cases there is the ^{117}La nucleus.

EXPERIMENT

At the Tandem + LINAC of the Laboratori Nazionali di Legnaro we recently measured the decay of ^{117}La nucleus. A first short run was done to get a energy calibration in the 1 MeV region from the ^{147}Tm proton decay; for this run we used a 1 pnA 261 MeV ^{58}Ni beam and a 1 mg/cm^2 thick ^{92}Mo target. We measured for a total of 10 hours and the RMS was set for M/q = 147/27$^+$ recoils with E_{rec} = 80 MeV. ^{147}Tm has two proton decaying levels with E_p = 1.051 and 1.119 MeV and $T_{1/2}$ = 560 ms and 360 μs respectively [5]. For these two peaks we got, respectively, the following halflives $T_{1/2}$ = 535(91) ms and $T_{1/2}$ = 295(86) μs in good agreement with [5].

Figure 1a shows the spectrum of all the decays detected in the case of ^{147}Tm nucleus; figure 1b refers to the decays which occurred in a time window of 1 second after the recoil implantation in a specific strip: the proton peak from ^{147}Tm ground

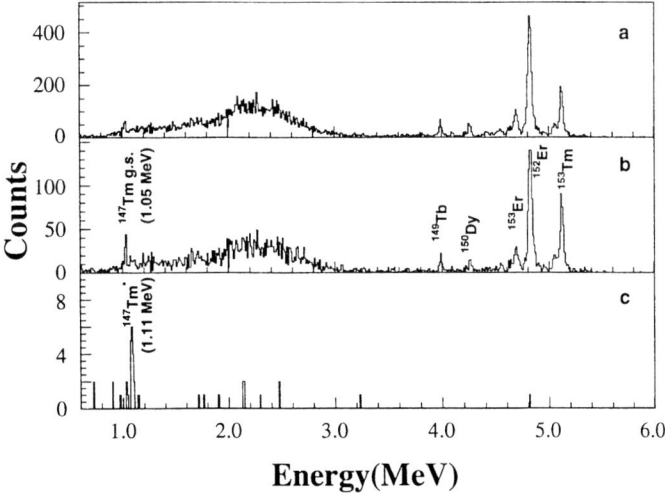

FIGURE 1. Decay spectra from ^{58}Ni + ^{92}Mo. a) All decays in DSSD; b) all decays occurring in a time interval of 1s after implantation of a recoil in a given strip; c) same as b) but with a time interval of 2 ms (see text for further details).

state decay is clearly visible together with some α decay lines from nuclei that reach the DSSD due to RMS M/q ambiguity. Figure 1c is same as 1b, but with a time window of 2 ms, enough to exhauste the short lived isomeric decay of ^{147}Tm: the 1.11 MeV proton line is the only peak present in this spectrum.

During the second part of the run a 310 MeV ^{58}Ni beam of 1.5 pnA bombarded a 1 mg/cm^2 ^{64}Zn target, the produced recoils were guided to the RMS focal plane (RMS was set for M/q = 117/30$^+$ and E$_{rec}$ = 118 MeV); recoils having M/q = 117/30 (or 121/31 or 113/29) were allowed to pass through a 14 mm wide slit and to leave the focal plane detector (a PPAC). After a flight path of 1 m they were implanted in a 60 μm thick DSSD with 40×40 strips each of 1 mm width. Before being implanted in the DSSD the recoils passed through a 1.8 mg/cm^2 thick Ni foil to reduce their energy by a factor of 2. The total measuring time was 36 hours.

Each event (either recoil or decay) is time-stamped by a 4 MHz clock. Signals from the DSSD are treated independently in the x-direction to get both position and energy information, while, in the y-direction, all strips are read together through a delay line with a 2 ns step between each strip, so that in this direction we have only position information (unfortunately two problems arise for this signal: a) the position resolution deteriorates for low energy events resulting in poor pixel definition for such events, b) a strong noise present in the experimental area

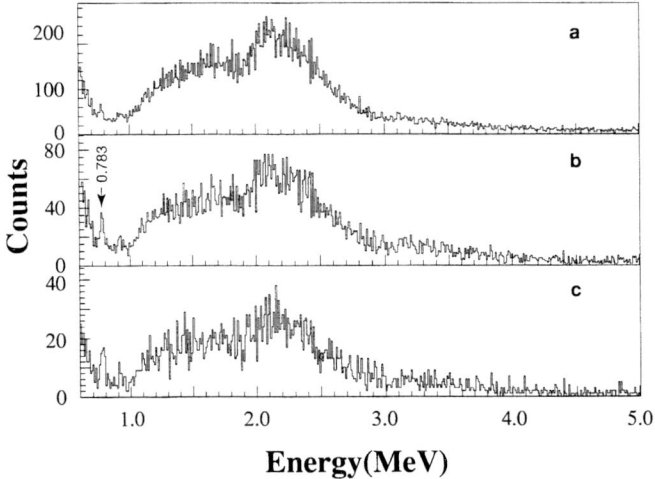

FIGURE 2. Decay spctra from ^{117}La: a) all decays in DSSD; b) all decays occurring in a 100 ms time interval after the recoil implantation in a given strip; c) same as b, but requiring M/q = 117/30$^+$.

could only be avoided by setting thresholds at about 1 MeV, therefore cutting almost all signals below this value).

RESULTS

Figure 2a shows decay spectra for all events in the case of ^{117}La; at low energy one can see a peak that becomes more prominent in figure 2b which includes only decays occurring in a time window of 100 ms after the recoil implantation in a strip; figure 2c shows decays occurring in the same time interval as in figure 2b, but in coincidence with recoils inthe central part of the M/q = 117/30$^+$ peak.

Since we did not use any detector downstream of the DSSD for vetoing β particles, the region between 1.5 and 4 MeV shows a lot of events due either to β particles scattered in the DSSD or to β delayed particles (most likely protons) coming, for example, from ^{117}Ba [6]. From the fit of the low energy peak we get a detected energy of $E_{det} = (775\pm3)$ keV, from which we deduce $E_p = (783\pm6)$ keV corresponding to $Q_p = (790\pm6)$ keV and to $Q_{p,nucl} = (800\pm6)$ keV (we have taken into account 10 keV for the screening effect).

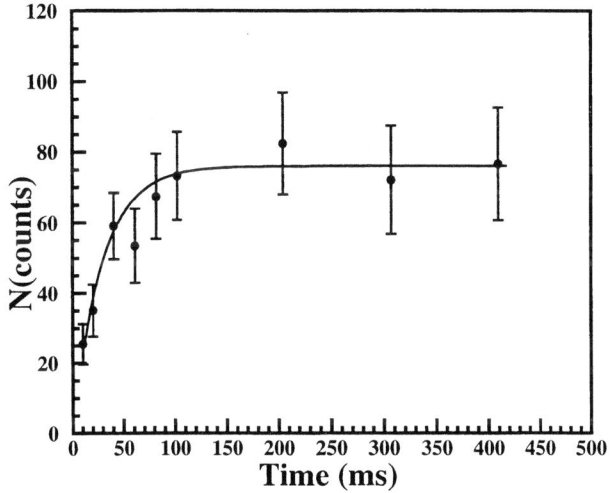

FIGURE 3. Experimental points for the 783 keV proton line halflife in ^{117}La and the fitted decay curve.

The time analysis of the line is given in figure 3 where the total number of the decays is shown as a function of time. The peak was always taken from the spectra corresponding to "strip condition": in fact, for the reasons given above, the y signal is very inefficient at low energy precluding the possibility of making a pixel analysis. The "strip condition" is nonetheless good since our total counting rate on the DSSD detector was about 1 count per 160 ms per strip (1 count per 1.48 s per strip in the ^{147}Tm case); fitting the points in figure 3 we get a halflife of $T_{1/2} = (20\pm5)$ ms and about 70 counts for the 783 keV proton peak.

Considering an efficiency of 60% for the DSSD and an efficiency between 5% and 10% for the RMS we get a cross section between 150 and 300 nb for ^{117}La, which is in good agreement with previously measured cross sections for (p,4n) evaporation channel in this region [3] [4].

If we now compare our results with existing mass predictions and calculations, we see that Audi and Wapstra give Q_p = (470±1026) keV [10], while Möller and Nix [7] predict Q_p = 501 keV, Liran and Zeldes [8] Q_p = 1011 keV and Jänecke and Masson [9] Q_p = 1021 keV. Our result lies in between these evaluations. From [7] the expected deformation parameters are β_2 = 0.29 and β_4 = 0.1. The nucleus should be strongly deformed and spherical calculations could be useful only for delimiting the range of the possible halflives. Using the code PREM [11], based on a simple WKB spherical model, with Q_p = 800 keV we get $T_{1/2}$ = 0.33 ms if the ground state is a $d_{5/2}$ and $T_{1/2}$ = 110 ms if the ground state is a $g_{7/2}$ level, these being the only candidates for the ground state configuration in ^{117}La.

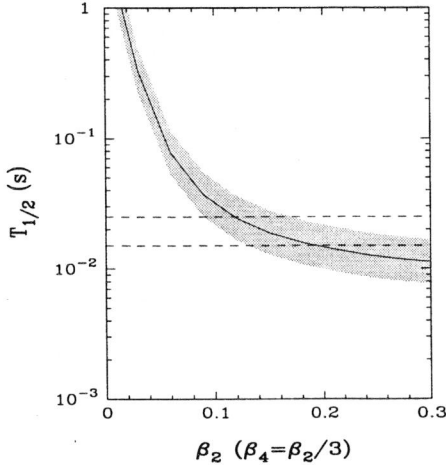

FIGURE 4. Halflife calculations for ^{117}La nucleus as a function of the deformation parameter. The gray band reflects the uncertainties in the experimental Q_p value. The dashed lines representthe measured $T_{1/2}$ and its experimental error.

At this point it is clear that we need a more sophisticated model to reproduce the experimental results and to get structure information for the ^{117}La nucleus. Using the approach of Maglione et al. [12] we can establish the Fermi surface for our nucleus: at large deformation (β_2 = 0.3 and β_4 = 0.1) the ground state is probably a K=3/2$^+$ from the $d_{5/2}$ level. The result for this level is shown in fig. 4.

The calculations agree with experimental result if one assumes a spectroscopic factor $S = 0.5$ (reasonable in the strong coupling limit) not included in figure 4. Our data do not show evidence of direct α emission from ^{117}La therefore we attribute a 100% branching to the proton decay.

Further investigations are needed in order to better understand ^{117}La decay and to possibly reveal its fine structure.

REFERENCES

1. Woods, P.J., and Davids, C.N., *Ann. Rev. Nucl. Part. Sci.* **47**, 541 (1997).
2. Hoffman, S., *Nuclear Decay Modes*, ed. Poenaru, D.N., IOP Publishing, 1996, 143.
3. Davids, C.N., Woods, P.J., Seweryniak, D., Sonzogni, A.A., Batchelder, J.C., Bingham, C.R., Davinson, T., Henderson, D.J., Irvine, R.J., Poli, G.L., Uusitalo, J., and Walters, W.B., *Phys. Rev. Lett.* **80**, 1849 (1998).
4. Rykaczewski, K., Batchelder, J.C., Bingham, C.R., Davinson, T., Ginter, T.N., Gross, C.J., Grzywacz, R., Karny, M., MacDonald, B.D., Mas, J.F., McConnel, J.W., Piechaczek, A., Slinger, R.C., Toth, K.S., Walters, W.B., Woods, P.J., Zganjar, E.F., Barmore, B., Ixaru, L.Gr., Kruppa, A.T., W.Nazarewicz, W., Rizea, M., and Vertse, T. *Phys. Rev.* **C60**, 011301 (1999).
5. Sellin, P.J., Woods, P.J., Davinson, T., Davies, N.J., Livingston, K., Page, R.D., Shotter, A.C., Hoffman, S., And James, A.N., *Phys. Rev.* **C47**, 1933 (1993).
6. Tideman-Peterson, P., Kirchner, R., Klepper, O., Roeckl, E., Schardt, D., Plochocki, A., Zylicz, J., *Nucl.Phys.* **A437**, 342 (1985).
7. Möller, P., Nix, J.R., Myers, W.D., Swiatecki, W.J., *At. Data Nucl. Data Tables* **59**, 185 (1995).
8. Liran, S., Zeldes, N., *At.Data Nucl. Data Tables* **17**, 431 (1976)
9. Jänecke, J., and Masson, P.J., *At.Data Nucl. Data Tables* **39**, 289 (1988).
10. Audi, G., and Wapstra, A.H., *Nucl. Phys.* **A595**, 409 (1995).
11. Poli, G.L., PhD thesis, University of Milan, Milan, Italy, 1998.
12. Maglione. E., Ferreira, L.S., Liotta, R.J., *Phys. Rev.* **C59**, R589 (1999) and these Proceedings.

Nuclear Structure Effects in the Proton Decay of Odd-Odd Nuclides

W.B. Walters

Department of Chemistry, University of Maryland, College Park, MD 20742, USA

Abstract. The systematic study of proton decay has provided considerable insight into the properties of nuclides beyond the proton drip line. In particular, the results have provided masses for odd-Z nuclides and spectroscopic factors that have been useful in charting the shifts of nuclear single-particle orbitals along the proton drip line. Most of the invesigations have, however, been focused on properties of odd-Z proton emitters whose daughter nuclides are even-even nuclides. In this paper, the relationship between the structure of odd-odd nuclides and their odd-neutron daughters will be discussed. These effects will be illustrated by specific examples for structure and decay of odd-odd Sb nuclides.

INTRODUCTION

Many of the aspects of proton decay were discussed in a recent review by Woods and Davids (1). The interpretation of the decay of odd-Z, odd-mass proton emitters reveals is relatively straightforward as the daughter state is usually the 0^+ ground state of an even-even nuclide. Hence, the energy and half-life can be used to determine parity and ground (isomeric)-state energy of the parent nuclide as well as make a good estimate of the spin of the parent and extract a spectroscopic factor for the transition.

For odd-odd proton emitters, the situation is considerably more complex as the final state is a low-energy, but not necessarily ground state level in an odd-neutron nuclide. In this paper, the structure and possible proton decay properties of the spherical odd-odd Sb nuclides with protons and neutrons in the $d_{5/2}$ and $g_{7/2}$ orbitals are discussed and used to illustrate how proton decay in other mass regions can be interpreted.

ODD-ODD NUCLIDES NEAR DOUBLE-MAGIC NUCLIDES

In Figure 1 are shown the structures of two well-studied odd-odd nuclides that lie near double magic nuclides. First, the levels of ^{210}Bi, with one $h_{9/2}$ proton and one $g_{9/2}$ neutron beyond double-magic ^{208}Pb are shown. The energies of adjacent spin states show large staggering when the multiplet is looked at as a whole. However, as can be observed, the T = 0 and T = 1 states form separate systems where the T = 0 states monotonically increase in energy and the T = 0 states exhibit a parabolic shape with minimum energy for the antialigned 1⁻ and aligned 9⁻ levels. Also shown are the levels of ^{48}Sc which has one $f_{7/2}$ proton particle and one $f_{7/2}$ neutron hole coupled to double-magic ^{48}Ca. Again, the two separate systems can be observed, a monotonic T = 1 system with the order inverted, and an inverted parabolic shape for the T = 0 states. These systems are discussed in some detail by Moinster et al., (2)

MULTIPLET INVERSION IN ODD-ODD NIOBIUM NUCLIDES

For nuclei away from double magic nuclides, the approaches described by Vladimir Paar (3) and by Van Maldeghem and Heyde (4) are of interest. Paar described the splitting of proton-neutron multiplets in odd-odd nuclei in an expression which includes a product of the terms $(U_p^2 - V_p^2) \cdot (U_n^2 - V_n^2)$ when plotted versus I(I+1). A well studied example of the I(I + 1) dependence of such splitting is shown in Figure 2 for the levels of ^{114}In. And, the occupancy dependence is illustrated in Fig. 3 where the structure for the odd-odd Nb nuclides just beyond ^{90}Zr are shown. Hence, ^{92}Nb has a single $g_{9/2}$ proton coupled to a single $d_{5/2}$ neutron and the resulting multiplet is seen to have the antialigned 2^+ and aligned 7^+ levels showing the lowest energies. The addition of a pair of neutrons for ^{94}Nb results in the term $(U_n^2 - V_n^2)$ becoming ~ 0 with the result that the positions of the members of the multiplet are nearly degenerate and their actual energies are a consequence of second order effects. Note that five of the six levels are squeezed into a very narrow energy range. Finally, with the addition of another pair of neutrons to make ^{96}Nb, the $d_{5/2}$ orbitals are now 5/6 full and a hole state. The result is an inversion of the parabola observed for ^{92}Nb with the state 6^+ = $9/2^+ + 5/2^+$ -1 being the ground state. Notice that for ^{94}Nb, the T = 1 states ARE inverted while the T = 0 states are still parabolic with the 5^+ state at highest energy.

ODD-ODD ANTIMONY NUCLIDES

The single-particle orbitals that are expected to be occupied by protons and neutrons beyond the double-magic nuclide ^{100}Sn are shown in Figure 4. As can be observed, the large split between the $g_{7/2}$ and $d_{5/2}$ proton orbitals suggests that the low energy multiplets in the odd-odd Sb nuclides will dominated by two multiplets, the $\pi d_{5/2} \cdot \nu d_{5/2}$ multiplet and the $\pi d_{5/2} \cdot \nu g_{7/2}$ multiplet as shown in Fig. 4. For ^{102}Sb it is possible to predict with some confidence that low-energy isomeric 0^+ and 5^+ states will be found. However, the structure of ^{104}Sb is less certain. With ~3 neutrons in the $d_{5/2}$ orbitals that multiplet could easily be quenched to such an extent that the 1^+ and 6^+ isomeric levels will lie at the lowest energies in ^{104}Sb. Finally, both a projection of expected levels and the known levels of ^{106}Sb are shown. It can be seen that many levels are nearly degenerate and that the increased occupancy of the $\nu g_{7/2}$ orbital has resulted in the quenching of the $\pi d_{5/2} \cdot \nu g_{7/2}$ multiplet with the effect that the 6^+ level is now significantly elevated above the lowest-energy levels of the now-inverted

$\pi d_{5/2} \cdot \nu d_{5/2}$ multiplet. The 7^+ member of the $\pi g_{7/2} \cdot \nu g_{7/2}$ multiplet is shown and seen to reflect the ~ 850 keV position of the $\pi g_{7/2}$ level relative to the $\pi d_{5/2}$ ground state.

Next the possible proton decay of these nuclides is discussed. For ^{102}Sb, the predicted proton decay would be easily be seen to be the emission of $\ell = 2$ $d_{5/2}$ protons from both the expected 0^+ and 5^+ isomers that would leave the nucleus in its expected $d_{5/2}$ neutron ground state. Much more interesting would be the possible proton decay of ^{104}Sb. If, indeed, 1^+ and 6^+ isomers are present, it can be seen that the configuration of the 1^+ level would be a mixture of the $(\pi d_{5/2} \cdot \nu d_{5/2})_{1+}$ and the $(\pi d_{5/2} \cdot \nu g_{7/2})_{1+}$ levels. As a consequence, that level could decay to BOTH the $d_{5/2}$ and $g_{7/2}$ levels in ^{103}Sn by emission of protons with $\ell = 2$ as shown. As can be seen by the half lives calculated by WKB methods, the $\ell = 2$ transition to the $d_{5/2}$ ground state would be dominant. On the other hand, it can be seen that the 6^+ level could decay to the $d_{5/2}$ ground state ONLY by emission of an $\ell = 4$ $g_{7/2}$ proton. The 6^+ level could also decay to the $g_{7/2}$ neutron level estimated to lie at 250 keV, with an $\ell = 2$ proton, but that half-life is 100 days for a 350-keV proton.

Shibata et al., (5) show half lives of 600(200) ms and 440(+150/-110) ms for ^{106}Sb, and ^{104}Sb, respectively. Subsequently, Sohler et al., (6) established a low spin (2+) for ^{106}Sb from which it is possible to estimate a minimum log ft value of 4.3 for that beta decay to the 2^+ level in ^{102}Sn. For ^{104}Sb, the increased Q_{EC} value estimated by Shibata et al., combined with a similar log ft value, would indicate a half-life of ~200 ms for decay of a low-spin isomer and 600 ms for a 6^+ level. The latter decay would have a reduced E_β as it would only be possible to populate levels above ~2 MeV. Notice that if the spin of the observed isomer is 5^+, then $\ell = 2$ proton decay would be possible and, for a 600-keV proton energy, a 7-ms half-life would be expected.

In a search for proton emission from ^{108}I, Page et al., (7) observed only alpha decay which would populate levels in ^{104}Sb. No proton emission from ^{104}Sb was found to correlate with the alpha decay of ^{108}I. As the ^{108}I was produced in a heavy ion reaction, it would be expected that if isomers exist in ^{108}I, the high-spin isomer would be most strongly populated. And that decay of such an isomer would populate the high-spin isomer in ^{104}Sb. Hence, the above analysis of the structure of ^{104}Sb is consistent with the results of the ^{108}I experiment, and could, therefore, be seen to be a consequence of the configuration of the state and the position of the daughter state, not just a consequence of the mass and the Q_p value, but does not exclude the possibility of the existence of a low-spin isomer in ^{104}Sb that undergoes proton decay with a half-life in the millisecond range. On the other hand, it can be expected that proton decay from a low-spin isomer would depend critically on the Q_p value as the beta decay half-life can be expected to be in the range near 200 ms. Indeed, the 440 ms value noted above with the large range of uncertainty could be a composite of approximately equal proportions of isomers with individual half lives of 200 and 600 ms. Were that to be the case, then Q_p would be limited to values below 500 keV.

DEFORMED ODD-ODD NUCLIDES

The unique aspects of the decay of highly deformed odd-odd nuclides is illustrated in Fig. 5 by reference to ^{80}Y that is not a possible proton emitter, but whose possible proton decay does show what can happen. The spins and parities for the ground state and isomers in ^{80}Y are consistent with the Gallagher-Moszkowski rules. (8) The key difference between proton decay from "spherical" and "highly deformed" odd-odd nuclides is the possible decay fine structure in the decay of a high-spin isomer that is not likely for a low-spin isomer, owing to the low energies of the rotational states that will lie close to the bandheads. For the 1$^-$ isomer, an $\ell = 0$ decay is possible to the 3/2$^-$ bandhead while decay to the 5/2$^-$ and 7/2$^-$ would take place with both LOWER energy and HIGHER $\ell = 2$ transitions. Hence, the transitions to the higher-spin levels would be unlikely to compete with the $\ell = 0$ transition to the ground state. On the other hand, the 4$^-$ isomer would require an $\ell = 2$ transition to populate the bandhead and first rotational state, but could decay via an $\ell = 0$ transition to the higher-energy levels. Thus, the lower energy for the transition to the 7/2$^-$ level would be offset by the lower ℓ value. Hence, it can be seen that considerable fine structure would be possible for the decay of the higher-spin isomer, whereas, little fine structure would be expected for the decay of the low-spin isomer. Similar behaviour is also noted for decay to the positive parity levels, the high-spin isomer can populate a wide range of levels with $\ell = 1$ transitions, but the low-spin isomer can only populate the bandhead with an $\ell = 1$ transition. In a recently observed case of fine structure for the decay of the 3/2$^+$ ground state of ^{131}Eu, the proton decay to the 2$^+$ excited state could compete as $\ell = 0$ components were possible, whereas only $\ell = 2$ decay was possible to the 0$^+$ level. The opposite can be seen for possible fine-structure for decay of the 1/2+ isomer of ^{141}Ho where the ground state decay would be $\ell = 0$, but decay to the higher-energy 2$^+$ level would be inhibited by both higher ℓ and lower energy.

ACKNOWLEDGEMENTS

Numerous helpful discussions with Professors Vladimir Paar and Kris Heyde are gratefully acknowledged. This work was supported by the U.S. Department of Energy.

REFERENCES

1. P. J. Woods and C. N. Davids, Ann. Rev. Nucl. Part. Sci. **47**, 541 (1997).
2. A. Molinari, *et al.*, Nucl. Phys. **A239**, 45 (1975).
3. V. Paar, Nucl. Phys. **A331**, 16 (1979).
4. J. Van Maldegham and K. Heyde, Phys. Rev. **C32**, 1067 (1985)

5. M. Shibata, *et al.*, Phys. Rev. **C55**, 1715 (1997).
6. D. Sohler, *et al.*, Phys. Rev. **C59**, 1324 (1999).
7. R. D. Page, *et al.*, Phys. Rev. **C49**, 3312 (1994).
8. C. J. Gallagher, Jr. and S. A. Moszkowski, Phys. Rev. **111**, 1282 (1958).
9. A. A. Sonzogni, *et al.*, Phys. Rev. Lett. **83**, 1116 (1999)

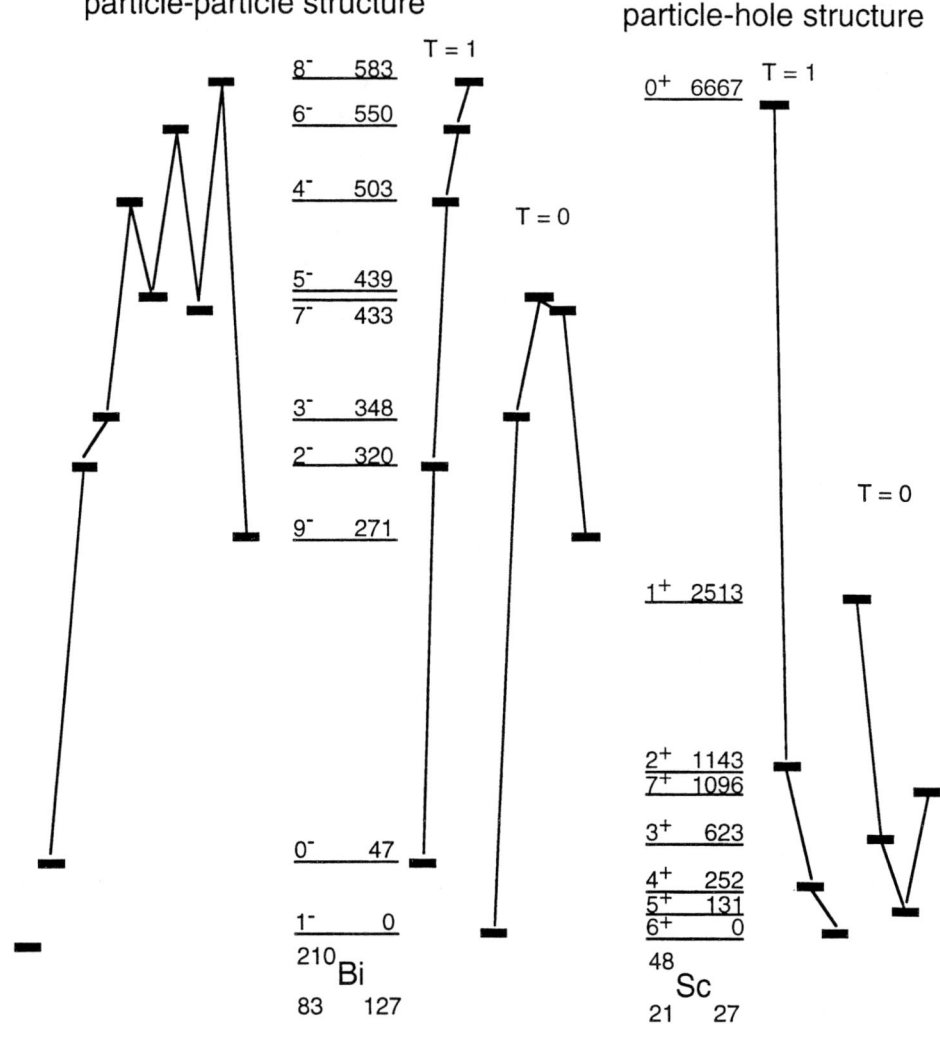

Figure 1. Odd-odd multiplets near double-magic nuclides.

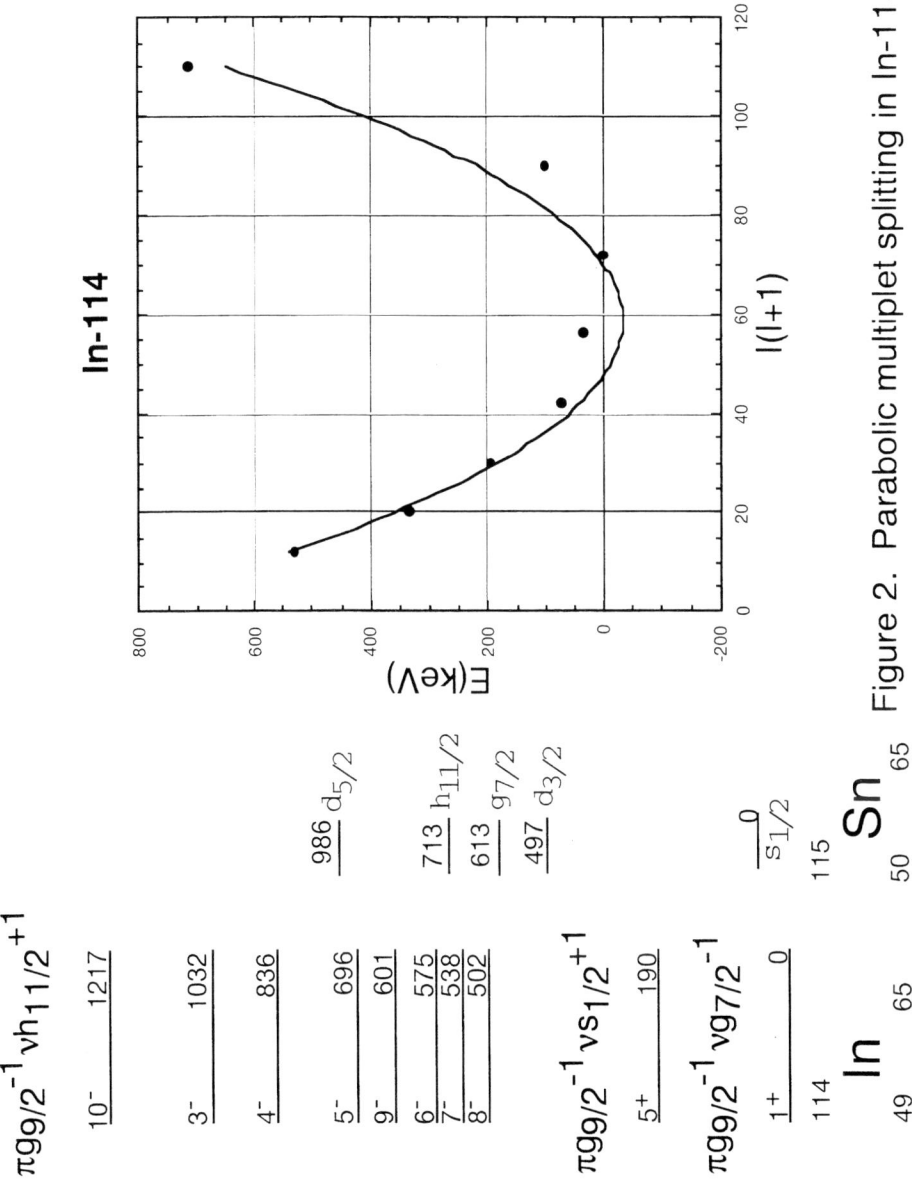

Figure 2. Parabolic multiplet splitting in In-114.

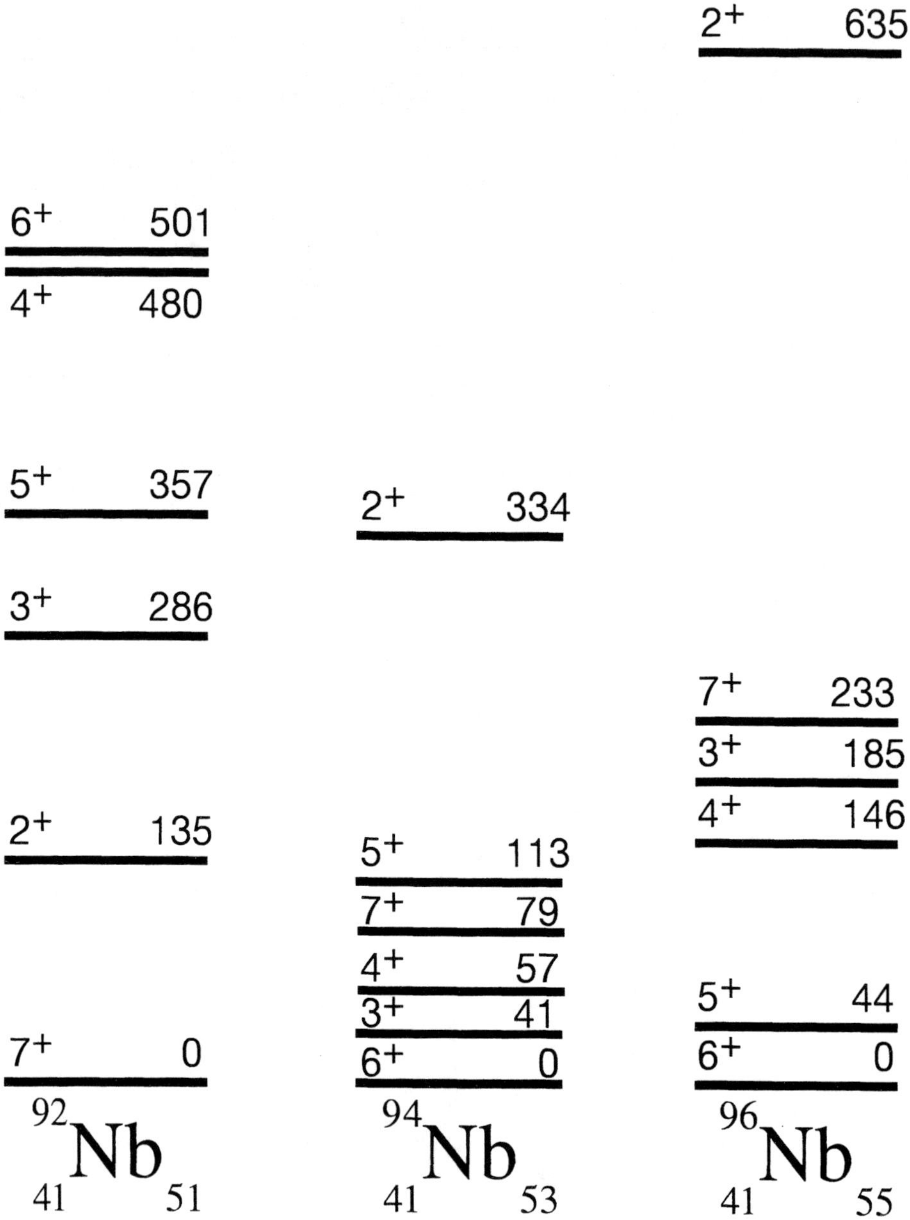

^{92}Nb (41, 51)
- 6^+ — 501
- 4^+ — 480
- 5^+ — 357
- 3^+ — 286
- 2^+ — 135
- 7^+ — 0

^{94}Nb (41, 53)
- 2^+ — 635
- 2^+ — 334
- 5^+ — 113
- 7^+ — 79
- 4^+ — 57
- 3^+ — 41
- 6^+ — 0

^{96}Nb (41, 55)
- 7^+ — 233
- 3^+ — 185
- 4^+ — 146
- 5^+ — 44
- 6^+ — 0

Figure 3. Inversion of the $\nu d_{5/2} \cdot \pi g_{9/2}$ multiplet in the odd-odd Nb nuclides.

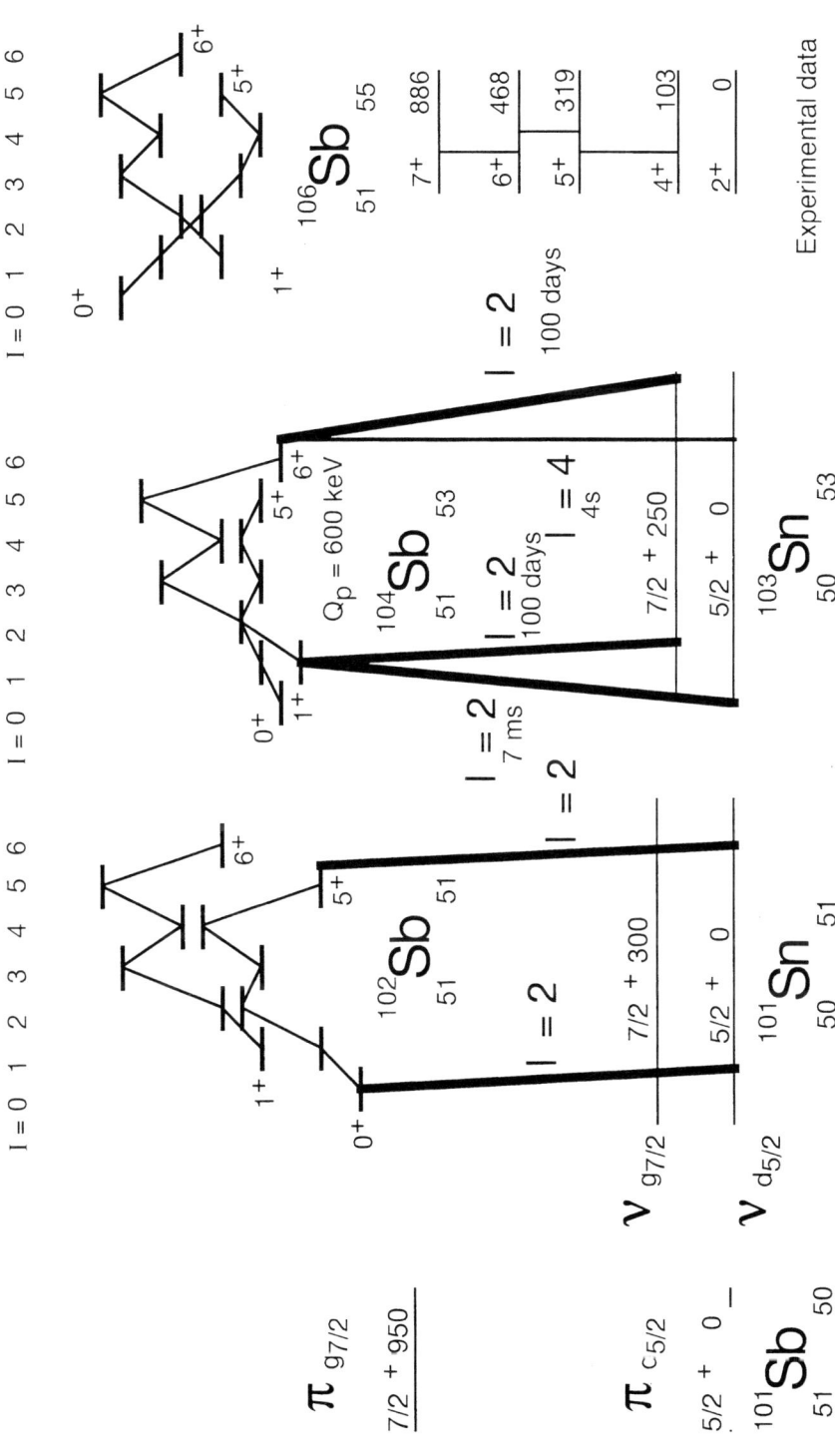

Figure 4. Structure and possible proton decay of odd-odd Sb nuclides.

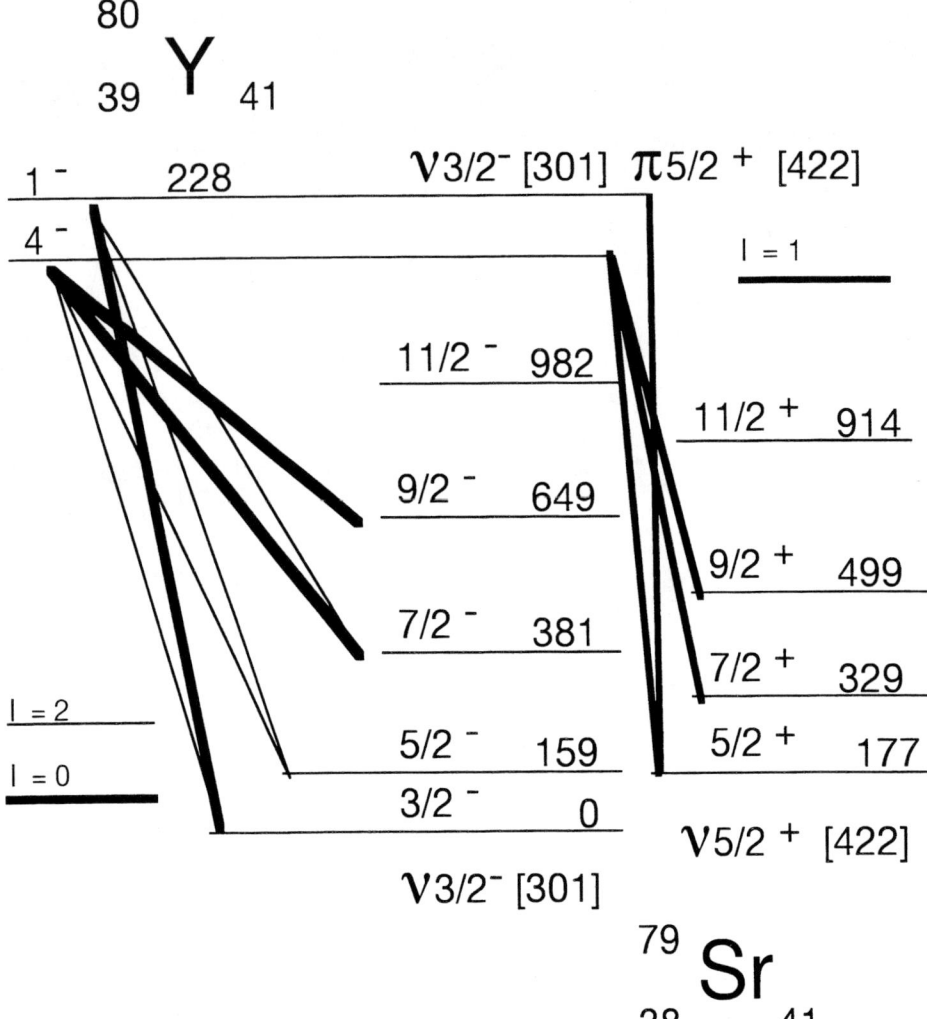

Figure 5. Structure and illustrative proton decay for Y-80 isomers.

A Search for Neutron Single-Particle States Populated Via Proton Emission from ^{146}Tm

T. N. Ginter[1], J. C. Batchelder[2], C. R. Bingham[3,4], C. J. Gross[4,5], R. Grzywacz[3,6], J. H. Hamilton[1], Z. Janas[6] A. Piechaczek[7], A. V. Ramayya[1], K. Rykaczewski[4,6], W. B. Walters[8], and E. F. Zganjar[7]

[1] *Department of Physics and Astronomy, Vanderbilt University, Nashville, Tennessee 37235*
[2] *UNIRIB, Oak Ridge Associated Universities, Oak Ridge, Tennessee 37831*
[3] *Department of Physics and Astronomy, University of Tennessee, Knoxville, Tennessee 37996*
[4] *Physics Division, Oak Ridge National Laboratory, Oak Ridge, Tennessee 37831*
[5] *Oak Ridge Institute for Science and Education, Oak Ridge, Tennessee 37831*
[6] *Institute of Experimental Physics, Warsaw University, PL-00681 Warsaw, Hoza 69, Poland*
[7] *Department of Physics and Astronomy, Louisiana State University, Baton Rouge, Louisiana 70803*
[8] *Department of Chemistry, University of Maryland, College Park, Maryland 20742*

Abstract. We studied the proton emission from $^{146}_{69}$Tm$_{77}$ and observed three new transitions. New transitions at 0.89 and 0.93 MeV have half-lives similar to that of the previously observed transition at 1.19 MeV, while a new transition at 1.02 MeV has a half-life similar to that of the previously observed transition at 1.12 MeV. These new transitions indicate the population of excited neutron single-particle states in $^{145}_{68}$Er$_{77}$.

INTRODUCTION

A previous study [1] of the proton emission from ^{146}Tm identified two proton transitions: one at 1119 ± 5 keV with a half-life of 235 ± 27 ms and the other at 1189 ± 5 keV with a half-life of 72 ± 23 ms. Both transitions are interpreted as originating from $h_{11/2}$ orbitals based on the comparison of their measured half-lives to predictions from simple WKB calculations.

In ^{145}Er, the proton decay daughter of ^{146}Tm, the three active neutron single-particle states (one of which is the ground state) are the $s_{1/2}$, $d_{3/2}$, and $h_{11/2}$ states. Calculations performed based on the microscopic-macroscopic model presented in Ref. [2] indicate that these three states lie close to each other — within an energy range of 200 keV. It is thus conceivable that these states can be populated in the decay of ^{146}Tm by proton emission. We have re-studied the proton emission from ^{146}Tm to search for proton transitions populating excited states in the daughter and to look for new proton emitting states with half-lives down to the microsecond time scale.

THE EXPERIMENT

We produced ^{146}Tm via the $p3n$ reaction channel using a beam of ^{58}Ni on a ^{92}Mo target of thickness 0.91 mg/cm^2. The beam was delivered at an energy of 292 MeV and with an intensity of 10 particle nA by the Holifield Radioactive Ion Beam Facility's 25 MV tandem accelerator located at Oak Ridge National Laboratory. The total beam-on-target time was about 72 hours.

We used the HRIBF Recoil Mass Spectrometer (RMS) to separate mass 146 ions for implantation into a double-sided silicon strip detector (DSSD) to study their subsequent decay by proton emission. A thin carbon foil was placed 10 cm downstream from the target position in front of the RMS to re-establish charge state equilibrium for any recoils that may have decayed by internal conversion before reaching the foil to prevent such recoils from being lost in the spectrometer. The RMS was scaled to accept recoils with a central energy of 90 MeV. The RMS was run in the converging mass mode to deliver two charge states of mass 146 recoils — 26$^+$ and 27$^+$ — into the DSSD. Baffles at the focal plane were used to prevent recoils from masses other than 146 from reaching the DSSD.

A multi-wire, gas-filled position sensitive avalanche counter (PSAC) was used at the focal plane in front of the DSSD. The PSAC not only provided mass identification of the recoils based on their observed positions, but it also was used to distinguish between decay and implantation events in the DSSD by whether or not these events were observed in coincidence with events from the PSAC.

Signals from the DSSD were processed using electronics provided by the University of Edinburgh. This system features Silena ADC's with FERA readout. The components of this system are discussed in detail in Ref. [3]. Use of this setup for observing proton activity at the RMS focal plane — particularly its effectiveness for observing short-lived activities with half-lives down to a few microseconds — has been discussed on several occasions [4–7].

The energy with which recoils were implanted into the DSSD was either well below 20 MeV or above 60 MeV depending on whether or not a 2.27 mg/cm^2 thick Cu foil was used between the PSAC and DSSD to reduce the energy of the recoils.

RESULTS

The overall gain in counts obtained in the two strong proton peaks from this experiment was about a factor of 20 over the previous work [1]. This gain resulted from extending the running time by a factor of four, from doubling the beam current used, from the collection of two charge states of mass 146 at the focal plane, and from use of a thicker target.

Although we obtained no evidence for new proton transitions with half-lives on the microsecond time scale, we did observe three new transitions at lower energies than the ones previously identified. Two transition at 0.89 and 0.93 MeV have half-lives in the range of 100 ms; the other transition at 1.02 MeV has a half-life in the range of 200 ms. This experiment will also provide more precise half-life values for the two previously identified transitions. Note that all new energy and half-life values stated here are preliminary.

To ensure that the new decay peaks were not somehow created by the way protons from the strong peaks at 1.12 and 1.19 MeV escaped from the surface of the DSSD on which the recoils were implanted, we varied the implantation depth of the recoils by using or not using the Cu foil to reduce the energy of the recoils. The new peaks remained under both experimental conditions.

Figure 1 shows the decay events observed in the DSSD within the first 100 ms after the arrival of a recoil when no Cu foil was used in front of the DSSD. The peaks above 4 MeV are from the α-decay of heavier nuclei. These nuclei arise from isotopic impurities within the target and reach the DSSD because they have mass-to-charge ratios similar to those of the mass 146 recoils.

As Fig. 1(a) illustrates, the background arising from escaping α-particles peaks at an energy well above the proton transitions near 1 MeV; very little of this background is present around the proton peaks. When the degrading foil was used, the α-escape background peaked on top of this crucial energy range. Deep implantation of recoils shifts the peak in the α-escape background to higher energies because the α-particles deposit a larger portion of their energy before they reach the surface of the DSSD to escape.

A disadvantage of deep implantation is the shift in energy observed for decay events which occur within a couple of hundred microseconds after the recoil is implanted into the DSSD. This shift occurs because of the extra time it takes the decay amplifiers to recover from the larger overload resulting from the higher implantation signal. For the case of ^{146}Tm, this overload effect turns out not to be an important issue since no short-lived proton transitions are present.

Figure 1(b) provides an expanded view of the energy range of interest. The three new peaks are clearly visible.

DISCUSSION

It should be noted that the previous work on ^{146}Tm is consistent with our observation of new proton transitions. Extra counts are visible in the region just below the proton peak at 1.12 MeV in Fig. 1 of Ref. [1].

The fact that the half-lives of the new transitions at 0.89 and 0.93 MeV seem to match that of the transition at 1.19 MeV suggests that all three transitions originate from the same state in ^{146}Tm; this implies that the new transitions are to

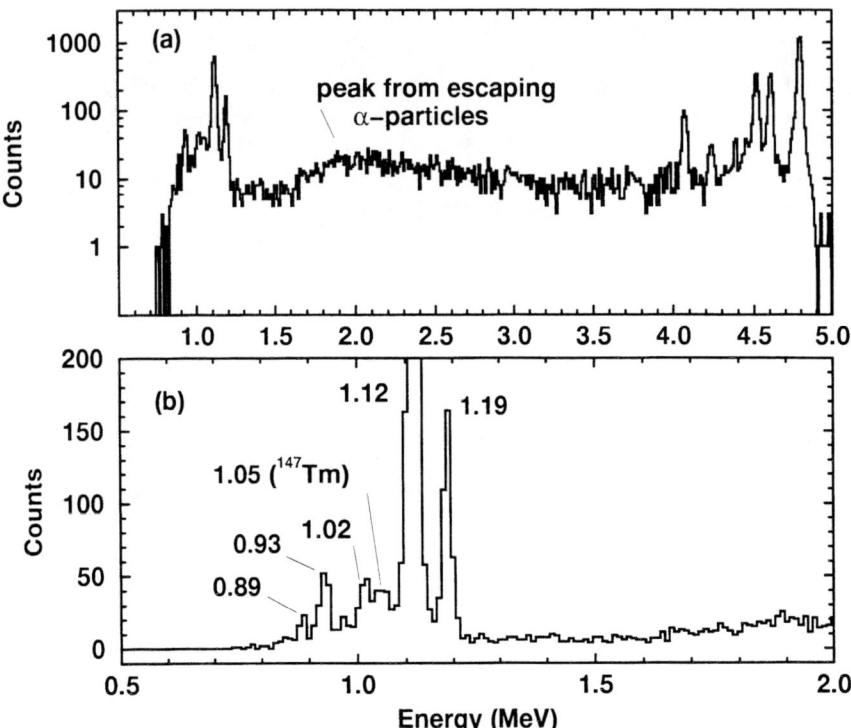

FIGURE 1. Decay events occurring within the first 100 ms after the arrival of a recoil at the DSSD. This data was generated with no Cu foil used in front of the DSSD to reduce the energy of the implanted recoils. The peaks above 4 MeV are from α-decay events, while those around 1 MeV are from proton emission events. The peak in the background from escaping α-particles is visible in (a) and clearly lies above the proton peaks. An expanded view of the energy range containing the ^{146}Tm proton peaks is given in (b).

TABLE 1. Simple WKB calculation of half-life as a function of angular momentum l for the five ^{146}Tm proton transitions. The new experimental energy and half-life values listed here are preliminary.

Energy (MeV)	Measured $t_{1/2}$ (ms)	WKB Half-Life Estimate (ms)		
		$\Delta l = 0$	$\Delta l = 2$	$\Delta l = 5$
0.89	~ 100 [a]	32	280	740,000
0.93	~ 100 [a]	7.2	63	160,000
1.02	~ 200 [a]	0.34	3.0	7,400
1.12	235 ± 27 [b]	0.018	0.16	370
1.19	72 ± 23 [b]	0.0029	0.025	57

[a] Preliminary half-life estimate.
[b] Value from Ref. [1]

excited states in ^{145}Er. The analogous argument implies that the new transition at 1.02 MeV is also to an excited state in ^{145}Er. The energy relationship that exists among the five transitions indicates that we are observing at least two new excited states in ^{145}Er.

If the new transitions are not to excited states in ^{145}Er, then they must be from previously unidentified states in ^{146}Tm. This would be the first instance in which a proton emitter has been observed with more than two proton emitting states.

Table 1 presents the half-lives as a function of angular momentum l as predicted using a simple WKB calculation for all four proton transitions. (The three proton orbitals active in this region of nuclei are $s_{1/2}$, $d_{3/2}$, and $h_{11/2}$.) The table suggests that the 0.93 MeV transition could arise from a state involving the $d_{3/2}$ proton orbital.

SUMMARY

Three new proton transitions have been identified in the decay of ^{146}Tm: ones at 0.89 and 0.93 MeV with half-lives of approximately 100 ms and one at 1.02 MeV with a half-life of approximately 200 ms. Whether these transitions arise from new excited states in ^{146}Tm or result from decays to new excited states in ^{145}Er, this work demonstrates that proton emission studies have advanced to the stage of multi-level decay spectroscopy.

ACKNOWLEDGMENTS

This work is supported by the U. S. Department of Energy under contract numbers DE-FG05-88ER40407 (Vanderbilt University), DE-AC05-76OR00033 (UNIRIB and ORISE), DE-FG02-96ER40983 (University of Tennessee), DE-FG02-96ER40978 (Louisiana State University), and DE-FG02-94ER40834 (University of Maryland). Oak Ridge National Laboratory is managed by Lockheed Martin Energy Research Corporation for the U. S. Department of Energy under contract No.

DE-AC05-96OR22464. UNIRIB is a consortium of universities, the state of Tennessee, Oak Ridge Associated Universities, and Oak Ridge National Laboratory and is partially supported by them.

REFERENCES

1. Livingston, K., et al., Phys. Lett. B **312**, 46 (1993).
2. Nazarewicz, W., Riley, M. A., and Garrett, J. D., Nucl. Phys. A **512**, 61 (1990).
3. Thomas, S. L., Davinson, T., and Shotter, A. C., Nucl. Instrum. Methods A **288**, 212 (1990).
4. Batchelder, J. C., et al., Phys. Rev. C **57**, R1042 (1998).
5. Bingham, C. R., et al., Phys. Rev. C **59**, R2984 (1999).
6. Rykaczewski, K., et al., Phys. Rev. C **60**, 011301 (1999).
7. Ginter, T. N., et al., Phys. Rev. C **61**, 014308 (1999).

Search for Two-Proton Emitters at FRS-GSI

M. Pfützner

Institute of Experimental Physics, Warsaw University, PL-00-861 Warsaw, Poland

Abstract. A project of studying proton drip-line nuclei in vicinity of ^{48}Ni, running at GSI Darmstadt, is shortly reviewed. Prospects for spectroscopy studies on ^{45}Fe, presently identified as the best candidate for the *2p* radioactivity, are briefly discussed.

One of the most interesting phenomena at the proton drip-line is the two-proton ground-state (*2p*) radioactivity which was predicted by Goldanskii already in 1960 [1]. The *2p* decay, if encountered, would proceed either by a simultaneous emission of two uncorrelated protons or by an emission of the diproton (^{2}He), where two protons are correlated, most probably in an $L = 0$ state. A sequential emission of protons in such a case is energetically forbidden or strongly suppressed.

The emission of two protons from very short lived ($T_{1/2} \approx 10^{-21}$s) ground state of ^{6}Be [2,3] and ^{16}O [4] has been observed, but data were found to be consistent with the sequential emission through tails of the broad intermediate states. In both cases, the decay energy (Q_{2p}) is larger then the Coulomb barrier, therefore the half-life is so extremely short. For the larger Z nuclei, where the Coulomb barriers are higher, much longer lifetimes are expected. Hence, the quest is for a *2p* candidate with Q_{2p} energy low enough, as compared to the Coulomb barrier, to keep the half-life of the *2p* decay measurably long but not much longer then the partial half-life of the β decay which is a competing disintegration mode. A search for such a candidate is going on since the prediction of the *2p* radioactivity. The main difficulty is that the *2p* decay rate is extremely sensitive to the decay energy Q_{2p}. Hence, a rather precise estimate of this value is needed to identify a good candidate.

Recent theoretical calculations in the framework of the nuclear shell model by Brown [5] and later by Ormand [6] focused on proton-rich nuclei in the titanium-to-nickel region. Several candidates were considered out of which two : ^{45}Fe and ^{48}Ni emerged as the most promising. Partial *2p* decay half-life, assuming diproton emission in the $L = 0$ state and the spectroscopic factor equal to one, were predicted to be in the range 10 ns — 100 μs and 10μs—4 s for ^{45}Fe and ^{48}Ni, respectively [6]. These ranges, spanning a few orders of magnitude are mostly due to 200 keV un-

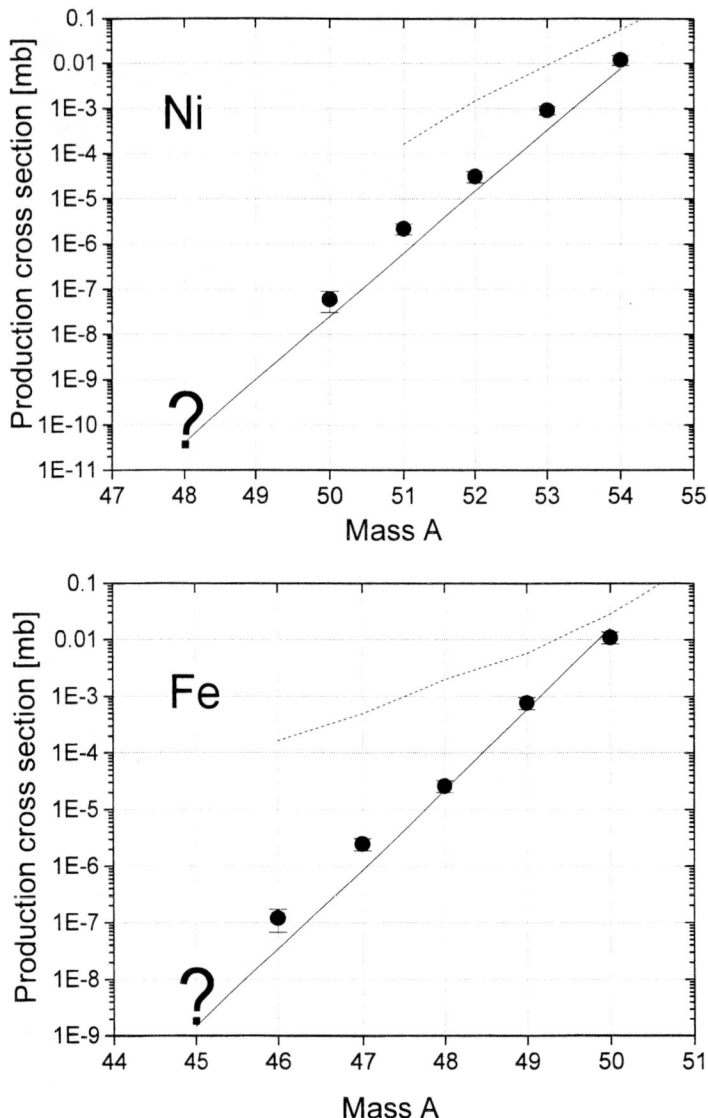

FIGURE 1. Experimental cross sections for nickel (upper panel) and iron (lower panel) isotopes produced by fragmentation of a ^{58}Ni beam at 600 MeV/nucleon on a ^{9}Be target. Solid line shows the values predicted by the EPAX formula [7], the dashed line represents the abrasion-ablation model [8].

certainty of the predicted decay energy, which illustrates the sensitivity mentioned above. The predicted β decay half-lifes were 7 ms and 9 ms, respectively. Prior to these theoretical calculations neither ^{45}Fe nor ^{48}Ni were known experimentally.

The technique of relativistic fragmentation followed by projectile fragment separation offers new possibilities of production and sensitive identification of exotic nuclei, far from the line of stability. The region of neutron-deficient nickel isotopes was studied at the Projectile FRagment Separator (FRS) at GSI Darmstadt [9]. In the first experiment, performed in 1992 [10], the primary beam of ^{58}Ni at 650 MeV/nucleon impinging on a ^{9}Be target (4 g/cm^2) was used. The average beam intensity was about 5×10^7 particles/s. With a few FRS settings, neutron-deficient nuclei between scandium and copper were selected and transmitted to the final FRS focus. They were identified in-flight, ion-by-ion, by the standard TOF–ΔE–$B\rho$ analysis. This experiment represented the first systematic survey of the production cross sections in this region of nuclidic chart for fragmentation reactions. Among the results were the first observation of ^{50}Ni and the evidence that ^{49}Co and ^{54}Cu are not bound. The extrapolation of the measured cross sections allowed an estimate of production rates of even more proton-rich isotopes. The systematics of the production cross sections for iron and nickel isotopes is shown in Fig. 1.

The next step towards more exotic nuclei was carried out at GSI in 1996. The ^{58}Ni beam of improved intensity of about 5×10^8 ions/s with energy of 600 MeV/nucleon bombarded a 4 g/cm^2 beryllium target [11]. The identification set-up was the same as in the previous experiment. After 72 h of counting with the FRS optimized for the transmission of ^{45}Fe, three new isotopes were identified : ^{49}Ni, ^{45}Fe and ^{42}Cr with 5, 3 and 10 events, respectively [11]. With ^{49}Ni and ^{45}Fe, the nuclei with $T_z = -7/2$ were reached experimentally for the first time.

Going for even more exotic species, like ^{48}Ni, and production of $T_z = -7/2$ nuclei in amounts sufficient for spectroscopy studies, requires further increase of beam intensity. An upgrade of GSI facilities has indeed been undertaken [12]. First, an electron cooling has been installed at the SIS synchrotron. Next, the Wideröe section of the linear accelerator UNILAC is being replaced by a high-current injector. Finally, it is expected that for the ^{58}Ni beam the maximum intensity of about 5×10^9 particles/s, given by the space charge limit of the synchrotron, will be reached.

Taking this beam intensity together with extrapolated cross sections and realistic simulations of the FRS transmission by means of MOCADI code [13], including losses due to secondary reactions, the rates of 16 ions/day and 0.4 ions/day were predicted for ^{45}Fe and ^{48}Ni at the final focus, respectively.

One should note that very recently a few atoms of ^{48}Ni have been identified at SISSI/LISE3 spectrometer at GANIL [14] among fragmentation products of 70 MeV/nucleon ^{58}Ni beam impinging on a ^{nat}Ni target. Yields of isotopes observed, including ≈ 50 ions of ^{45}Fe, were found to be consistent with the cross section systematics obtained at the FRS.

From all these experimental findings and from expected beam developments one can conclude that ^{45}Fe is the only *2p* decay candidate established for which there is

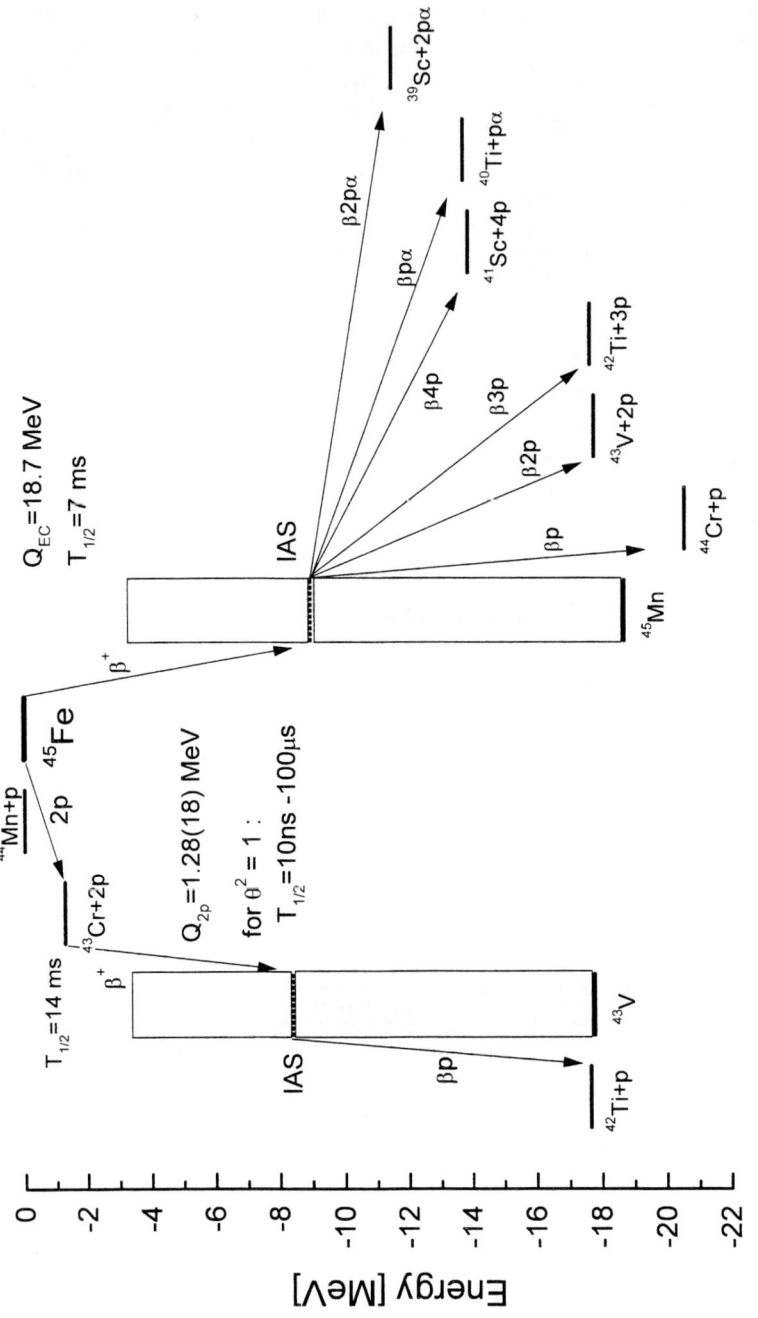

FIGURE 2. Expected decay scheme of ^{45}Fe based on the mass predictions from Ref. [6].

a chance for spectroscopy studies in the near future. The predicted decay channels of ^{45}Fe, assuming mass values calculated by Ormand [6], are sketched in Fig. 2. The value of Q_{2p} energy determines the principal decay mode. The mere fact that this isotope was identified at the final focus of FRS and LISE spectrometers, surviving the flight time of about 300 ns and about 1 μs, respectively, indicates that its half-life is longer then a few hundred nanoseconds. Hence, the Q_{2p} value must be smaller than about 1.5 MeV. If it is well below 1 MeV, then the beta decay will dominate. In such a case, because of the large Q_{EC} value (18.7 MeV) many exotic channels of beta-delayed particle emission would be opened. Detailed studies of them would be an interesting endeavor in itself. Only when the Q_{2p} value is between these two limits, the emission of two protons will proceed faster than β decay.

The range straggling of relativistic projectile-like fragments at the final focus of a spectrometer is typically large. For ^{45}Fe at the FRS, in the conditions optimized for the spectroscopy of this isotope, it amounts to about 60 mg/cm^2 (1σ value) in silicon. Thus, ions stopped in the a stack of typical silicon detectors of thickness between 100 and 500 μm, will be implanted in most cases deeply in the detector material. Since the range of 1 MeV protons in silicon is about 17 μm, one will not be able to track two emitted protons separately. Such a set-up allows a measurement of the total energy released and of the half-life.

In the next experiment at the FRS it is planned to surround a silicon implantation stack by a segmented NaI detector for efficient detection of 511 keV photons emitted back to back which will help to distinguish ground-state particle emission from β^+ decays. A signature of the *2p* decay in such conditions will be a release of about 1 MeV energy in one of Si detectors, a few microseconds after an implantation of identified ^{45}Fe ion, followed (with $T_{1/2} \approx 14$ ms) by the β^+ decay of ^{43}Cr which will be accompanied by two 511 keV quanta and by a beta-delayed proton emitted from an exited ^{43}V state. Such a sequence of events would clearly indicate a decay by particle emission. Once this is proved to occur, a different experimental set-up will have to be designed which would allow the detection of two protons separately in order to measure their energies and the angle between their final velocities. Only then the detailed mechanism of the process could be investigated and the basic question on the correlation of the two protons addressed.

This work was partly supported by the Polish Committee of Scientific Research under grant KBN 2 P03B 036 15. The support from Oak Ridge Associated Universities is gratefully acknowledged.

REFERENCES

1. Goldanskii, V.I., *Nucl. Phys.* **19**, 482 (1960).
2. Geesaman, D.F. et al., *Phys. Rev. C* **15**, 1835 (1977).
3. Bochkarev, O.V. et al., *Sov. J. Nucl. Phys.* **55**, 955 (1992).
4. Kryger, R.A. et al., *Phys. Rev. Lett.* **74**, 860 (1995).

5. Brown, B.A., *Phys. Rev. C* **43**, R1513 (1991).
6. Ormand, W.E., *Phys. Rev. C* **53**, 214 (1996).
7. Sümmerer, K. and Blank, B. submitted to *Phys. Rev. C* (1999).
8. Gaimard, J.-J. and Schmidt, K.-H., *Nucl. Phys.* **A531**, 709 (1991) and Schmidt, K.-H. et al., *Phys. Lett. B* **300**, 313 (1993).
9. Geissel, H. et al., *Nucl. Instrum. Methods* **B 70**, 286 (1992).
10. Blank, B. et al., *Phys. Rev. C* **50**, 2398 (1994).
11. Blank, B. et al., *Phys. Rev. Lett.* **77**, 2893 (1996).
12. Böhne, D., *GSI Scientific Report 1995* **96-1**, 166 (1996).
13. Schwab, Th., *GSI Report* **91-10** (1989).
14. Giovinazzo, J. et al., *contribution to this symposium*.

Recent Studies of Proton Drip-Line Nuclei Using the Berkeley Gas-Filled Separator

M. W. Rowe[1], J. C. Batchelder[2], V. Ninov[1], K. E. Gregorich[1], K. S. Toth[3], C. R. Bingham[3,4], A. Piechaczek[5], X. J. Xu[1,6], J. Powell[1], R. Joosten[1] and Joseph Cerny[1,7]

[1] *Lawrence Berkeley National Laboratory, MS 88, 1 Cyclotron Rd, Berkeley, California 94720*
[2] *UNIRIB, Oak Ridge Associated Universities, Oak Ridge, Tennessee 37831*
[3] *Department of Physics, Oak Ridge National Laboratory, Oak Ridge, Tennessee 37831*
[4] *The University of Tennessee, Knoxville, Tennessee 37996*
[5] *Lousiana State University, Baton Rouge, Louisiana 70803*
[6] *Institute of Modern Physics, Lanzhou, 730000, China*
[7] *Department of Chemistry, University of California, Berkeley, CA 94720*

Abstract. The Berkeley Gas-filled Separator provides new research opportunities at Lawrence Berkeley National Laboratory's 88-Inch Cyclotron. The use of this apparatus for the study of proton drip-line nuclides is discussed. Preliminary results of ^{78}Kr bombardments of ^{102}Pd targets at mid-target energies of 360, 375 and 385 MeV are presented. Improvements planned partially as a result of this measurement are also discussed.

INTRODUCTION

The properties of proton drip-line nuclides have long been a focus of study at Lawrence Berkeley National Laboratory's 88-Inch Cyclotron. This work includes the confirmation of the first example of direct proton emission, from an isomer of ^{53}Co, in 1970 [1]. However, until recently Berkeley has lacked the optimal tools for the study of proton emitters in the region above Sn, where all subsequent examples of direct proton decay have been observed [2]. Virtually all of these later studies have relied on some form of mass separation or analysis to identify the nuclides of interest and reduce unwanted background from other species produced simultaneously. In fact, it is the development of high resolution mass analyzers and velocity filters, as well as high-granularity silicon detector arrays, which has fueled the recent explosion in knowledge of proton drip-line isotopes in the region between Sn and Bi.

During the Fall of 1998, commissioning of the Berkeley Gas-filled Separator (BGS) began. Although this apparatus was designed primarily as a tool for research on the heaviest elements, its capabilities also make it well suited for the study of proton drip-line isotopes, particularly in the region near lead. After a discussion of the BGS's design and detector systems, preliminary results of the first measurement of very proton-rich nuclides using the BGS will be presented. The primary goal of this experiment was to observe proton emission from ^{177}Tl [3] and ^{176}Tl. In the context of the results of this measurement, improvements to the BGS which are planned or have already been implemented, will be discussed.

THE BERKELEY GAS-FILLED SEPARATOR

In order to successfully study proton drip-line nuclei, the experimental apparatus used must meet three requirements. Because such isotopes are produced in low yield, the target must be capable of withstanding large beam currents and/or the transport of the products from the target to the detector system must be very efficient. Second, this transport must be accomplished rapidly, since the half-lives of proton emitters are typically on the order of a few tens of milliseconds or shorter [2]. Finally, there must be some way of identifying the proton-emitter in order to distinguish it from other nuclides produced in much higher yields.

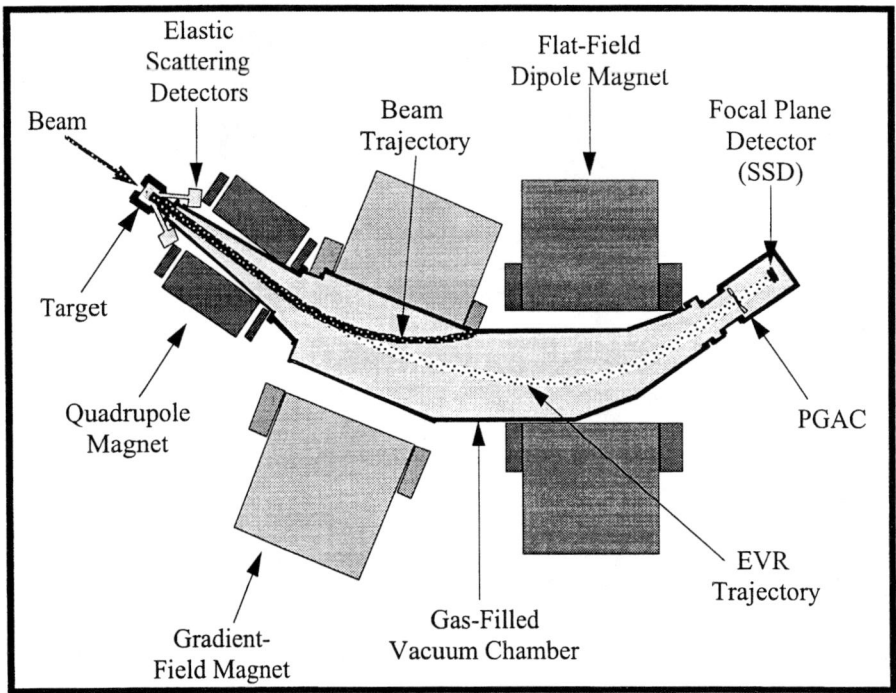

FIGURE 1. A schematic diagram of the Berkeley Gas-filled Separator. The beam enters the apparatus from the left. The detector systems are shown as arranged during the ^{78}Kr + ^{102}Pd measurement described herein. Recent modifications of the system will be described later in this paper.

Gas-filled separators can fulfill these requirements. The residues of fusion-evaporation reactions emerge from the target in a variety of charge states. Since the magnetic rigidity, upon which the separation of the various products is based, varies as the inverse of the charge, this can substantially reduce the percentage of a given nuclide transported through the separator to the detector system. To reduce this problem, the magnet of a gas-filled separator is filled with a dilute inert gas. According to Bohr's approximation, collisions with the atoms in the gas will leave the reaction products in an average charge state that is proportional to the cube root of their atomic number. To first order, the resulting magnetic rigidity is proportional to the mass over the cube root of the atomic number. This significantly improves the transport efficiency of a gas-filled separator relative to conventional mass analyzers and velocity filters. Unfortunately, the collisions in the gas cause much of the kinematic information to be lost such that the resolution is only sufficient to separate fusion-evaporation residues from incident beam and direct-reaction products. In effect, the gas-filled separator acts as a purifying beam dump. One must rely entirely on the detector system for identification of the implanted nuclides.

A schematic diagram of the Berkeley Gas-filled Separator is shown in Fig. 1. The beam enters from the left through a 100 $\mu g/cm^2$ carbon window. The entire apparatus downstream of this window is filled with He gas at a pressure of ~ 1.3 Torr. The gas helps to cool the target, which may either consist of a single foil on a target ladder, or a series of foils mounted on a continuously rotating target wheel. Two PIN-diode detectors behind the target monitor beam elastically scattered from the target. Beam and reaction products exit the target and enter a quadrupole magnet followed by a gradient-field dipole magnet that provide vertical and horizontal focussing, respectively. The latter magnet also increases the separation between beam and fusion-evaporation residues, as does the flat-field dipole that follows. The incident beam is bent such that it is dumped between the two dipole magnets. The focal plane of the system is located inside the detector chamber, approximately 1 m from the exit of the second dipole. The operating parameters of the BGS are presented in Table 1. Note that many of these numbers will depend on the kinematics of the specific reaction employed.

TABLE 1. Operating Parameters of the Berkeley Gas-filled Separator.

Parameter	Value
Maximum beam current:	5×10^{12} ions/s
Target thickness:	$0.1 - 1.0$ mg/cm^2 optimum
Angular acceptance:	±75 mrad horizontal
	±150 mrad vertical
Momentum acceptance:	> 50%
Charge acceptance:	~ 100%
Transport efficiency:	20 – 70% typical
Total Bend:	70°
Path Length:	4.6 m
Transport time:	< 2 µs
Maximum rigidity:	2.5 T m
Overall background rejection:	> 10^{12}
Dispersion:	18 mm / %Bρ
Focal plane image size:	50 mm vertical
	80 – 150 mm horizontal

Inside the detector chamber reaction, products pass through a parallel-grid avalance counter (PGAC) and are subsequently implanted into a 300 µm-thick single-sided silicon strip detector (SSD) located at the focal plane. The SSD is 80 mm wide by 35 mm high. It is divided into 16 vertical strips which give crude position sensitivity in the horizontal plane. Signals are measured from both ends of each strip. The sum of these signals is proportional to the implantation or decay energy. Typical resolution for alpha-decay events was 70 keV full-width at half maximum (FWHM). The ratio of either signal to the sum is used to determine the vertical position of an event along the strip due to resistive charge division. One may think of a particular event occuring within a pixel, defined by the strip and the vertical-position resolution of 900 μm FWHM. The total number of pixels for the focal plane detector is ~620.

Positive identification of a particular reaction product may be accomplished by observing its decay chain if the decays of the daughter nuclides are known. A high-energy signal in the SSD, observed in coincidence with a signal from the PGAC, indicates that a fusion-evaporation residue has been implanted in the silicon detector. One may then look for decays occurring within the same pixel. By time-stamping each event, it is trivial to determine the half-life for a particular decay. Since the residues are implanted near the surface of the detector, the detection efficiency for subsequent decays is approximately 50%. Due to the large number of pixels, the average time between implantations in a given pixel is on the order of seconds, thus permitting the decay chain to be observed through several generations. The decay-correlation technique is particularly well suited for work near the proton drip line where the half-lives of the nuclides of interest are typically less than 50 ms, making false coincidences unlikely. Unfortunately, this technique is not well suited for the region of the proton drip-line immediately below the N=82 closed shell, since these nuclides decay primarily by electron capture rather than alpha emission.

A SEARCH FOR LIGHT THALLIUM ISOTOPES

In a first experiment to search for new proton emitters using the BGS, three ^{78}Kr bombardments of a 1.5 mg/cm^2 70%-enriched ^{102}Pd target were performed at mid-target energies of 360, 375 and 385 MeV. Analysis of the data from this experiment is still in progress, and all results presented herein should be considered preliminary. This paper will focus on the results of the 375 MeV bombardment. The goal of this measurement was to reproduce the results of the earlier study of ^{177}Tl performed at Argonne National Laboratory, both to provide energy calibration points and to determine if the BGS was working as expected. Poli, et al.[3], produced ^{177}Tl in its ground state using the ^{102}Pd(^{78}Kr,p2n) reaction at 370 MeV with a cross section of 10 nb; an isomer was also produced with a 30 nb cross section. The ground state [$T_{1/2}$ = 67(37) ms] decayed by emission of either an 1156(20) keV proton [27(13)%] or a 6907(7) keV alpha particle. The isomer decayed by emission of a 1958(10) keV proton [51(8)%] or a 7487(13) keV alpha particle with a half life of 451(106) μs.

Figure 2 shows the alpha spectrum measured at the focal plane of BGS during a 17 hour 375 MeV bombardment at an average beam current of approximately 10 pnA. Decay events were differentiated from implantations by requiring that they be anti-coincident with signals from the PGAC; no other conditions were applied to generate this spectrum. The energy resolution for alpha decays was 70 keV. The peaks have been assigned to various Hg, Au and Pt isotopes on the basis of their energies.

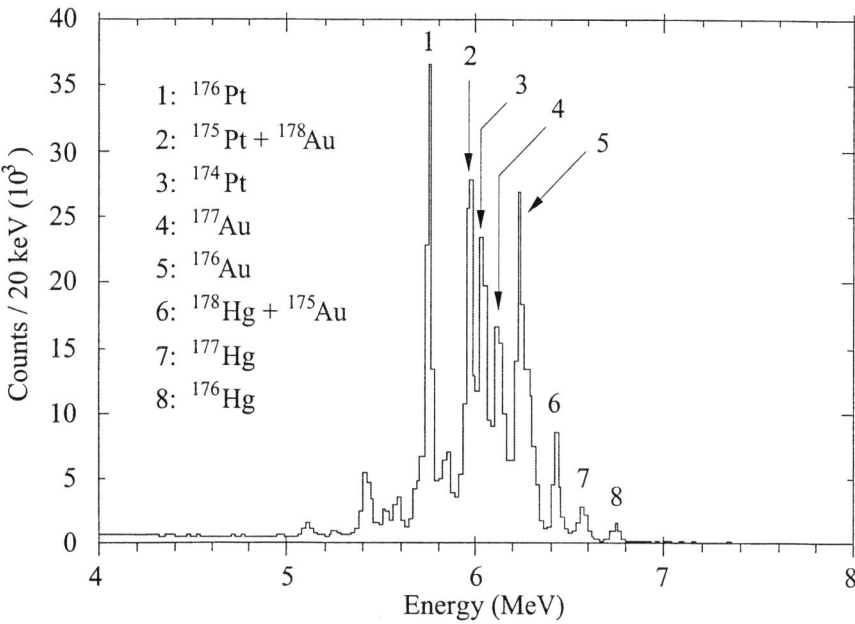

FIGURE 2. Alpha decays observed during the 375 MeV ^{78}Kr bombardment of ^{102}Pd. All events shown were detected in anti-coincidence with signals from the PGAC. Preliminary assignments have been made for some of the peaks based on their energies.

When compared to the work of Poli and collaborators [3], one sees that the distribution of isotopes produced was slightly different. This may be due in part to the somewhat higher beam energy employed. Based on the yield of two key isotopes, ^{177}Hg and ^{177}Au, it would appear that the transport efficiency of the BGS was significantly lower than had been expected. Several factors may have contributed. In an effort to increase the suppression of the primary beam at the focal plane, a stack of carbon degrader foils, with a total thickness that was varied from 1.09 to 0.76 mg/cm^2, was placed immediately behind the target. This may have adversely affected the BGS optics and lowered transmission. Subsequent measurements suggest that these foils were not necessary to achieve adequate beam suppression. These foils also blocked the elastic scattering detectors used to measure beam current, so it is unclear how accurate the beam intensity cited above is. Finally, much of the time was spent attempting to optimize the tune of the separator. At the time of this measurement, the analysis code was still in development, which complicated tuning due to lack of on-line diagnostics. However, the yield of alpha emitters was within a factor of two of that measured using the Fragment Mass Analyzer at Argonne [3], and this result was achieved in a shorter time period (17 hours vs. 65 hours).

In order to positively identify the various isotopes produced, correlations were sought between the implantation of evaporation residues and the subsequent chain of decays. A time stamp for each event was generated using independent MHz, kHz and Hz clocks. During analysis, all events coincident with signals in the PGAC were assumed to be implantations and stored in a correlation buffer. When a decay event was observed, a search was made for earlier implants or decays within 0.9 mm on the same strip of the SSD. The maximum preceding time interval that was searched for correlations was determined by the energy of the event in accordance with typical half-lives; for alpha decays of greater than 6.7 MeV, this was set to 150 ms. When correlations were found between alpha decays and earlier events, the decays were added to the correlation buffer as well.

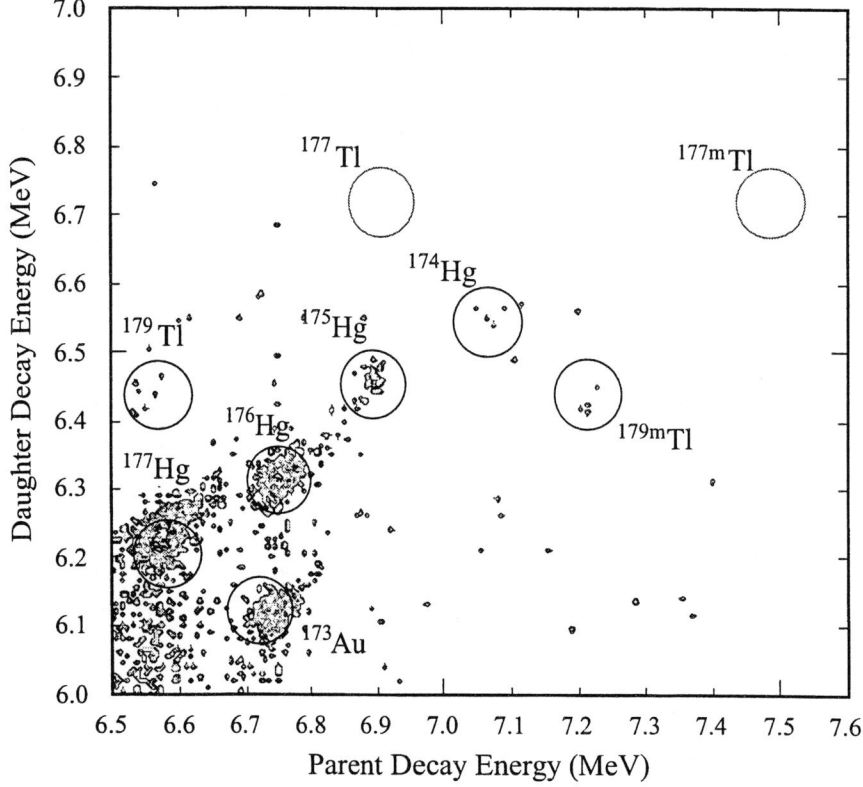

FIGURE 3. A parent-daughter alpha-decay correlation plot generated from the data collected during the 375 MeV ^{78}Kr bombardment. See text. The contour lines increase logarithmically; black indicates the 8-count countour.

Figure 3 shows a correlation plot between successive generations of alpha decays from the same data set that generated Fig. 2. The decay energy from the parent lies along the x-axis and the subsequent decay is plotted on the y-axis. The circles, which have a diameter encompassing 100 keV, indicate the positions of various parent-daughter decay series, as based on their literature decay energies [4]. Each is labelled with the parent isotope. The ^{177}Hg-^{173}Pt, ^{173}Au-^{169}Ir, ^{176}Hg-^{172}Pt and ^{175}Hg-^{171}Pt parent-daughter decay pairs are clearly seen. Several events are observed which correspond to the decay energies for the ground state and isomer of ^{179}Tl and their daughter ^{175}Au. No evidence is observed of the decays of ^{177}Tl or ^{177m}Tl as indicated in the grey circles. However, several correlated events are observed which have the same decay energies as ^{174}Hg and ^{170}Pt. Though additional analysis will be required to confirm this assignment, this is very encouraging, since only a few ^{174}Hg decay events have been observed in previously published studies [5, 6].

A search was also made for protons correlated with alpha decays. For these purposes, all events with energies between 1 and 2 MeV that were anti-coincident with signals from the PGAC were considered to be protons. Since proton emission from high-spin isomers can occur with a relatively long half-life, a maximum correlation-search time of 500 ms was used. A position resolution of 0.9 mm was also used again as the condition for spatial correlation. Using these criteria, only six event pairs were found in the 375 MeV data set with a proton corrrelated to an alpha decay with an energy greater than 6.5 MeV.

Table 2 shows the data for two of these events which are ^{177}Tl candidates. The first three columns from the left show the type of event observed, followed by the time interval and difference in position between an event and the correlated event which preceded it. The next three columns show the nuclide to which the decay is tentatively assigned, and the decay energies and half-lives, taken from the literature [4]. The first event is assigned to the decay chain of the ^{177}Tl isomer. There is excellent agreement between the energies observed and previously measured values of the ^{177m}Tl decay chain, and all four events (including the implantation) occur very close together on a single silicon strip. However, the time intervals between the implantation and the first two decays are much longer than would be expected from the literature half-lives. It should be noted that the scalars used to generate the time stamps for each event would occasionally give bad values. A great deal of effort was required to correct for this problem in the analysis, and thus the timing information is somewhat suspect. However, there is virtually no doubt that all of these four events occurred within a very short time period. The second candidate is assigned to the ^{177}Tl ground-state decay. Again, there is excellent agreement between decay energies, and the events occur very near to one another. Also, the decay times are in better agreement with the half-lives for this chain, though the above stipulations still apply.

TABLE 2. Candidate ^{177}Tl Events Compared to Literature Values.

Decay Chain	Δt	Δy	Nuclide	Literature Decay	Half-Life
(EVR implanted)					
1930 keV proton	53 ms	+0.21 mm	^{177m}Tl	1958 keV proton	451 μs
6760 keV alpha	358 ms	-0.23 mm	^{176}Hg	6740 keV alpha	30 ms
6310 keV alpha	73 ms	-0.29 mm	^{172}Pt	6310 keV alpha	100 ms
(EVR implanted)					
1110 keV proton	1.5 ms	+0.09 mm	^{177g}Tl	1156 keV proton	67 ms
6710 keV alpha	22 ms	-0.14 mm	^{176}Hg	6740 keV alpha	30 ms

OUTLOOK FOR THE FUTURE

These preliminary results are quite encouraging. However, this experiment also made it clear that the BGS, in its original configuration, had several shortcomings. In the months since, a variety of improvements have been planned and, in most cases, implemented. Figure 4 is provided to illustrate some of the problems with the original configuration of the separator and its detectors. It shows energy spectra measured in the SSD at the focal plane during the 360 MeV ^{78}Kr bombardment. The black line shows the sum of all events measured at the focal plane. The dark grey line shows the subset of those events which were coincident with signals from the PGAC, and thus interpreted as implantations of fusion-evaporation residues. The light grey line shows events that were anticoincident with PGAC signals, and therefore interpreted as decays.

It is seen that the majority of events above 2 MeV in Fig. 4 are coincident with the PGAC. However, between 5 and 7 MeV alpha decay signals dominate. Above, 7 MeV, virtually no alpha decays should be observed, so it is clear that some implantation events are passing through the PGAC without producing a signal. To improve the discrimination between implant- and decay-type events, a second PGAC was added to the BGS upstream of the focal plane shortly after this experiment. This greatly improves the discrimination.

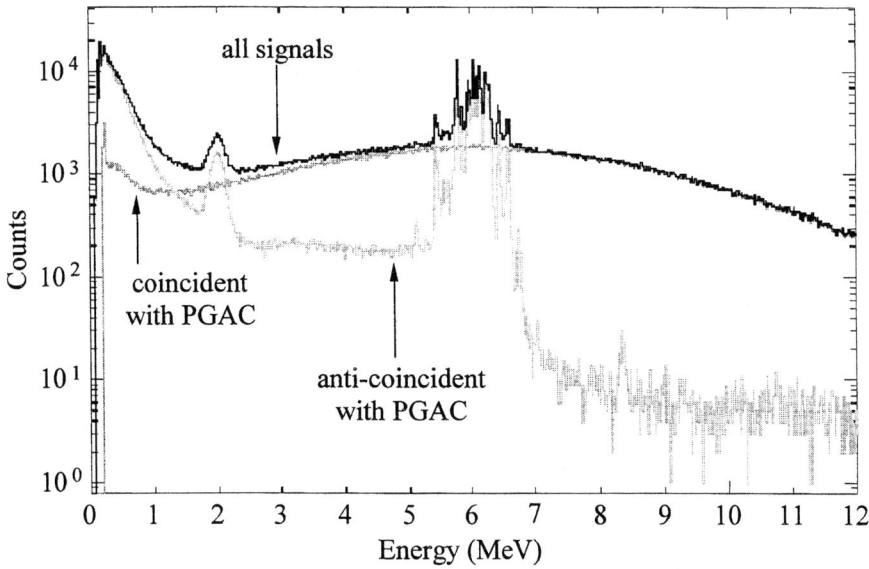

FIGURE 4. Energy spectra of events observed in the silicon strip detector (SSD) located at the BGS focal plane. The black line indicates the spectrum of all events. The heavy and light grey lines indicate the subsets of those events which were either coincident or anticoincident with PGAC signals, respectively.

A broad peak which is mostly anticoincident with the PGAC is seen in the spectrum near 2 MeV. This is due to high-energy He ions resulting from collisions between heavy ions and the He gas in the BGS. Because they are high-energy and low-Z, they produce minimal signals in the PGAC and are not stopped in the SSD. Soon after this experiment, a second silicon strip detector was put in place behind the focal plane SSD. Events observed in coincidence with signals > 1 MeV from this detector are discarded. The small peak near 8 MeV is of unknown origin, but may have been caused by a similar mechanism. It does not seem to be correlated with alpha decay chains.

As noted earlier, the energy and position resolution of the SSD average approximately 70 keV and 0.9 mm, with some strips behaving much worse than others. This was due to radiation damage sustained during the commissioning of the BGS. This was later improved somewhat by annealing the detector. It is planned that this SSD will be replaced early in 2000 by two 60 mm x 60 mm 16-strip silicon detectors placed side-by-side at the focal plane. Besides improving the energy and position resolution, this will also provide more focal plane coverage. The enhanced position sensitivity will increase the average time between implantations at a given position by a factor of approximately 8, decreasing the chances of false coincidences between different decay chains. This should permit more decay generations to be observed, or higher event rates to be tolerated.

Most of the decay events observed below 5 MeV in Fig. 4 are due to alpha particles which decay out of the detector, so that only a partial energy signal is observed. To improve this situation, six additional SSD's will be placed around the four sides of the focal plane SSD, perpendicular to its surface. These will detect approximately 60% of the alpha particles or protons that escape from the focal plane detector. In addition to increasing the odds of observing decay chains through multiple generations, this will reduce the "background" in the region between 1 and 2 MeV where proton decays will be observed. These side detectors will also be installed in early 2000.

During this experiment, the dead time of the data acquisition system after each event was about 200 µs. This would prevent the observation of isotopes with very short half-lives. The CAMAC-based system which was used is being replaced by a VME-based system, which should significantly reduce this problem. This is especially important due to the large number of additional channels which will be needed for the new detectors described above. There have also been improvements in the data analysis code that will improve on-line diagnostics.

Finally, experience with tuning the separator will also prove important. In particular, recent measurements have indicated that suppression of the primary beam is sufficient in this reaction without degrading the beam and recoils after the target. In Fig. 4, the evaporation residues have been degraded to the same energy range as the alpha decays. Without the carbon degrader foils, the occasional implant event that does not produce a signal in the PGAC is unlikely to be mistaken for a decay. Also, the elastic scattering detectors behind the target will no longer be blocked by degrader foils, so the beam intensity may be accurately assessed.

If necessary, we plan to repeat this experiment in order to take advantage of these improvements to the BGS. In addition, several other measurements of proton and alpha-particle emitters in this region of the Chart of the Nuclides are planned. These studies will provide important decay, mass and structure information about nuclei at the extreme limits of stability.

CONCLUSION

Although yields observed in this early BGS experiment were not as high as had been expected, the results of the preliminary analysis of the 375 MeV data are intriguing. Much more work still remains to be done on the analysis at this time; only about half of the 375 MeV data has been examined. Improvements in the correlation algorithms may also yield additional results which were missed. Analysis of the data taken at 360 MeV and 385 MeV has only been cursory at this point.

Partially as a result of this experiment, a series of improvements to the BGS are being implemented. Some of these improvements were in place during the element 118 measurements [7], and proved to be critical to the success of those experiments. The present study has indicated that the BGS will be an excellent tool for the investigation of proton drip-line nuclei near the $Z=82$ shell closure. In particular, its high transport efficiency and its ability to utilize high beam intensities will permit isotopes produced in very low yields to be observed and identified. This will be especially true once all of the planned improvements are in place.

ACKNOWLEDGMENTS

We wish to acknowledge the excellent work David Ruiz, Ron Oort and the rest of the 88" Cyclotron machine shop and technical staff did in building and installing the BGS. We also thank Sigurd Hofmann and GSI for the silicon strip detector used in this measurement, and the 88" Cyclotron operations staff for their assistance with this experiment. This work was supported by the U. S. Department of Energy under contracts DE-AC03-76SF00098 (Lawrence Berkeley National Laboratory), DE-AC05-76OR00033 (UNIRIB), DE-AC05-96OR22464 (Oak Ridge National Laboratory, managed by Lockheed Martin Energy Research Corporation), DE-FG02-96ER40983 (University of Tennessee), and DE-FG02-96ER40978 (Louisiana State University).

REFERENCES

1. Cerny, J., Esterl, J. E., Gough, R. A. and Sextro, R. G., Phys. Letters 33B, 284-286 (1970).
2. Woods, P. J., and Davids, C. N., "Nuclei Beyond the Proton Drip-Line," in Ann. Rev. Nucl. Part. Sci., 47, 541-90 (1997).
3. Poli, G. L., et al., Phys. Rev. C 59, R2979-R2983 (1999).
4. Pfennig, G., Klewe-Nebenius, H., Seelmann-Eggebert, W., Karlsruher Chart of the Nuclides, 6th ed., revised reprint, Institute fuer Instrumentelle Analytik, 1998.
5. Uusitalo, J., et al., Z. Phys. A 358, 375-376 (1997).
6. Seweryniak, D., et al., Phys. Rev. C 60, 031304-1—031304-4 (1999).
7. Ninov, V. et al., Phys. Rev. Lett. 83, 1104-7 (1999).

Two-Proton Decay Experiments at MSU

M. Thoennessen, M. J. Chromik* and P. G. Thirolf*

*National Superconducting Cyclotron Laboratory and
Department of Physics & Astronomy, Michigan State University
East Lansing, MI 48824, USA*

Abstract. First evidence for direct two-proton radioactivity has been observed in the decay of the first excited state of ^{17}Ne. The decay is in competition with the γ-decay back to the ground state of ^{17}Ne and from the branching ratio the lifetime for the two-proton decay is estimated to be 0.9 ps. The proton-proton angular distribution is statistically not significant to observe any correlations. The first excited state was populated with relativistic Coulomb excitation of the exotic beam of ^{17}Ne. This method can also be used to study the inverse reaction (2p, γ). Although the present reaction ^{15}O(2p,γ)^{17}Ne is not important for astrophysical purposes two-poton capture reactions on heavier proton rich nuclei can have a large impact on the path of the rp-process.

INTRODUCTION

So far all experimental attempts to identify two-proton radioactivity at or near the proton dripline have been unsuccessful (e.g. [1]). Two different experimental approaches have been pursued to search for two-proton decay. In medium mass nuclei where the predicted lifetimes can be on the order of nanoseconds or longer [2,3] the exotic isotopes are produced in fragmentation reactions and then implanted in a detector array where the time correlated decay is observed. The most recent successful observation of ^{48}Ni offers the opportunity to search for two-proton emission in this doubly magic nucleus [4,5]. In the lighter mass region the Coulomb and centrifugal barriers are much smaller and the lifetimes are significantly shorter. These nuclei can be studied by kinematic reconstruction of the decay products (daughter and two protons) following the decay in flight. Again, the exotic isotopes of interest are produced in high-energy fragmentation reactions [1]. ^{12}O was the first candidate studied at MSU. An upper limit of 7% for correlated two-proton emission was determined and the dominant decay process was the sequential decay via the broad intermediate ground state of ^{11}N [1,6-8]. A similar situation is expected for the next heavier candidate ^{16}Ne, where the decay most likely proceeds via the broad ground state of ^{15}F. However, another promising candidate is ^{17}Ne, where the first excited state ($J^+ = 3/2^-$, E* = 1.288 MeV) is bound by 168 keV with respect to one-proton emission but unbound with respect to two-proton emission by 344 keV (for details see [9]). Therefore this state can decay via a simultaneous emission of two protons to ^{15}O, because the widths of the low-lying states in ^{16}F are too small (~ 40 keV) for a

---

*Present address: Luwig Maximilian Universität München, D-85748 Garching, Germany

sequential decay through their tails. The two-proton decay is in competition with the γ-decay to the ground state of ^{17}Ne. In a recent intermediate energy Coulomb excitation experiment the γ-decay from the first excited state to the ground state ($J^+ = 1/2^-$) has been measured and the experimental yield has been compared to the theoretically expected cross section. The measured γ-ray yield accounts for only 43% of the predicted one, thus encouraging the investigation of a potential two-proton decay branch [9].

Intermediate energy or relativistic Coulomb excitation (as for example ^{17}Ne(γ,2p)^{15}O) is not only suitable to study the two-proton decay, it is also useful to measure reaction rates for the inverse process. The inverse reaction ^{15}O(2p, γ)^{17}Ne has been suggested as a possible breakout reaction of the hot CNO cycle [10]. The detailed level structure can have a significant influence on the reaction rates.

In the following we will discuss the latest results of the two-proton decay experiment, followed by reaction rate calculations based on the measured level structure of ^{17}Ne and future possible experiments in heavier nuclei relevant for the astrophysical rp-process.

THE ^{17}NE(γ,2p)^{15}O REACTION

The experiment to search for the two-proton decay of ^{17}Ne was performed at the National Superconducting Cyclotron Laboratory at Michigan State University. A 60 MeV/u radioactive ^{17}Ne beam was produced from a primary ^{20}Ne beam using the A1200 fragment separator. A Wien filter was used to further purify the secondary beam and a 90% pure beam with an intensity of ~5000 ^{17}Ne particles/s was achieved. In order to identify the two-proton decay from the first excited state in ^{17}Ne a complete reconstruction of the decay kinematics in the center-of-mass system (CM) was necessary. Thus the interaction point on the target as well as the energies and directions of all outgoing decay particles were measured. Details of the experimental setup can be found elsewhere [11,12].

The invariant mass and thus the excitation energy spectrum of ^{17}Ne was reconstructed from the energies and angles of the outgoing two protons and ^{15}O. The bottom part of Figure 1 shows the data together with fits from simulations that include in addition to the first

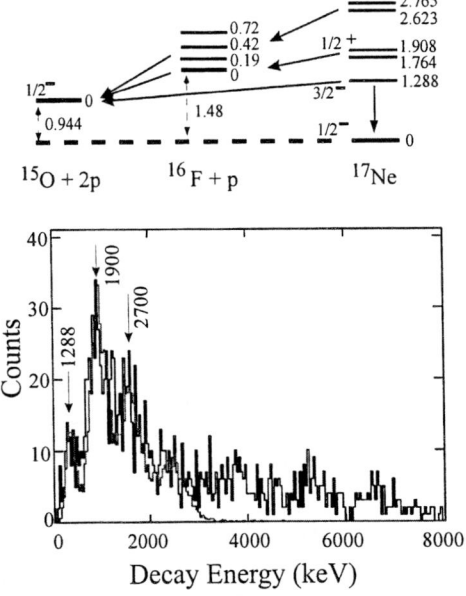

FIGURE 1. Partial level scheme (top) and invariant mass spectrum of ^{17}Ne (bottom).

excited state also high lying states which decay sequentially via intermediate states in ^{17}F. The partial level scheme and possible decay paths are indicated in the upper panel of the figure. Note that the decay energy is measured relative to the (^{15}Opp) separation energy, whereas the indicated peaks correspond to the excitation energy of ^{17}Ne. The lowest peak at a decay energy of 340±40^{stat} ±50^{syst} keV [13] corresponds to an excitation energy of 1284 keV in ^{17}Ne. This agrees within the uncertainties with the first excited state of ^{17}Ne at 1288 keV [14,15] and is first evidence for simultaneous two-proton decay of this state in ^{17}Ne. The peaks at higher decay/excitation energies also correspond to known states in ^{17}Ne [14,15]. The energy resolution is on the order of 250 keV, mainly dominated by the error in the determination of the interaction

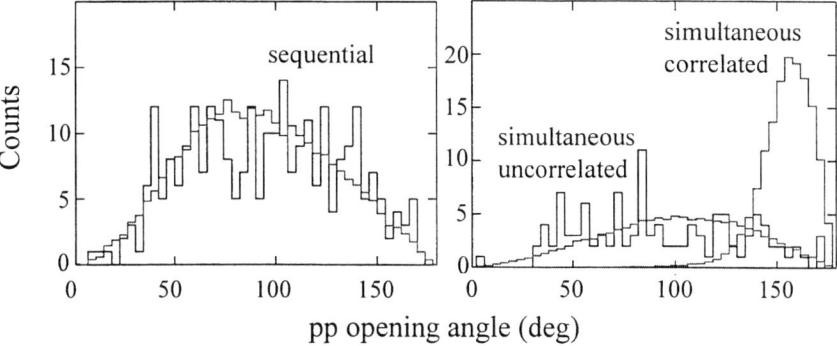

FIGURE 2. Proton-proton angular distribution for the sequential decay of the 2.7 MeV state in ^{17}Ne (left) and the direct decay of the first excited 1.29 MeV state (right). Included are results of sequential and (un)correlated simultaneous simulations.

point on the target.

Thus it was not possible to resolve the doublets around 1.9 MeV and 2.7 MeV in ^{17}Ne. Figure 2 shows the angular distributions of the two protons with respect to each other. The simulations include the experimental efficiencies and acceptances. The statistics for the most interesting first excited state is very limited and it is not possible to draw conclusions about any correlations. However, there is no evidence for highly correlated di-proton emission or emission from opposite sides of the fragment. For comparison simulations for simultaneous correlated and simultaneous uncorrelated emissions are also shown. A recent experiment to improve the statistics for the angular distribution was performed and is currently being analyzed [16].

From the calculated excitation cross section and the measured γ-decay branch it is possible to extract the lifetime for the two-proton decay. In the γ-ray experiment only 43% of the excitation cross section was detected for the γ-decay branch [9]. However, attributing all of the missing strength to two-proton decay would result in a lifetime of γ_{2p} = 0.08 ps. This would be substantially faster than estimates of ~360 ps for the barrier penetration of a diproton. From the γ-ray yield and the number of presently observed two-proton events from the first excited state the branching ratio $\Gamma\gamma/\Gamma_{2p}$ is approximately 7.5. This translates into a 6% decay branch for the two-proton decay. Thus the two-proton yield does not account for all of the missing excitation to the first excited state. Although secondary excitations to higher excited states could account for the remainder of the missing strength, it seems most likely that the B(E2)-value is overpredicted by the shell model calculations. The branching ratio of $\Gamma_\gamma/\Gamma_{2p} \sim 7.5$ corresponds to a two-proton lifetime of $\tau_{2p} = 0.9$ ps, which is still longer than the barrier penetration calculations. These calculations assumed the emission of a $l = 2$ diproton. However, it is conceivable that there is a sizable fraction of $d_{5/2} \otimes s_{1/2}$ contribution to the 3/2⁻ first excited state of ^{17}Ne which would reduce the barrier significantly. For example, the barrier for an $l = 0$ transition is reduced by a factor of ~40 compared to an $l = 2$ transition. However, detailed three-body model calculations are necessary to calculate the lifetime for the two-proton decay from a $d_{5/2} \otimes s_{1/2}$ configuration. This reduction, in connection with the possible halo configuration which has been suggested [17], and the uncertainty of 50 keV in the mass of ^{17}Ne [18] can result in lifetimes on the order of picoseconds which would be consistent with the present experiment.

THE ^{15}O(2p,γ)^{17}NE REACTION

The measurement of the ^{17}Ne(γ,2p)^{15}O reaction can also be an interesting tool to study the inverse reaction ^{15}O(2p, γ)^{17}Ne, which could have important implications for astrophysical processes. This reaction in particular has been pointed out to be a possible break-out reaction from the hot CNO cycle [10]. The solid and dot-dashed lines of figure 3 show the calculated reaction rates (left) and the regions where different processes dominate as a function of density and temperature (right). The hot

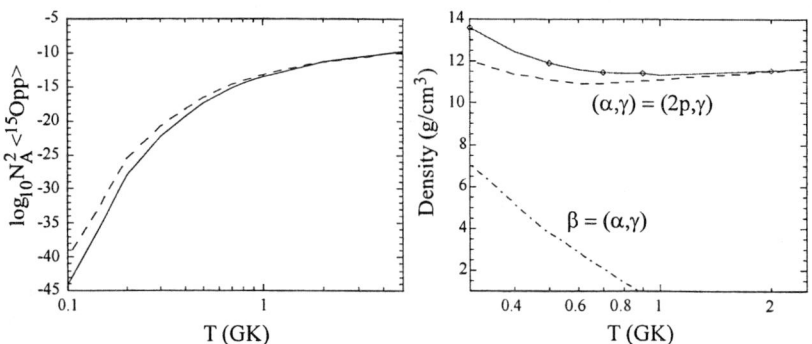

FIGURE 3. Calculated 2p-capture rate ^{15}O(2p,γ)^{17}Ne as a function of temperature (left). The dividing lines where different decay processes dominant for given density and temperature region are shown on the right. The dot-dashed line indicates equal timescales for β-decay and the (α,γ) break-out reaction. The solid and dashed lines correspond to equal timescales for the (α,γ) and (2p,γ) reactions based on levels from the ^{17}N mirror and the measured for the ^{17}Ne levels, respectively..

CNO cycle proceeds through ^{15}O via β-decay to ^{15}N. At temperatures and densities above the β = (α,γ) line the reaction ^{15}O(α,γ)^{19}Ne begins to dominate and serves as a break-out from the cycle (dot-dashed). Only at very high (unrealistic) densities begins the ^{15}O(2p, γ)^{17}Ne reaction to dominate. Thus the two-proton capture reaction is not an important reaction for the astrophysical network calculations.

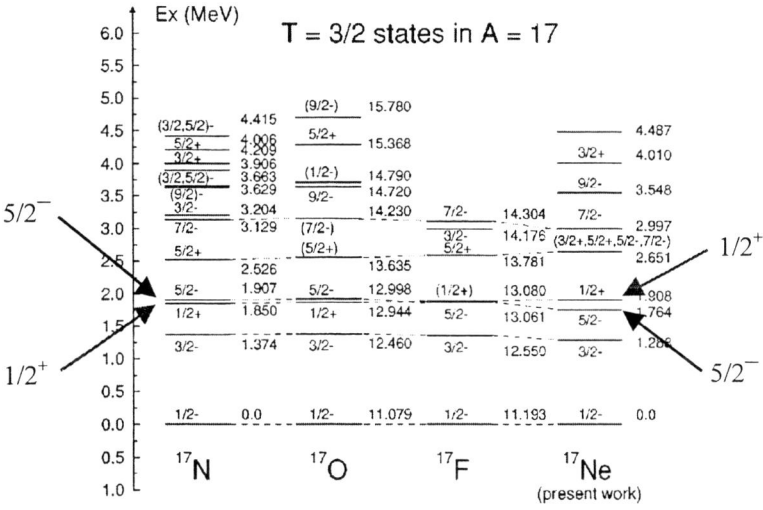

FIGURE 4. Level schemes of A = 17 nuclei. The level inversion of the 5/2⁻ and the 1/2⁺ states is indicated. Adapted from reference [15].

These calculations were based on levels in ^{17}Ne deduced from the mirror nucleus ^{17}N [10]. The recent measurement of ^{17}Ne levels found a level inversion of the 5/2⁻ and 1/2⁺ level in ^{17}Ne compared ^{17}N as shown in Figure 4 [15]. Including this level inversion in the reaction rate calculations result in a large increase of the 2p proton capture reaction at low temperatures as shown as the dashed curve in the left part of Figure 3. However, although it also reduced the boundary between the (α,γ) and the $(2p,\gamma)$ reactions in these temperature region to lower densities, it is still not sufficient to make a significant contribution to the breakout. The break-out of the hot CNO cycle is still dominated by the (α,γ) reaction.

Although in this particular case the detailed structure had no influence on the network calculations this example was meant to demonstrate that even small changes of individual levels can have a large influence on reaction cross sections.

Two-proton capture reactions are predicted to have a significant impact in the medium mass region of the rp-process [19]. For example ^{68}Se is a waiting point of the rp-process which can be bypassed via the reaction ^{68}Se$(2p,\gamma)^{70}$Kr. Coulomb excitation reaction like ^{17}Ne$(\gamma,2p)^{15}$O discussed above can give important information for the inverse reaction. Thus measuring ^{70}Kr$(\gamma,2p)^{68}$Se would be an important reaction which should be feasible with the next generation radioactive beam facilities. Of course the mass (and structure of excited levels) of the intermediate (unbound) nucleus ^{69}Br is also important and can be studied with the reaction ^{9}Be$(^{70}$Br,^{69}Br) similar to the methods used to study ^{12}O and ^{11}N [1,7,8].

CONCLUSIONS

In conclusion, we observed evidence for two-proton radioactivity of the first excited state in ^{17}Ne. The preliminary lifetime of 0.9 ps indicates a significant contribution of $d_{5/2} \otimes s_{1/2}$ configuration of the $3/2^-$ state. The angular distribution of the two protons in the decay is statistical not significant to make definite conclusions, although it does not seem to be strongly correlated (diproton) nor strongly anti-correlated (two protons on the opposite site of the fragment). Using an improved experimental setup with optimized efficiency and energy resolution will help to clarify the remaining uncertainties in the context of the reported first evidence for two-proton radioactivity.

In addition, it has been shown that relativistic (or intermediate) energy Coulomb excitation of exotic nuclei along the proton dripline can be important for astrophysical processes. The ground state contribution to two proton capture reactions can be deduced from the inverse Coulomb excitation reaction.

Acknowledgments

One of us (MJC) acknowledges the support of the "Studienstiftung des Deutschen Volkes". We acknowledge the help of A. Azhari, M. Fauerbach, T. Glasmacher, R. Ibbotson, R. A. Kryger, H. Scheit, P. J. Woods and S. Yokoyama during the experiment and thank J. Brown, D. J. Morrissey and M. Steiner for producing the radioactive ^{17}Ne beam. This work was supported by the National Science Foundation under grant PHY95-28844.

References

1. R. A. Kryger *et al.*, Phys. Rev. Lett. **74**, 860, (1995).
2. B. Alex Brown, Phys. Rev. C **43**, R1513 (1991).
3. W. E. Ormand, Phys. Rev. C **55**, 2407 (1997).
4. B. Blank *et al.*, Phys. Rev. Lett., accepted for publication (1999).
5. J. Giovinazzo *et al.*, these proceedings.
6. L. Axelsson *et a.,l* Phys. Rev. C **54**, R1511 (1996).
7. A. Azhari, et al., Phys. Rev. C **57**, 628 (1998).
8. A. Azhari, R. A. Kryger, and M. Thoennessen, Phys. Rev. C 58, 2568 (1998).
9. M. J. Chromik, B. A. Brown, M. Fauerbach, T. Glasmacher, R. Ibbotson, H. Scheit, M. Thoennessen, and P. Thirolf, Phys. Rev. C **55**, 1676 (1997).
10. Joachim Görres, Michael Wiescher and Friedrich-Karl Thielemann, Phys. Rev. C **51**, 392 (1995)
11. M. J. Chromik, P. G. Thirolf, M. Thoennessen, M. Fauerbach, T. Glasmacher, R. Ibbotson, R. A. Kryger, H. Scheit, and P. J. Woods, Proceedings of the 2nd International Conference on Exotic Nuclei and Atomic Masses, edited by B. M. Sherrill, D. J. Morrissey, and C. N. Davids, AIP Conference Proceedings **455**, p. 286 (1998).
12. M. J. Chromik, P. G. Thirolf, M. Thoennessen, M. Fauerbach, T. Glasmacher, R. Ibbotson, R. A. Kryger, H. Scheit, and P. J. Woods, Proceedings of the International Conference on Nuclear Structure '98, edited by C. Baktash, AIP Conference Proceedings **481**, p. 187 (1999).
13. P. G. Thirolf, M. J. Chromik, M. Thoennessen, M. Fauerbach, T. Glasmacher, R. Ibbotson, R. A. Kryger, H. Scheit, and P. J. Woods, Proceedings of the 7th International Conference on Clustering Aspects of Nuclear Structure and Dynamics, to be published in World Scientific (1999).
14. V. Guimarães *et al.*, Z. Phys. A **353**, 117 (1995).

15. V. Guimarães *et al.*, Phys. Rev. C **58**, 116 (1998).
16. M. J. Chromik *et al.*, to be published.
17. M. V. Zhukov and I. J. Thompson, Phys. Rev. C **52**, 3505 (1995).
18. *Table of Isotopes*, 8th ed., edited by R. B. Firestone and V. S. Shirley (Wiley, New York 1996).
19. H. Schatz *et al.*, Phys. Rep. **294**, 167 (1998).

In-beam Studies of Proton Emitters using the Recoil-Decay Tagging Method [1]

D. Seweryniak[1,2], P.J. Woods[3], J.J. Ressler[2], C.N. Davids[1],
A. Heinz[1], A.A. Sonzogni[1], J. Uusitalo[1], W. B. Walters[2],
J.A. Caggiano[1], M.P. Carpenter[1], J.A. Cizewski[4], T. Davinson[3],
K.Y. Ding[4], N. Fotiades[4], U. Garg[5], R.V.F. Janssens[1], T.-L. Khoo[1],
F.G. Kondev[1], T. Lauritsen[1], C.J. Lister[1], P. Reiter[1], J. Shergur[2],
I. Wiedenhoever[1]

[1] *Argonne National Laboratory, Argonne, Illinois, USA*
[2] *University of Maryland, College Park, Maryland, USA*
[3] *University of Edinburgh, Edinburgh, United Kingdom*
[4] *Rutgers University, New Brunswick, New Jersey, USA*
[5] *University of Notre Dame, Notre Dame, Indiana, USA*

Abstract. The last five years have witnessed a rapid increase in the volume of data on proton decaying nuclei. The path was led by decay studies with recoil mass separators equipped with double-sided Si strip detectors. The properties of many proton decaying states were deduced, which triggered renewed theoretical interest in the process of proton decay.

The decay experiments were closely followed by in-beam γ-ray studies which extended our knowledge of high-spin states of proton emitters. The unparalleled selectivity of the Recoil-Decay Tagging method combined with the high efficiency of large arrays of Ge detectors allowed, despite small cross sections and overwhelming background from strong reaction channels, the observation of excited states in several proton emitters.

Recently, in-beam studies of the deformed proton emitters ^{141}Ho and ^{131}Eu have been performed with the GAMMASPHERE array of Ge detectors and the Fragment Mass Analyzer at ATLAS. Evidence was found for rotational bands in ^{141}Ho and ^{131}Eu. The deformations and the single-particle configurations proposed for the proton emitting states from the earlier proton-decay studies were confronted with the assignments deduced based on the in-beam data. It should be noted that the cross section for populating ^{131}Eu is only about 50 nb, and it represents the weakest channel ever studied in an in-beam experiment.

---

[1] This work was supported by the U.S. Department of Energy, Nuclear Physics Division, under contracts No. W-31-109-ENG-38 and No. DE-FG02-94-ER40834

INTRODUCTION

The domain of nuclei situated far from the line of β stability has always been an arena of numerous experimental pursuits and a testing ground for new theoretical models of nuclear behavior. With the recent advances in experimental techniques, and with radioactive beams on the horizon, the physics of nuclei with an excess of neutrons or protons has become one of the focal points of nuclear physics. In particular, nuclei at the drip lines are expected to draw a lot of attention. They define the very limits of nuclear existence and will be susceptible more than any other nuclei to phenomena associated with low binding energy such as halos, skins or mixing with the continuum. Because protons are expected to be kept in check by the Coulomb barrier, these effects are expected to play a more important role for neutron-rich nuclei. However, the neutron drip line can be accessed experimentally only for a handful of light elements. On the other hand, the proton drip line can be studied experimentally now, as shown by recent rapid progress in proton-decay studies (see Ref. [1] for the most recent review, and several other papers in these proceedings for the latest results). In fact, more detailed studies of proton emitters are already possible. Their excited states have been studied, providing independent information on the structure of the proton emitters and elucidating their behavior at high spin. This has been made possible by combining in-beam spectroscopic techniques with the selectivity of proton and α decay studies. The experimental techniques used in these experiments will be described in more detail in the following section.

RECOIL-DECAY TAGGING

Nuclei at the proton drip line are in general produced in heavy-ion fusion-evaporation reactions using the most neutron deficient beam-target combinations available. For these reactions, the compound system emits mainly protons, rarely α particles and very seldom neutrons. The cross sections drop rapidly with the number of emitted neutrons. Typically, about 20-30 channels are open. As a result, a highly efficient and selective detection system is necessary to observe weak proton-rich evaporation channels. It is especially true for in-beam γ-ray studies where 10-20 γ-ray transitions are emitted per reaction channel, and weak γ-ray transitions are buried under Compton scattered events originating from strong reaction channels. One can reach cross sections as low as 10μb with conventional methods, such as the detection of light evaporation particles, or the measurement of the mass and the atomic number of recoils using a recoil mass separator. The cross sections for producing proton emitters do not exceed 100μb. For the deformed proton emitters discussed in this paper the cross sections are below 1μb and in the case of the isomer in ^{141}Ho only 50 nb. The required selectivity was achieved using the Recoil-Decay Tagging method. In this method, the proton or α decays detected using a recoil mass separator, equipped with a double-sided Si strip detector, are used to tag prompt γ-ray transitions. This technique was first implemented at GSI using the SHIP separator combined with NaI detectors to detect prompt γ rays [2], and independently at Daresbury using for the first time an array of Ge detectors in combination with a recoil mass separator [3]. Subsequently, the method was implemented at the Argonne National Laboratory, the University of Jyväskylä, and the Oak Ridge National Laboratory. Already in the first RDT experiment some preliminary results were reported on γ-ray transitions in the proton emitter ^{109}I [3]. In an experiment at ATLAS with the Aye-Ball array of Ge detectors and the Argonne fragment mass analyzer a ground-state band in ^{147}Tm was observed and was interpreted as a rotationally aligned $h_{11/2}$ band, and a moderate deformation of β=0.13 was deduced for the ground state [4]. The availability of even more efficient arrays of Ge detectors led to further advances in in-beam studies of proton emitters. Gamma-ray transitions in ^{151}Lu were identified [5]. Excited $h_{11/2}$ bands were found in the moderately deformed proton emitters ^{109}I [6] and ^{113}Cs [7], but the decay of these bands to the ground state was not determined. A complex level scheme including a decoupled $h_{11/2}$ band was also constructed for ^{167}Ir [8]. Finally, after the discovery of the first highly deformed proton emitters ^{141}Ho and ^{131}Eu [9], and the observation of proton fine structure in ^{131}Eu [10], attempts were made to find evidence for rotational bands in these nuclei. The following chapter reports on the results of these experiments.

ROTATIONAL BANDS IN ^{141}HO AND ^{131}EU

Experiments

The Recoil-Decay Tagging method was used to study excited states in the proton emitters ^{141}Ho and ^{131}Eu. Fig. 1 shows the implementation of the Recoil-Decay Tagging at ATLAS. The prompt γ rays were detected using the array of 101 Compton suppressed HPGe detectors GAMMASPHERE. The recoiling evaporation residues were dispersed in the Argonne fragment mass analyzer (FMA) according to their mass-to-charge-state ratio. Behind the focal plane of the FMA the recoils were implanted into a double-sided Si strip detector (DSSD) where they subsequently decayed. The front and back side of the 60 μm thick, 40 mm by 40 mm DSSD were divided into 40 horizontal and 40 vertical strips, respectively, forming 1600 quasi-pixels. Using spatial and time correlation, the decays were associated with their parent nuclei and the prompt γ rays emitted from their excited states. This allowed the assignment of γ rays to particular reaction channels based on the characteristic proton decays of the implants.

Two proton lines have been observed in ^{141}Ho. The 1169(8)-keV line with a half-life of 4.4(4) ms, corresponding to the ground-state proton decay [9], and the 1230(20)-keV line with a half-life of 8$\pm$3 μs associated with an isomer [12]. In

^{131}Eu the protons have an energy of 932(7) keV and a half-life of 17.8(19) ms, and a 24(5)% decay branch to the 2^+ state in the daughter nucleus was found [10].

In the first experiment a ^{54}Fe beam at 292 MeV from ATLAS impinged on a 0.7 mg/cm^2 ^{92}Mo target to produce ^{141}Ho as the p4n evaporation channel. In order to increase the statistics, a second experiment was performed using inverse kinematics. Thanks to the increase in the efficiency of the FMA, a factor of about 4 was gained in the proton yield. In Fig. 2(a) the γ-ray spectrum tagged by the proton decay of the ground state in ^{141}Ho is shown. Fig. 2(b) shows the sum of selected γ-γ coincidence gates. The transitions marked with stars in Fig. 2(b) are in coincidence with each other and form a regular sequence, most likely the ground-state rotational band. In Fig. 3 the γ-ray spectrum tagged by the proton decay of the isomeric state in ^{141}Ho is shown. Only 4 relatively strong γ-ray transitions are present in this spectrum. Although the statistics were not sufficient to obtain coincidence relationships between these transitions, it is plausible that they are in coincidence with each other and form a rotational band built on the isomer. Assuming transport efficiency of 10% for the FMA, cross sections of 250 nb and 50 nb were deduced for populating the ground state and the isomeric state in ^{141}Ho.

In the second experiment a 0.75 mg/cm^2 thick ^{58}Ni target was bombarded with a 402-MeV ^{78}Kr beam to study the p4n channel leading to ^{131}Eu. In Fig. 4 γ rays which were correlated with the ground-state proton decay of ^{131}Eu are shown. The γ-ray spectrum assigned to ^{131}Eu is much more complex than those obtained for ^{141}Ho. Already a simple inspection of Fig. 4 suggests that more than one γ-ray

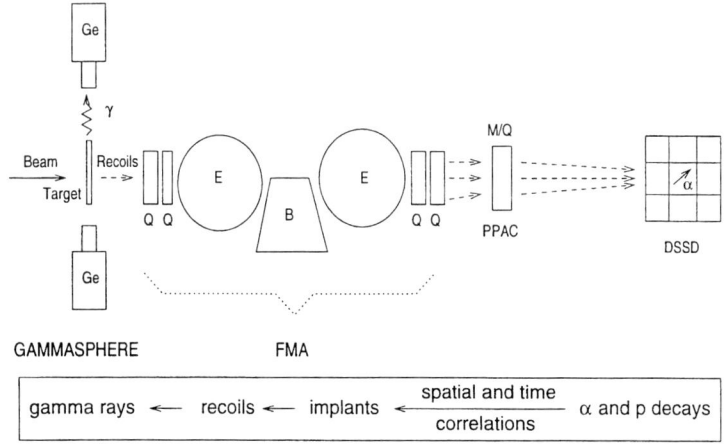

FIGURE 1. The implementation of the Recoil-Decay Tagging at ATLAS.

sequence must be present. It is worth noting that the γ-ray spectrum tagged by the proton decay to the 2^+ excited state resembles the γ-ray spectrum correlated with the ground state-to-ground state proton decay. This confirms unambiguously that both proton lines in ^{131}Eu are emitted from the same state. This is in contrast to the situation in ^{141}Ho, where there is no overlap between the γ-ray spectra correlated with the two proton lines. Assuming an efficiency of 10% for the FMA a cross section of 90 nb was deduced for populating the ground state of ^{131}Eu.

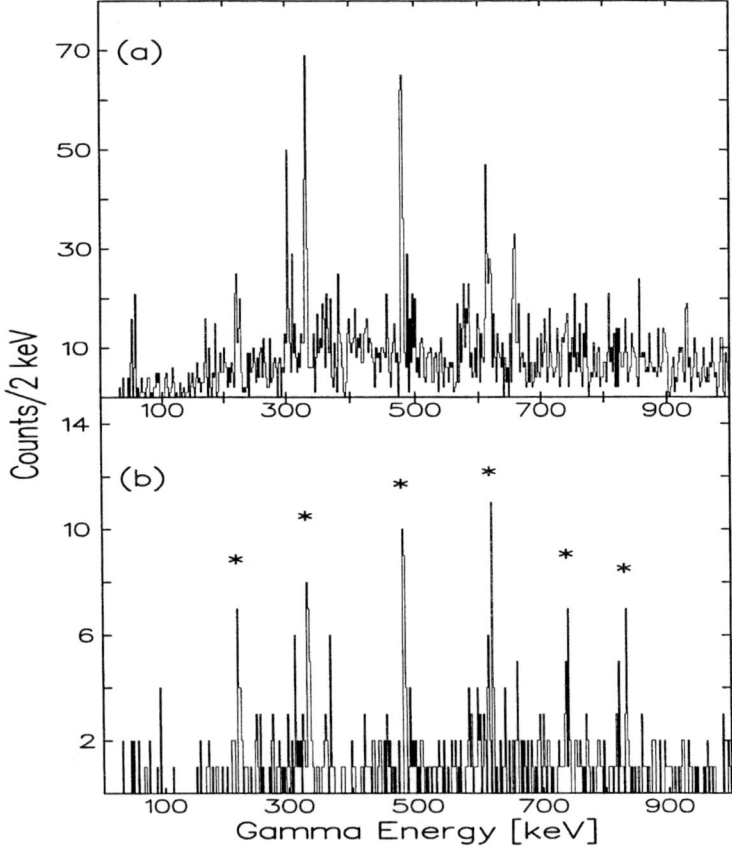

FIGURE 2. (a) The spectrum of γ rays tagged with the protons emitted from the ground state in ^{141}Ho. (b) The sum of the proton tagged γ-γ coincidence gates. The transitions used as gates are marked with stars.

Discussion

A deformation of $\beta=0.29$ was calculated for the ground state of ^{141}Ho [11]. At this deformation the $7/2^-[523]$ Nilsson orbital originating from the $h_{11/2}$ spherical state is predicted to be the ground state. The $1/2^+[411]$ $d_{3/2}$ state is expected to be located close in energy to the ground state. In addition, several orbitals originating from the $g_{7/2}$ and $d_{5/2}$ spherical states should lie at higher excitation energies. The observed proton-decay rates from the ground state and the isomeric state in ^{141}Ho are in agreement with calculations only if the $7/2^-[523]$ and $1/2^+[411]$ configurations are assigned to the ground state [9] and to the isomeric state [12], respectively. In Ref. [9] a deformation of $\beta=0.30$ was found to best fit the data.

In ^{131}Eu, the $3/2^+[411]$ $d_{5/2}$ state and $5/2^+[413]$ $g_{7/2}$ states are predicted to be located close to the Fermi surface at the calculated deformation of $\beta=0.33$ [11]. The ground-state proton decay rate is consistent with the theory for both the $3/2^+[411]$ and $5/2^+[413]$ configurations and deformations of about $\beta=0.35$ [9]. However, the observed proton-decay branch to the excited 2^+ state in the daughter nucleus unambiguously favors the $3/2^+[411]$ assignment [10].

There are no data available on the excited states of the proton-decay daughter nuclei, neighboring odd-Z isotones, and even-N isotopes of both ^{141}Ho and ^{131}Eu.

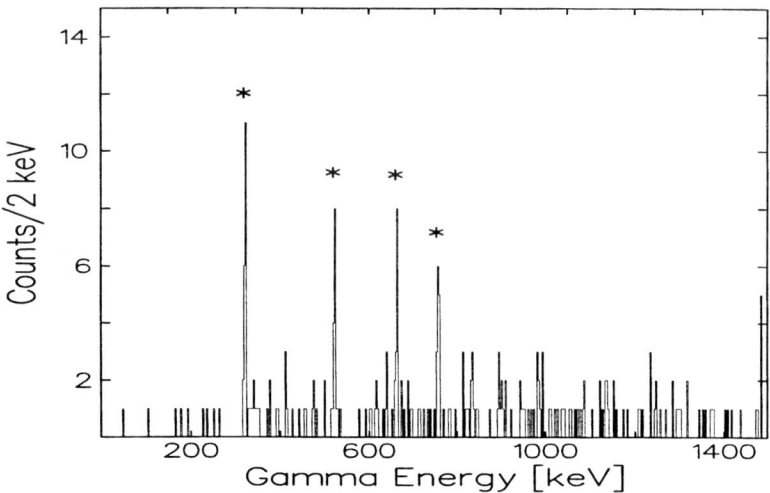

FIGURE 3. The spectrum of γ rays tagged with the protons emitted from the isomeric state in ^{141}Ho.

The data on nuclei situated further away is limited to a few transitions in ground-state rotational bands for even-even systems, and in bands based on the $h_{11/2}$ orbital for odd-Z nuclei. This situation makes any systematic comparison difficult, especially in a region where deformation changes rapidly with the number of valence nucleons. In nuclei with Z>50 and N<82 which have enough valence protons and neutrons to develop deformation, strongly populated bands based on the $h_{11/2}$ orbital are a common occurrence. In most of these cases the $h_{11/2}$ proton is aligned with the axis of rotation because a low-K orbital is involved (above Z=50) or because of the small deformation (along the N=82 shell gap). One expects the $h_{11/2}$ bands to be strongly populated in ^{141}Ho and ^{131}Eu as well. However, due to the larger deformation and the Fermi surface moving toward the medium-K $h_{11/2}$ orbitals, strong coupling might be energetically favored. Strongly coupled bands based on the medium-K $h_{11/2}$ orbitals have been observed on the other side of the N=82 shell gap at comparable deformation (see, for example, ^{157}Ho [13]).

The differences between transition energies in the $h_{11/2}$ band marked in Fig. 2(b) suggest that it is a cascade of stretched quadrupole transitions. Fig. 5 shows the dynamic moment of inertia $\mathcal{J}^{(2)}$ as a function of rotational frequency ω for the rotational bands in ^{141}Ho and ^{131}Eu. $\mathcal{J}^{(2)}$ increases gradually up to $\omega \approx 0.4 MeV$

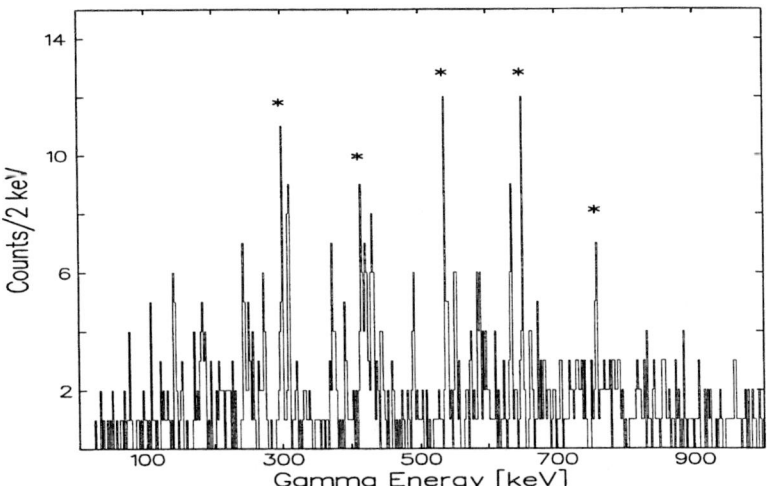

FIGURE 4. The spectrum of γ rays tagged with the protons emitted from the ground state of ^{131}Eu. The transitions marked with stars could possibly form an $h_{11/2}$ band.

for the ground-state band. The first crossing is expected to be due to the alignment of a pair of $h_{11/2}$ protons and was observed to take place at a rotational frequency of about $0.25 MeV$ in this region. The dynamic moment of inertia of the ground-state band does not exhibit any crossing at such a low rotational frequency. This suggests that the band is built on the $h_{11/2}$ orbital since, as a result, the crossing is blocked. The observed transitions could form the favored sequence $27/2^- \to 23/2^- \to 19/2^- \to 15/2^- \to 11/2^- \to 7/2^-$ of the $7/2^-[523]$ band. Contrary to expectations the unfavoured signature partner does not seem to be populated. Triaxiality could explain a larger than expected signature splitting.

It was shown by Mueller [14] that deformation is correlated with the J_0 parameter in the Harris expansion of the dynamic moment of inertia as a function of the rotational frequency: $\mathcal{J}^{(2)} = J_0 + 3J_1\omega^2$. Using this approach a deformation of $\beta=0.28\pm0.04$ is deduced for the ground state band in ^{141}Ho.

The transitions correlated with the proton decay of the ^{141}Ho isomer have similar energy spacings as seen in the ground-state band at low energies, but there is a compression at higher energies. As can be seen in Fig. 5, the dynamic moment of inertia for this band starts to increase at $\omega \approx 0.2 MeV$, indicative of a low-lying band crossing. In fact, a careful inspection of Fig. 3 reveals several weak transitions in the region where the backbending takes place. Since we observed only one signature partner for this band it must have a significant signature splitting. Among the non-$h_{11/2}$ orbitals which are located near the Fermi surface only the $1/2^+[411]$

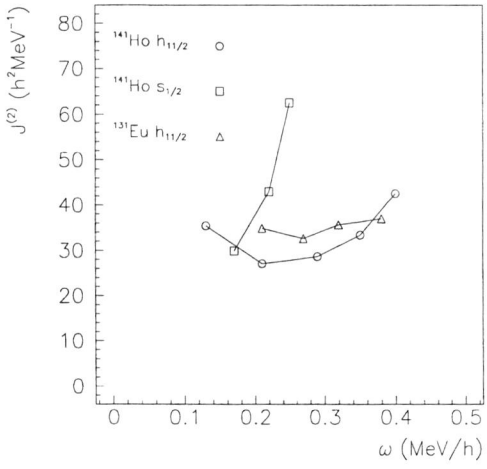

FIGURE 5. The dynamic moments of inertia as a function of rotational frequency for the rotational bands in ^{141}Ho and ^{131}Eu.

band is expected to exhibit a large signature splitting. Thus, the observed γ-ray transitions could form the $15/2^+ \to 11/2^+ \to 7/2^+ \to 3/2^+$ favored signature band. The low energy transition to the $1/2^+$ band-head and the unfavored signature band have so far remained unobserved.

Due to lower statistics and the complex γ-ray spectrum 4, band assignments in ^{131}Eu are more difficult. The γ-ray transitions marked in Fig. 4 could be good candidates for an $h_{11/2}$ band. Their regular spacing indicates that the first crossing is blocked. The dynamic moment of inertia for this band is larger than that for the analogous $h_{11/2}$ band in ^{141}Ho. As a result, a slightly higher deformation of $\beta=0.34\pm0.05$ was obtained using the same method as for ^{141}Ho. The remaining transitions in Fig. 4 most likely form a strongly coupled band. They could correspond to the members of the $3/2^+[411]$ $d_{5/2}$ band which was observed in ^{133}Pm [15]. In particular, transitions below 120 keV could be the lowest lying M1 transitions connecting the two signature-partner bands. At the moment the structure of this band is not known, but the data analysis is still in progress.

OUTLOOK

To draw more firm conclusions on the structures observed in ^{141}Ho and ^{131}Eu one needs data with better statistics. In particular, γ-γ coincidence relations are needed for ^{131}Eu. This would require a detection system with higher efficiency and/or a system which can handle more intense beams. In principle, several other deformed proton emitters could be studied in-beam. ^{117}La [16] and ^{145}Tm [17] nuclei are the most promising cases for in-beam studies since their production cross sections are comparable to those of ^{141}Ho and ^{131}Eu. Other, not yet discovered, Pr, Pm and Tb proton emitters could be another possibility, although the cross sections might be prohibitively low if p6n reaction channels have to be used. Finally, one could search for light deformed proton emitters between ^{56}Ni and ^{100}Sn, provided that their half-lives lie within the observation window of the existing detection systems. Obviously, intense radioactive beams will help to study all the cases listed above by means of much larger cross sections.

SUMMARY

The Recoil-Decay Tagging method has established itself as a powerful technique for in-beam studies of proton emitters. With state-of-the-art detection systems, reaction channels with cross sections as low as 50 nb have been observed. In ^{141}Ho and ^{131}Eu rotational bands were observed and deformations of $\beta \approx 0.28$ and $\beta \approx 0.34$, respectively, were deduced in agreement with the values obtained from experimental proton-decay rates and theoretical calculations. In addition, the dynamic moments of inertia extracted for the bands in ^{141}Ho support the configuration assignments proposed from the analysis of the proton-decay rates. Because of low statistics and a more complex γ-ray spectrum in ^{131}Eu, firm conclusions on band assignments have not yet been made. The study of excited states in deformed proton emitters has already proven to be a source of valuable information on the structure of the proton decaying states, and has shed light on the response of proton emitters to the stress of rotation.

REFERENCES

1. P.J. Woods and C.N. Davids, *Annu. Rev. Nucl. Part. Sci.* **47**, 541 (1997).
2. R.S. Simon et al., *Nucl. Phys.* **A325**, 197 (1986).
3. E.S. Paul et al., *Phys. Rev.* **C55**, R2137 (1997).
4. D. Seweryniak et al., *Phys. Rev.* **C56**, R723 (1997).
5. C.-H. Yu et al., *Phys. Rev.* **C58**, R3042 (1998).
6. C.-H. Yu et al., *Phys. Rev.* **C59**, R1834 (1999).
7. C.J. Gross et al., in *Proceedings of the 2nd International Conference on Exotic Nuclei and Atomic Masses*, Shanty Creek, USA, 1998, edited by B.M. Sherrill, D.J. Morrissey and C.N. Davids (AIP, New York, 1998), p. 444.
8. M.P. Carpenter et al., *Acta Physica Polonica* **30**, 581 (1999).
9. C.N. Davids et al., *Phys. Rev. Lett.* **80**, 1849 (1998).
10. A.A. Sonzogni et al., *Phys. Rev. Lett.* **83**, 1116 (1999).
11. P. Möller et al., *At. Nucl. Data Tables* **59**, 185 (1995).
12. K. Rykaczewski et al., *Phys. Rev.* **C60**, 011301 (1999).
13. D.C. Radford et al., *Nucl. Phys.* **A545**, 665 (1992).
14. W.F. Mueller, PhD thesis, University of Tennessee, Knoxville, May 1997.
15. A. Galindo-Uribarri et al., *Phys. Rev.* **C54**, 1057 (1996).
16. F. Soramel, these proceedings.
17. J.C. Batchelder et al., *Phys. Rev.* **C57**, R1042 (1998).

Spectroscopic Factors
and
Orbital Mixing

Spherical Proton Emitters and Spectroscopic Factors

Paul B. Semmes

Physics Department, Tennessee Technological University, Cookeville, TN 38505

Abstract. The most commonly used implementation of the WKB method for calculating spherical decay rates is shown to be unsatisfactory, and some improvements are suggested. Also, the smaller than expected experimental spectroscopic factor for the $1d_{3/2}$ state in ^{151}Lu can be explained in a particle-core coupling calculation.

INTRODUCTION

In a typical experimental study of a proton-emitting isotope, the energy of the emitted protons and the total half-life of the decaying state are measured. The partial half-life for proton emission, $t_{1/2,p}$, can be deduced from the total half-life once the branching fraction for proton emission is known. Ideally, the branching fraction is measured by observing the competing decay modes of the same parent state (i.e., β^+/EC decay), but in many cases, the partial half-life for β^+/EC decay is simply estimated from calculations. Given a suitable method for calculating the expected proton emission rate, the experimental spectroscopic factor is defined as the ratio of the theoretical and experimental proton partial half-lives,

$$S_p^{\exp} = \frac{t_{1/2,p}^{th}}{t_{1/2,p}^{\exp}}. \qquad (1)$$

The calculated decay rates depend very sensitively on the decay energy (Q_p) and on the orbital angular momentum ℓ of the emitted proton. Although a number of different methods and parameter sets have been used to calculate $t_{1/2,p}^{th}$, the uncertainties in these calculated values are relatively small, compared to the typical uncertainties that result from the experimental decay energy, half-life and branching fraction. Consequently, at least for spherical nuclei, the experimental spectroscopic factor can be used to identify the quantum numbers n, ℓ, j of the proton in the parent state (1-3).

One of the important developments in the last few years has been the progression from the *qualitative* assignments of orbital quantum numbers to a more *quantitative* analysis of the spectroscopic factors which then allows a comparison with nuclear structure models through the theoretical expression (4-6)

$$S_p^{th} = \frac{1}{2I_i+1}\left|\langle I_i \| a_{n\ell j}^+ \| I_f \rangle\right|^2. \qquad (2)$$

However, in order to extract an experimental spectroscopic factor, the theoretical $t_{1/2,p}^{th}$ value must be obtained with some care. In this contribution, two issues will be considered: First, a critical evaluation of the usual semiclassical estimate of the decay rate based on the WKB method is presented. Second, an estimate of the effect of <u>weak</u> quadrupole collectivity on the theoretical spectroscopic factors of some nuclei in the range $71 \leq Z \leq 75$ is given.

SPHERICAL DECAY RATE CALCULATIONS

All of the spherical decay rate calculations are based on a one-body model in which the quasibound proton moves in an effective potential

$$V_{eff}(r) = V_{WS}(r) + V_{\ell s}(r) + V_{Coul}(r) + V_\ell(r). \qquad (3)$$

The first three terms in this equation are the spherical Woods-Saxon potential, the spin-orbit interaction, and the Coulomb interaction. Each of these potentials contains some parameters (nuclear radius, diffuseness, charge radius, strength parameters) which are usually taken from the Becchetti-Greenlees optical model potential (7) or something similar. The fourth term is the "centrifugal potential"

$$V_\ell(r) = \frac{\hbar^2 \ell(\ell+1)}{2\mu r^2}. \qquad (4)$$

The most commonly used method of calculating $t_{1/2,p}^{th}$ is a version of the semiclassical WKB approach, in which the decay rate λ is written as the product of a frequency factor ν and a barrier penetration P

$$\lambda = \frac{\ln 2}{t_{1/2}} = \nu P. \qquad (5)$$

The barrier penetration is

$$P = \exp\left\{-2\sqrt{\frac{2\mu}{\hbar^2}} \int_{r_1}^{r_2} \sqrt{V(r) - Q_p}\, dr\right\} \qquad (6)$$

where the integral in Eq. 6 is taken over the forbidden region between the classical turning points r_1 and r_2. The frequency factor is obtained from the normalization of the WKB wavefunction in the classically allowed region

$$\nu = \frac{N\hbar}{4\mu} \qquad (7)$$

where

$$N^{-1} = \frac{1}{2}\sqrt{\frac{\hbar^2}{2\mu}} \int_{r_0}^{r_1} \frac{dr}{\sqrt{Q_p - V(r)}}. \qquad (8)$$

Thus, if all the parameters of the potential are known, this provides one method of estimating the decay rate. The most common implementation of the WKB method for proton emission uses a much simpler expression for the frequency factor (1,2,6), i.e.,

$$v = \sqrt{2}\pi^2\hbar^2\left[\mu^{3/2}R_c^3\sqrt{(zZe^2/R_c - Q_p)}\right]^{-1}. \quad (9)$$

However, this expression is borrowed from the theory of α decay where the total potential was simply assumed to be a square well inside the charge radius R_c and the Coulomb potential beyond this radius. Using this expression for the frequency factor effectively replaces the potential $V_{eff}(r)$ inside the classically allowed region with a simple square well and by ignoring angular momentum effects (s-wave emission only). Such a drastic approximation cannot be expected to provide more than a qualitative estimate of the decay rate.

How much error is introduced into the WKB calculations by this extreme simplification of the frequency factor? How accurate is the WKB method, when compared to a better calculation? A partial answer can be found in Ref.(5), where calculations were made using the Distorted Wave (DW) method and the Two-Potential Approach (TPA) and compared with WKB results. The DW and TPA calculations are both perturbation methods, and should be essentially exact since the decay width is so small. Both begin by solving an eigenvalue problem, for the quasistationary state in the DW method, and for a bound state approximation in the TPA. However, instead of finding the energy eigenvalue corresponding to a given potential $V_{eff}(r)$, one of the parameters of the potential must be adjusted so that the eigenvalue matches the desired decay energy Q_p. In practice, the depth of the Woods-Saxon potential was adjusted, and so this parameter was not exactly at the Becchetti-Greenlees value (7). Also note that the strength of the spin-orbit interaction was chosen to be proportional to the Woods-Saxon strength, so that it was also adjusted in the fitting process. Furthermore, the Coulomb radius, R_c, was taken to be the same as the Woods-Saxon radius R_0, rather than a separate value (5).

One of the results from Ref. (5) is that the decay rates obtained from the TPA and DW calculations were shown to be practically identical. Furthermore, the WKB calculations using the normalization given in Eq. (8), and using the same parameters for the potential as in the DW or TPA (including the adjusted depth of the Woods-Saxon potential) agreed with the DW and TPA values within ~10% or less. (Note that two different normalizations were considered in the WKB calculations in Ref. (5), and the labels were inadvertently interchanged there.)

A comparison of these results with the WKB calculations using the simplified frequency factor and strictly using the Becchetti-Greenlees parameters (with no adjustment of the depth of the potential) can be made by examining Table IV of Ref. (5) and Table 1 of Ref. (6). This comparison has been made by calculating the ratios of the WKB and DW proton half-lives, i.e. $t_{1/2,p}(WKB)/t_{1/2,p}(DW)$, for all cases that use the same experimental decay energy. The ratios tend to cluster around different values for different orbitals, and so the averages of these ratios are presented separately for the different orbitals in Table 1.

TABLE 1. Comparing WKB and DW half-lives

Orbital	$t_{1/2,p}(WKB)/t_{1/2,p}(DW)$
$1d_{5/2}$	0.59
$0h_{11/2}$	1.24
$1d_{3/2}$	0.86
$2s_{1/2}$	0.77

Thus, on average, the WKB half-life calculated for a $1d_{5/2}$ orbital is only 59% of the DW (or TPA) result, while the WKB half-life for a $0h_{11/2}$ orbital is 24% larger than the DW value. Such large discrepancies are unacceptable, and the clear variation with the orbital considered is particularly troubling. This orbital sensitivity is largely due to the inappropriate approximation for the frequency factor, and the next most important shortcoming is that the Becchetti-Greenlees value for the potential depth is not equally appropriate for all orbitals. The other differences in the parameters of the potential used in Ref. (5) and the Becchetti-Greenlees values are essentially negligible.

The WKB calculations have the advantage of their simplicity, but they must be improved in order to be useful at a quantitative level. One obvious improvement is to replace the oversimplified frequency factor (Eq. 9) with the normalization condition (Eq. 8). A second improvement is to adjust the depth of the central potential V_0 to satisfy the semiclassical Bohr-Sommerfeld quantization condition

$$\int_{r_1}^{r_2} \left[\frac{2\mu}{\hbar^2} (Q_p - V(r)) \right]^{1/2} = (2n+1)\frac{\pi}{2} \qquad (10)$$

A third possible improvement is to use the Langer modified form of the centrifugal barrier, where the $\ell(\ell+1)$ term is replaced by $(\ell+1/2)^2$ (8,9).

To investigate these possible improvements in the WKB calculations, calculations have been made with five different WKB versions for four representative cases, and compared with the DW half-lives. The four cases considered are the $1d_{5/2}$ orbital for ^{113}Cs (DW $t_{1/2,p}$ = 5.40 × 10^{-7} sec), the $0h_{11/2}$ orbital for ^{147}Tm (DW $t_{1/2,p}$ = 2.60 sec), the $1d_{3/2}$ orbital for ^{166}Ir (DW $t_{1/2,p}$ = 2.10 × 10^{-2} sec), and the $2s_{1/2}$ orbital for ^{161}Re (DW $t_{1/2,p}$ = 1.90 × 10^{-4} sec). The five different versions of the WKB calculations are as follows: WKB1 is the form used in Refs.(1,2,6) with the simplified frequency factor and the Becchetti-Greenlees value for V_0; WKB2 is the form used in Ref. (5), with the normalization from Eq. (8) and the depth V_0 adjusted to the DW value; WKB3 uses the normalization from Eq. (8) but keeps V_0 at the Becchetti-Greenlees value; WKB4 uses the normalization from Eq. (8) and fits V_0 according to the Bohr-Sommerfeld condition; and finally WKB5 uses the normalization from Eq. (8), fits V_0 according to the Bohr-Sommerfeld condition and uses the Langer modified centrifugal term. The ratios of the WKB and DW proton half-lives, i.e. $t_{1/2,p}(WKB)/t_{1/2,p}(DW)$, are given in Table 2 for the four different cases considered.

The best overall agreement is achieved for WKB2, but this is somewhat misleading since the depth V_0 was adjusted to the fitted value from the DW (or TPA) eigenvalue problem. Clearly the DW or TPA calculations are preferred over any version of the WKB method, so one must assume that those results are not available. The WKB3 version gives good results for all but the $0h_{11/2}$ orbital; the problem here is that the Becchetti-Greenlees value for V_0 differs from the fitted depth by about 2 or 3 MeV.

TABLE 2. Comparing WKB and DW half-lives: $t_{1/2,p}$ (WKB) / $t_{1/2,p}$ (DW)

	$1d_{5/2}$ (^{113}Cs)	$0h_{11/2}$ (^{147}Tm)	$1d_{3/2}$ (^{166}Ir)	$2s_{1/2}$ (^{161}Re)
WKB1	0.59	1.23	0.87	0.75
WKB2	0.95	1.04	1.01	0.95
WKB3	1.01	1.16	0.94	0.92
WKB4	0.94	1.00	1.00	1.14
WKB5	1.01	1.07	1.06	0.99

The version WKB4 works well for all cases except the $2s_{1/2}$ orbital. This is because the Bohr-Sommerfeld quantization condition as written in Eq. (10) is not appropriate if the first turning point is at $r_1 = 0$. This deficiency is remedied by using the Langer modified form of the centrifugal barrier, as shown by WKB5. In fact, this version is essentially the same as the WKB calculations presented in Ref (3), except that the radius parameter was fitted through the quantization condition instead of V_0 in that work. Note that the largest discrepancy in the WKB5 calculations is for the $0h_{11/2}$ orbital, for which the WKB half-life is 7% larger than the DW value. Since the main assumption in the WKB approximation is that the potential should vary slowly over a wavelength, it is expected that this method will be less accurate for the low n orbitals. Thus it seems that using a WKB method to calculate $t_{1/2,p}^{th}$ introduces an uncertainty in the extracted experimental spectroscopic factor of about 5%. In most of the currently known proton emitters, this is small compared to the uncertainty from the decay energy Q_p, and/or the branching fraction.

WHY ARE THE $1d_{3/2}$ SPECTROSCOPIC FACTORS QUENCHED?

For spherical nuclei, assuming that the odd-Z parent state can be treated as a one-quasiproton state coupled to an inert core, then the theoretical spectroscopic factors can be calculated from the BCS expression

$$S_p^{th} = u_j^2 \tag{11}$$

where u_j^2 is the probability that the spherical orbital ($n\ell j$) is empty in the even-Z daughter nucleus. For most of the proton emitters that are expected to be spherical or nearly spherical in shape, the experimental spectroscopic factors agree very well with the values calculated from the BCS expression (5). Recent experiments for ^{151}Lu (10) found a $1d_{3/2}$ isomeric proton emitting state, as indicated by the decay energy and half-life. An experimental spectroscopic factor of $0.34^{+0.12}_{-0.08}$ was extracted using TPA calculations. This result is significantly lower than the BCS u_j^2 value of 0.73. In fact, as shown in Ref. (10), four of the five $1d_{3/2}$ proton emitters currently known have spectroscopic factors below the BCS value. The one point apparently in good agreement (^{156}Ta) should be regarded as an upper limit since the proton partial half-life was extracted from the experimental total half-life assuming a 100% proton branch (i.e., that β^+/EC decay is negligible).

One possible explanation for the reduction in the spectroscopic factor is that the 3/2+ parent state is not a pure $1d_{3/2}$ quasiproton state but is mixed with other low-lying configurations, e.g., $s_{1/2} \otimes 2^+$, or $d_{3/2} \otimes 2^+$. In order to test this suggestion, calculations have been made in the framework of the Core-Quasiparticle Coupling Model (11). In this approach, the Hamiltonian for the odd-mass system is written

$$H_{tot} = H_{s.p.} + H_{core} + H_{pair} - \kappa q \cdot Q. \tag{12}$$

where H_{core} is the Hamiltonian for the collective core, H_{pair} is the pairing interaction, and the last term is a quadrupole-quadrupole coupling between the particle and core. The single-particle Hamiltonian $H_{s.p.}$ is just the energies of the spherical $(n\ell j)$ orbitals included in the calculation. The collective core is regarded as "known" and is specified by the set of core energies E_R and reduced quadrupole matrix elements $\langle R' \| Q \| R \rangle$. The Hamiltonian matrix is diagonalized within a weak coupling basis

$$|I\rangle = |\tilde{j} \otimes R\rangle \tag{13}$$

where $\tilde{j}$ indicates a quasiparticle with particle and hole amplitudes u and v for each (j,R) basis state. In this approach, the theoretical spectroscopic factor for the decay to the ground state of the even-even daughter is just

$$S_p^{th} = u^2(j, R=0). \tag{14}$$

Calculations have been made for ^{151}Lu using the spherical Woods-Saxon single particle energies for the proton $0g_{7/2}$, $1d_{5/2}$, $2s_{1/2}$ and $1d_{3/2}$ orbitals, and the Fermi level and pairing gap parameters taken from the BCS calculation. The core was treated as an oblate rotor with $\beta = -0.15$, as suggested in Ref. (12).

The result of this calculation is that the mixing of the $d_{3/2} \otimes 0^+$ configuration with the other I = 3/2+ configurations reduces the spectroscopic factor $u^2(j, R = 0)$ to 0.45, which is just within the experimental limits of $0.34^{+0.12}_{-0.08}$. This calculation is robust in the sense that small changes the deformation, or considering a more vibrational description of the collective core produces very little change in the calculated spectroscopic factor.

Furthermore, similar calculations for the $s_{1/2}$ proton emitters ^{157}Ta, ^{161}Re and ^{167}Ir reduce their calculated spectroscopic factors very little. For example, for ^{161}Re, the experimental spectroscopic factor is 0.51 ± 0.10 and the spherical BCS u_j^2 value of 0.59 is in good agreement (5). The predicted deformation (12) is β = 0.08, which suggests a core structure that is more vibrational than rotational. When a variety of reasonable core descriptions are considered, the calculated $u^2(j, R = 0)$ varies between 0.55 to 0.50, and thus the agreement with the experimental result is not spoiled.

Finally, similar calculations for the $0h_{11/2}$ state in ^{151}Lu using the oblate rotor description for the core with $\beta = -0.15$ reduces the calculated spectroscopic factor very little. The experimental spectroscopic factor is 0.5 ± 0.4 and the spherical BCS u_j^2 value of 0.54 is in good agreement (5), although the error bars are large. When the rotor core description is included, the calculated $u^2(j, R = 0)$ is 0.45, which still agrees very well with the experimental value. Note that this calculation includes the Coriolis mixing between Nilsson levels, but it does not include any effects of the deformation on the decay process. Consequently, although these results might not be considered as definitive, they are very encouraging.

ACKNOWLEDGMENTS

Partial support for this work was provided by the U. S. Department of Energy under Grant No. DE-FG02-92ER40694.

REFERENCES

1. Hofmann, S., in *Particle Emission from Nuclei*, Boca Raton, CRC, 1989, ch. 2, pp. 5-72.
2. Hofmann, S., in *Nuclear Decay Modes*, Bristol, IOP, 1996, ch. 3, pp. 143-203.
3. Buck, B., Merchant, A. C., and Perez, S. M., *Phys. Rev. C* **45**, 1688-1692 (1992).
4. Davids, C. N., Woods, P. J., Batchelder, J. C., Bingham, C. R., Blumenthal, D. J., Brown, L. T., Busse, B. C., Conticchio, L. F., Davinson, T., Freeman, S. J., Henderson, D. J., Irvine, R. J., Page, R. D., Penttilm, H. T., Seweryniak, D., Toth, K. S., Walters, W. B., and Zimmerman, B. E., *Phys. Rev. C* **55**, 2255-2266 (1997).
5. Åberg, S., Semmes, P. B., and Nazarewicz, W., *Phys. Rev. C* **56**, 1762-1773 (1997); *Phys. Rev. C* **58**, 3011 (1998).
6. Woods, P. J., and Davids, C. N., *Annu. Rev. Nucl. Part. Sci.* **47**, 541-590 (1997).
7. Becchetti, F. D., and Greenlees, G. W., *Phys. Rev.* **182**, 1190-1209 (1969).
8. Langer, R. E., *Phys. Rev.* **51**, 669-676 (1937).
9. Morse, P. M., and Feshbach H., *Methods of Theoretical Physics, Part II*, New York, McGraw-Hill, 1953, p. 1101.
10. Bingham, C. R., Batchelder, J. C., Rykaczewski, K., Toth, K. S., Yu, C.-H., Ginter, T. N., Gross, C. J., Grzywacz, R., Karny, M., Kim, S. H., MacDonald, B. D., Mas, J., McConnell, J. W., Semmes, P. B., Szerypo, J., Weintraub, W., and Zganjar, E. F., *Phys. Rev. C* **59**, R2984-R2988 (1999).
11. Dönau, F., and Frauendorf, S., *Phys. Lett. B* **71**, 263-266 (1977).
12. Möller, P., Nix, J. R., Myers, W. D., and Swiatecki, W. J., *At. Data Nucl. Data Tables* **59**, 185-381 (1995).

Ground-State Properties of Deformed and Transitional Proton Emitters in the Relativistic Hartree-Bogoliubov Model

G.A. Lalazissis[1], D. Vretenar[1,2], and P. Ring[1]

[1] *Physik-Department der Technischen Universität München, D-85748 Garching, Germany*
[2] *Physics Department, Faculty of Science, University of Zagreb, Croatia*

Abstract. The Relativistic Hartree Bogoliubov (RHB) model is applied in the description of ground-state properties of proton-rich odd-Z nuclei in the region $53 \leq Z \leq 73$. For nuclei close to the drip-lines the RHB framework provides a unified and self-consistent description of mean-field and pairing correlations. The location of the proton drip-line, the ground-state quadrupole deformations and one-proton separation energies at and beyond the drip-line, the deformed single-particle orbitals occupied by the odd valence proton, and the corresponding spectroscopic factors are compared with available experimental data, and with predictions of the macroscopic-microscopic mass model.

INTRODUCTION

The structure and decays modes of nuclei beyond the proton drip-line present one of the most active areas of experimental and theoretical studies of exotic nuclei with extreme isospin values. In the last few years many new data on ground-state and isomeric proton radioactivity have been reported. In particular, detailed studies of odd-Z ground-state proton emitters in the regions $51 \leq Z \leq 55$ and $69 \leq Z \leq 83$ have shown that the systematics of spectroscopic factors is consistent with half-lives calculated in the spherical WKB or distorted-wave Born (DWBA) approximations [1,2]. More recent data [3,4] indicate that the missing region of light rare-earth nuclei contains strongly deformed systems at the drip-lines.

Two essentially complementary theoretical approaches have been used for the description of ground-state and isomeric proton radioactivity. One possibility is to start from a spherical or deformed phenomenological single-particle potential, a Woods-Saxon potential for instance, and to adjust the parameters of the potential well in order to reproduce the experimental one-proton separation energy. The width of the single-particle resonance is then determined by the probability of tunneling through the Coulomb and centrifugal barriers. Since the probability strongly depends on the valence proton energy and on its angular momentum, the calculated half-lives provide direct information about the spherical or deformed orbital occupied by the odd proton. For a spherical proton emitter it is relatively simple to calculate half-lives in the WKB or DWBA approximations [2]. It is much more difficult, however, to quantitatively describe the process of three-dimensional quantum mechanical tunneling for deformed proton emitters. Modern reliable models for calculating proton emission rates from deformed nuclei have been developed only recently [4,5]. A shortcoming of this approach is that it does not predict proton separation energies, i.e. the models do not predict which nuclei are likely to be proton emitters. In fact, if they are used to calculate decay rates for proton emission from excited states, the depth of the central potential has to to be adjusted for each proton orbital separately. In addition, such a description does not provide any information about the spectroscopic factors of the proton orbitals. Instead, experimental spectroscopic factors are defined as ratios of calculated and measured half-lives, and the deviation from unity is attributed to nuclear structure effects.

In Refs. [6-9] we have used the relativistic Hartree Bogoliubov (RHB) theory to calculate properties of proton-rich spherical even-even nuclei with $14 \leq Z \leq 28$, and to describe odd-Z deformed ground-state proton emitters in the region $53 \leq Z \leq 73$. RHB presents a relativistic extension of the Hartree-Fock-Bogoliubov theory, and it provides a unified framework for the description of relativistic mean-field and pairing correlations.Such a unified and self-consistent formulation is especially important in applications to drip-line nuclei. The RHB framework has been used to study the location of the proton drip-line, the ground-state quadrupole deformations and one-proton separation energies at and beyond the drip line, the deformed single particle orbitals occupied by the odd valence proton, and the corresponding spectroscopic factors. The results of fully self-consistent calculations have been compared with experimental data on ground-state proton emitters.

THE RELATIVISTIC HARTREE-BOGOLIUBOV MODEL

Models based on the relativistic mean-field approximation have been very successfully applied in the description of a variety of nuclear structure phenomena over the whole periodic table, from light nuclei to superheavy elements [10]. When also pairing correlations are included in the self-consistent Hartree-Bogoliubov framework, the relativistic mean-field theory can be applied to the physics of exotic nuclei at the drip-lines. The RHB model with finite range pairing interactions has been applied in studies of the halo phenomenon in light nuclei [11], properties of light nuclei near the neutron-drip [12], reduction of the spin-orbit potential in nuclei with extreme isospin values [13], ground-state properties of Ni and Sn isotopes [14], properties of Λ-hypernuclei with a large neutron excess [15], the deformation and shape coexistence phenomena that result from the suppression of the spherical N=28 shell gap in neutron-rich nuclei [16], properties of proton-rich nuclei and the phenomenon of ground-state proton radioactivity [6–9].

The relativistic mean field theory describes the nucleons as Dirac point particles. Inside a nucleus, protons and neutrons move independently in the mean fields which originate from the nucleon-nucleon interaction. Conditions of causality and Lorentz invariance impose that the interaction is mediated by the exchange of point-like effective mesons, which couple to the nucleons at local vertices. The single-nucleon dynamics is described by the Dirac equation

$$\left\{-i\boldsymbol{\alpha}\cdot\boldsymbol{\nabla} + \beta(m+g_\sigma\sigma) + g_\omega\omega^0 + g_\rho\tau_3\rho_3^0 + e\frac{(1-\tau_3)}{2}A^0\right\}\psi_i = \varepsilon_i\psi_i. \quad (1)$$

The spinor ψ denotes the nucleon with mass m. σ, ω, and ρ are the meson fields, and A denotes the electromagnetic potential. g_σ g_ω, and g_ρ are the corresponding coupling constants for the mesons to the nucleon. The lowest order of the quantum field theory is the *mean-field* approximation: the meson field operators are replaced by their expectation values. The A nucleons, described by a Slater determinant $|\Phi\rangle$ of single-particle spinors ψ_i, $(i = 1, 2, ..., A)$, move independently in the classical meson fields. The sources of the meson fields are defined by the nucleon densities and currents. The ground state of a nucleus is described by the stationary self-consistent solution of the coupled system of Dirac and Klein-Gordon equations. Due to time reversal invariance, there are no currents in the static solution for an even-even system, and therefore the spatial vector components $\boldsymbol{\omega}$, $\boldsymbol{\rho_3}$ and $\mathbf{A}$ of the vector meson fields vanish.

In addition to the self-consistent mean-field potential, pairing correlations have to be included in order to describe ground-state properties of open-shell nuclei. And while for strongly bound systems pairing can be included in the simple BCS scheme in the valence shell, exotic nuclei with extreme isospin values require a careful treatment of the asymptotic part of the nucleonic densities, and therefore a unified description of mean-field and pairing correlations. In the framework of the relativistic Hartree-Bogoliubov model, the ground state of a nucleus $|\Phi >$ is represented by the product of independent single-quasiparticle states. These states are eigenvectors of the generalized single-nucleon Hamiltonian which contains two average potentials: the self-consistent mean-field $\hat{\Gamma}$ which encloses all the long range particle-hole (ph) correlations, and a pairing field $\hat{\Delta}$ which sums up the particle-particle (pp) correlations. The single-quasiparticle equations result from the variation of the energy functional with respect to the hermitian density matrix ρ and the antisymmetric pairing tensor κ. In the Hartree approximation for the self-consistent mean field, the relativistic Hartree-Bogoliubov equations read

$$\begin{pmatrix} \hat{h}_D - m - \lambda & \hat{\Delta} \\ -\hat{\Delta}^* & -\hat{h}_D + m + \lambda \end{pmatrix} \begin{pmatrix} U_k(\mathbf{r}) \\ V_k(\mathbf{r}) \end{pmatrix} = E_k \begin{pmatrix} U_k(\mathbf{r}) \\ V_k(\mathbf{r}) \end{pmatrix}. \quad (2)$$

where $\hat{h}_D$ is the single-nucleon Dirac Hamiltonian (1), and m is the nucleon mass. The chemical potential λ has to be determined by the particle number subsidiary condition in order that the expectation value of the particle number operator in the ground state equals the number of nucleons. The column vectors denote the quasi-particle spinors and E_k are the quasi-particle energies. The pairing field $\hat{\Delta}$ in (2) is defined

$$\Delta_{ab}(\mathbf{r},\mathbf{r}') = \frac{1}{2}\sum_{c,d} V_{abcd}(\mathbf{r},\mathbf{r}') \sum_{E_k>0} U^*_{ck}(\mathbf{r})V_{dk}(\mathbf{r}'), \tag{3}$$

where a,b,c,d denote quantum numbers that specify the Dirac indices of the spinors, $V_{abcd}(\mathbf{r},\mathbf{r}')$ are matrix elements of a general two-body pairing interaction. The RHB equations are solved self-consistently, with potentials determined in the mean-field approximation from solutions of Klein-Gordon equations

$$[-\Delta + m_\sigma^2]\,\sigma(\mathbf{r}) = -g_\sigma\,\rho_s(\mathbf{r}) - g_2\,\sigma^2(\mathbf{r}) - g_3\,\sigma^3(\mathbf{r}) \tag{4}$$
$$[-\Delta + m_\omega^2]\,\omega^0(\mathbf{r}) = g_\omega\,\rho_v(\mathbf{r}) \tag{5}$$
$$[-\Delta + m_\rho^2]\,\rho^0(\mathbf{r}) = g_\rho\,\rho_3(\mathbf{r}) \tag{6}$$
$$-\Delta\,A^0(\mathbf{r}) = e\,\rho_p(\mathbf{r}). \tag{7}$$

for the sigma meson, omega meson, rho meson and photon field, respectively. The source terms in equations (4) to (7) are sums of bilinear products of baryon amplitudes.

The input parameters of the RHB model are the coupling constants and the masses for the effective mean-field Lagrangian, and the effective interaction in the pairing channel. In most applications we have used the NL3 effective interaction [17] for the RMF Lagrangian. Properties calculated with NL3 indicate that this is probably the best effective interaction so far, both for nuclei at and away from the line of β-stability. For the pairing field we employ the pairing part of the Gogny interaction

$$V^{pp}(1,2) = \sum_{i=1,2} e^{-((\mathbf{r}_1-\mathbf{r}_2)/\mu_i)^2}\,(W_i\,+\,B_i P^\sigma - H_i P^\tau - M_i P^\sigma P^\tau), \tag{8}$$

with the set D1S [18] for the parameters μ_i, W_i, B_i, H_i and M_i ($i=1,2$). This force has been very carefully adjusted to the pairing properties of finite nuclei all over the periodic table. In particular, the basic advantage of the Gogny force is the finite range, which automatically guarantees a proper cut-off in momentum space.

GROUND-STATE PROTON EMITTERS

Proton-rich nuclei display many interesting structure phenomena which are important both for nuclear physics and astrophysics. These nuclei are characterized by exotic ground-state decay modes such as direct emission of charged particles and β-decays with large Q-values [1]. The properties of many proton-rich nuclei should

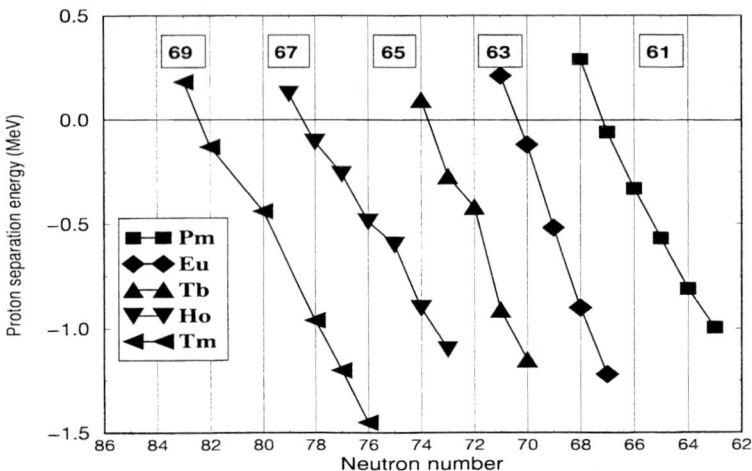

FIGURE 1. Calculated one-proton separation energies for odd-Z nuclei $61 \le Z \le 69$ at and beyond the drip-line.

also play an important role in the process of nucleosynthesis by rapid-proton capture. In addition to decay properties (particle emission, β-decay), of fundamental importance are studies of atomic masses and separation energies, and especially the precise location of proton drip-lines.

Proton radioactivity from the ground-state is determined by the Coulomb and centrifugal terms of the effective potential. For Z≤50, nuclei beyond the proton drip-line exist only as short lived resonances, and ground-state proton decay probably cannot be observed directly. On the other hand, the relatively high potential energy barrier enables the observation of ground-state proton emission from medium-heavy and heavy nuclei. At the drip-lines proton emission competes with β^+ decay; for heavy nuclei also fission or α decay can be favored. Experimental studies of ground-state proton radioactivity in odd-Z nuclei 51≤Z≤55 and 69≤Z≤ 83, have shown that the systematics of spectroscopic factors is consistent with half-lives calculated in the spherical WKB or DWBA approximations. On the other hand, recently reported proton decay rates indicate that the missing region of light rare-earth nuclei contains strongly deformed systems at the drip-lines. The lifetimes of deformed proton emitters should provide direct information on the Nilsson configuration occupied by the odd proton, and therefore on the shape of the nucleus. In Refs. [7,8] we have applied the RHB model in a detailed study of deformed odd-Z proton-rich nuclei with 53≤Z≤69. Ground-state properties of isotopes at the proton drip-line have been calculated with the NL3 + D1S combination of effective interactions in the ph and pp channels. In Figs. 1 and 2 the one-proton separation energies are displayed for the odd-Z nuclei $53 \le Z \le 69$, as function of the number of neutrons. The model predicts the drip-line nuclei: ^{110}I, ^{115}Cs, ^{118}La, ^{124}Pr, ^{129}Pm, ^{134}Eu, ^{139}Tb, ^{146}Ho, and ^{152}Tm.

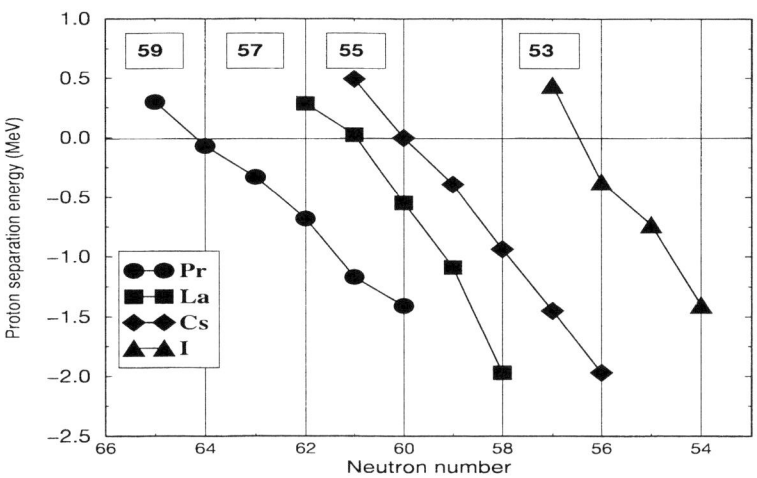

FIGURE 2. Same as in Fig. 1, but for odd-Z isotopes $53 \leq Z \leq 59$.

For the ground-state proton emission to occur, the valence proton must penetrate the wide Coulomb and centrifugal potential barriers, and this process competes with the β^+ decay. The half-life of the decay strongly depends on the energy of the odd proton and on its angular momentum. For a typical rare-earth nucleus the energy window in which ground-state proton decay can be directly observed is about 0.8 – 1.7 MeV. For the most probable proton emitters, in Table 1 we enclose the ground-state properties calculated in the RHB model. For each nucleus we include the one-proton separation energy S_p, the quadrupole deformation β_2, the deformed single-particle orbital occupied by the odd valence proton, and the corresponding theoretical spectroscopic factor. The spectroscopic factor of the deformed odd-proton orbital is defined as the probability that this state is found empty in the daughter nucleus with even number of protons. The results of RHB calculations are compared with the predictions of the finite-range droplet mass (FRDM) model: the projection of the odd-proton angular momentum on the symmetry axis and the parity of the odd-proton state Ω_p^π [19], the one-proton separation energy [19], and the ground-state quadrupole deformation [20]. The separation energies are also compared with available experimental data on proton transition energies. While an excellent agreement is found between quadrupole deformations calculated in the two theoretical models, the predictions for the deformed orbitals occupied by the odd proton differ in many cases. The models predict similar values for the one-proton separation energies of proton emitters with $Z \geq 59$. The differences are largest for I and Cs: the RHB predicts stronger binding for I, and weaker binding for Cs. When compared with available experimental proton energies, both models fail to reproduce the observed anomaly in the one-proton separation energies of ^{112}Cs and ^{113}Cs.

The theoretical separation energies are also compared with recently reported

TABLE 1. Odd-Z ground-state proton emitters in the region of nuclei with $53 \leq Z \leq 69$. Results of the RHB calculation for the one-proton separation energies S_p, quadrupole deformations β_2, and the deformed single-particle orbitals occupied by the odd valence proton, are compared with predictions of the macroscopic-microscopic mass model, and with the experimental transition energies. All energies are in units of MeV; the RHB spectroscopic factors are displayed in the sixth column.

	N	S_p	β_2	p-orbital	u^2	Ω_p^π [19]	S_p [19]	β_2 [20]	E_p exp.
^{107}I	54	-1.40	0.15	$3/2^+[422]$	0.84	$1/2^+$	-2.14	0.14	
^{108}I	55	-0.73	0.16	$3/2^+[422]$	0.79	$1/2^+$	-1.12	0.15	
^{109}I	56	-0.37	0.16	$3/2^+[422]$	0.81	$1/2^+$	-0.95	0.16	0.8126(40) [23]
^{111}Cs	56	-1.97	0.20	$1/2^+[420]$	0.74	$3/2^+$	-1.43	0.19	
^{112}Cs	57	-1.46	0.20	$1/2^+[420]$	0.74	$3/2^+$	-0.76	0.21	0.807(7) [24]
^{113}Cs	58	-0.94	0.21	$1/2^+[420]$	0.73	$3/2^+$	-0.76	0.21	0.9593(37) [21]
^{115}La	58	-1.97	0.26	$1/2^+[420]$	0.20	$3/2^+$	-1.50	0.27	
^{116}La	59	-1.09	0.30	$3/2^-[541]$	0.73	$3/2^+$	-0.67	0.28	
^{119}Pr	60	-1.40	0.32	$3/2^-[541]$	0.39	$3/2^-$	-1.41	0.31	
^{120}Pr	61	-1.17	0.33	$3/2^-[541]$	0.33	$3/2^-$	-0.66	0.32	
^{124}Pm	63	-1.00	0.35	$5/2^-[532]$	0.72	$5/2^-$	-1.34	0.33	
^{125}Pm	64	-0.81	0.35	$5/2^-[532]$	0.74	$5/2^-$	-1.24	0.33	
^{130}Eu	67	-1.22	0.34	$5/2^-[532]$	0.44	$3/2^+$	-1.17	0.33	
^{131}Eu	68	-0.90	0.35	$5/2^+[413]$	0.44	$3/2^+$	-1.01	0.33	0.950(8) [3]
^{135}Tb	70	-1.15	0.34	$3/2^+[411]$	0.62	$3/2^+$	-1.15	0.33	
^{136}Tb	71	-0.90	0.32	$3/2^+[411]$	0.65	$3/2^+$	-0.55	0.31	
^{140}Ho	73	-1.10	0.31	$7/2^-[523]$	0.61	$7/2^-$	-0.81	0.30	
^{141}Ho	74	-0.90	0.32	$7/2^-[523]$	0.64	$7/2^-$	-0.89	0.29	1.169(8) [3]
^{145}Tm	76	-1.43	0.23	$7/2^-[523]$	0.47	$1/2^+$	-1.0	0.25	1.728(10) [21]
^{146}Tm	77	-1.20	-0.21	$7/2^-[523]$	0.50	$7/2^-$	-0.60	-0.20	1.120(10) [22]
^{147}Tm	78	-0.96	-0.19	$7/2^-[523]$	0.55	$7/2^-$	-0.56	-0.19	1.054(19) [23]

experimental data on proton radioactivity from ^{131}Eu, ^{141}Ho [3], ^{145}Tm [21], ^{146}Tm [22], and ^{147}Tm [23]. The ^{131}Eu transition has an energy $E_p = 0.950(8)$ MeV and a half-life 26(6) ms, consistent with decay from either $3/2^+[411]$ or $5/2^+[413]$ Nilsson orbital. For ^{141}Ho the transition energy is $E_p = 1.169(8)$ MeV, and the half-life 4.2(4) ms is assigned to the decay of the $7/2^-[523]$ orbital. The calculated RHB proton separation energy, both for ^{131}Eu and ^{141}Ho, is -0.9 MeV. In the RHB calculation for ^{131}Eu the odd proton occupies the $5/2^+[413]$ orbital, while the ground state of ^{141}Ho corresponds to the $7/2^-[523]$ proton orbital. This orbital is also occupied by the odd proton in the calculated ground states of ^{145}Tm, ^{146}Tm and ^{147}Tm. For the proton separation energies we obtain: -1.43 MeV in ^{145}Tm, -1.20 MeV in ^{146}Tm, and -0.96 MeV in ^{147}Tm. These are compared with the experimental values for transition energies: $E_p = 1.728(10)$ MeV in ^{145}Tm, $E_p = 1.120(10)$ MeV in ^{146}Tm, and $E_p = 1.054(19)$ MeV in ^{147}Tm. Calculations also predict possible proton emitters ^{136}Tb and ^{135}Tb with separation energies -0.90 MeV and -1.15 MeV, respectively. In both isotopes the predicted ground-state

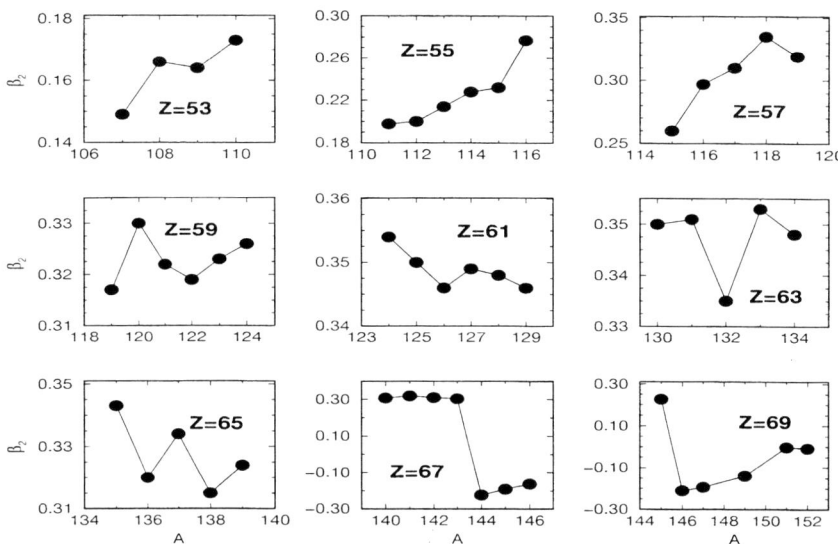

FIGURE 3. Self-consistent ground-state quadrupole deformations for odd-Z nuclei $53 \leq Z \leq 69$.

proton configuration is $3/2^+[411]$. Another possible proton emitter is ^{130}Eu with separation energy -1.22 MeV and the last occupied proton orbital $5/2^-[532]$ or $5/2^+[413]$.

The calculated mass quadrupole deformation parameters for the odd-Z nuclei $53 \leq Z \leq 69$ at and beyond the drip line are displayed in Fig. 3. While prolate deformations $0.15 \leq \beta_2 \leq 0.20$ are calculated for most of the I and Cs nuclei, the proton-rich isotopes of La, Pr, Pm, Eu and Tb are strongly prolate deformed ($\beta_2 \approx 0.30 - 0.35$). By increasing the number of neutrons, Ho and Tm display a transition from prolate to oblate shapes. The absolute values of β_2 decrease as we approach the spherical solutions at $N = 82$.

A detailed analysis of single proton levels, including spectroscopic factors, can be performed in the canonical basis which results from the fully microscopic and self-consistent RHB calculations. For the Eu isotopes this is illustrated in Fig. 4, where we display the proton single-particle energies in the canonical basis as function of the neutron number. The dashed line denotes the position of the Fermi level. The proton energies are the diagonal matrix elements of the Dirac Hamiltonian h_D in the canonical basis. The phase-space of positive-energy states should not be confused with the continuum of scattering states which asymptotically behave as plane waves. The RHB ground-state wave function can be written either in the quasiparticle basis as a product of independent quasi-particle states, or in the *canonical basis* as a highly correlated BCS-state. In the *canonical basis* nucleons occupy single-particle states. The canonical states are eigenstates of the RHB density matrix. The eigenvalues are the corresponding occupation numbers. In particular, we notice that for the proton emitter ^{131}Eu, the ground-state corresponds to the

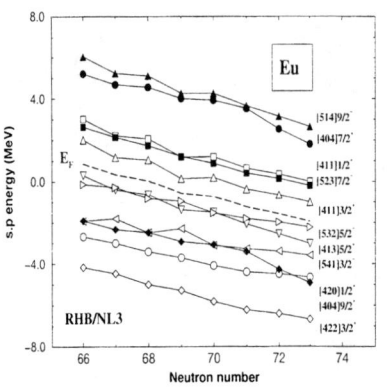 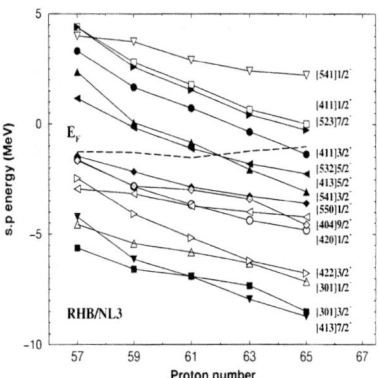

FIGURE 4. The proton single-particle levels for the Eu isotopes. The dashed line denotes the position of the Fermi level.

FIGURE 5. The proton single-particle levels for the isotopes at the proton drip-line $57 \leq Z \leq 65$.

odd valence proton in the $5/2^+[413]$ orbital. For ^{130}Eu the states $5/2^+[413]$ and $5/2^-[532]$ are almost degenerate, and this is an example of a situation in which only the comparison of calculated and measured half-lives can decide which is the last occupied proton orbital. Both for ^{130}Eu and ^{131}Eu the macroscopic-microscopic model predicts the last proton orbital to be $\Omega^\pi = 3/2^+$, which corresponds to the Nilsson orbital $3/2^+[411]$. In the energy diagram of Fig. 4 the states $5/2^+[413]$ and $5/2^-[532]$ are closer to the Fermi level for all calculated Eu isotopes.

In Fig. 5 we plot the proton single-particle energies in the canonical basis for the nuclei at the drip-line: ^{118}La, ^{124}Pr, ^{129}Pm, ^{134}Eu, ^{139}Tb. The levels are shown as functions of the number of protons, and again the dashed line denotes the position of the Fermi level. From this diagram one can easily deduce which levels are most likely to be occupied by the odd-proton, at and beyond the drip-line. Figure 5 should be compared with Table 1 for the odd-proton orbitals occupied in the proton emitters. Of course, by further decreasing the number of neutrons, i.e. by going beyond the drip-line, the proton levels and the Fermi level are shifted upwards in energy. However, as can be seen from Fig. 4 for the Eu isotopes, this increase in energy is very smooth and dramatic changes should not be expected for the occupation of the proton orbitals beyond the drip-line.

In Ref. [9] we have analyzed the recent data on proton emission from the closed neutron shell nucleus ^{155}Ta [25], and the proposed experiment to search for direct proton emission from ^{149}Lu [26]. The one-proton separation energies for the Lu and Ta isotopes are displayed in Fig. 6, as function of the number of neutrons. The predicted drip-line nuclei are ^{154}Lu and ^{159}Ta. The calculated separation energies are compared with experimental transition energies for ground-state proton emission in ^{150}Lu, ^{151}Lu [23], ^{155}Ta [25], ^{156}Ta [27], and ^{157}Ta [28]. In all five cases an excellent agreement is observed between model predictions and experimental data. In addition to ^{151}Lu, which was the first ground-state proton emitter to be dis-

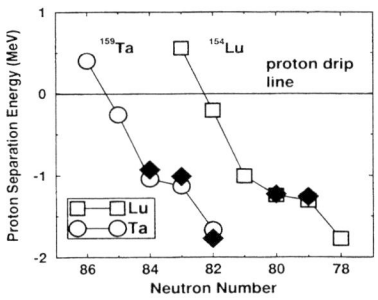
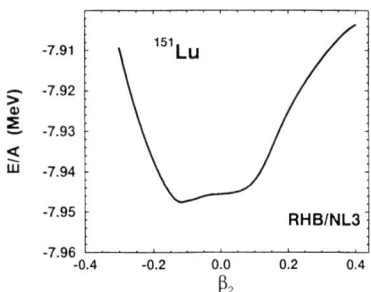

FIGURE 6. Proton separation energies for Lu and Ta isotopes at and beyond the drip-line.

FIGURE 7. Potential energy curve for ^{151}Lu.

TABLE 2. Lu ground-state proton emitters. The results of the RHB calculation for the one-proton separation energies S_p, qudrupole deformations β_2, and the deformed single-particle orbitals occupied by the odd valence proton, are compared with predictions of the macroscopic-microscopic mass model, and with the experimental transition energies. All energies are in units of MeV; the RHB spectroscopic factors are displayed in the sixth column.

	N	S_p	β_2	p-orbital	u^2	Ω_p^π [19]	S_p [19]	β_2 [20]	E_p exp.
^{149}Lu	78	-1.77	-0.158	7/2$^-$[523]	0.60	5/2$^-$	-1.51	-0.175	
^{150}Lu	79	-1.31	-0.153	7/2$^-$[523]	0.61	5/2$^-$	-1.00	-0.158	1.261(4) [23]
^{151}Lu	80	-1.24	-0.151	7/2$^-$[523]	0.58	7/2$^-$	-0.99	-0.150	1.233(3) [23]

covered [29], and ^{150}Lu, the self-consistent RHB calculation predicts ground-state proton decay in ^{149}Lu. The calculated one-proton separation energy -1.77 MeV corresponds to a half-life of a few μs, if one assumes that the nucleus is spherical. Direct proton emission with a half-life of the order of few μs is just above the lower limit of observation of current experimental facilities. An experiment to search for direct proton emission from ^{149}Lu has been proposed recently [26]. For the Lu ground-state proton emitters, in Table 2 the results of the RHB model calculation are compared with the predictions of the FRDM model. Both models predict oblate shapes for the Lu proton emitters, and similar values for the ground-state quadrupole deformations. On the other hand, while the FRDM assigns spin and parity 5/2$^-$ to the deformed single-particle orbitals occupied by the odd valence proton in all three proton emitters, the RHB model predicts the 7/2$^-$[523] Nilsson orbital to be occupied by the odd proton. We also notice that the RHB separation energies are much closer to the experimental values. The spectroscopic factors of the 7/2$^-$[523] orbital are displayed in the sixth column of Table 2.

The detailed analysis of odd-Z proton emitters ($53 \leq Z \leq 69$) has shown that,

TABLE 3. Spherical Ta ground-state proton emitters. RHB results for the proton separation energies, the single particle orbitals occupied by the odd proton, and the corresponding spectroscopic factors are compared with the predictions of the finite-range droplet model (FRDM) and with experimental data.

		RHB/NL3	FRDM [19]	EXP.
^{155}Ta	S_p	-1.677	-1.09	-1.765(10) [25]
	J^π	h 11/2$^-$	9/2$^-$	11/2$^-$
	spectr. factor	0.60		0.58(20)
^{156}Ta	S_p	-1.129	-0.60	-1.007(5) [27]
	J^π	d 3/2$^+$	3/2$^-$	3/2$^+$
	spectr. factor	0.51		
^{157}Ta	S_p	-1.040	-0.48	-0.927(7) [28]
	J^π	h 11/2$^-$	9/2$^-$	1/2$^+$
	spectr. factor	0.42		0.56(24)

while the proton-rich isotopes of La, Pr, Pm, Eu and Tb are all strongly prolate deformed ($\beta_2 \approx 0.30 - 0.35$), Ho and Tm isotopes at the proton drip-line display a transition from prolate to oblate shapes. Spherical shapes are expected as the nuclei with unbound protons approach the $N = 82$ neutron shell. The Lu proton emitters are found in the transitional region between oblate and spherical shapes. This is illustrated in Fig. 7, where we plot the binding energy curve for ^{151}Lu as function of the quadrupole deformation parameter. The binding energies result from self-consistent calculations performed by imposing a quadratic constraint on the quadrupole moment. A very shallow minimum is found at $\beta \approx -0.15$, but otherwise the potential is rather flat with a shoulder at $\beta = 0$.

The proton-rich Ta isotopes are spherical. In Fig. 6 we compare the calculated one-proton separation energies with experimental transition energies for ^{155}Ta [25], ^{156}Ta [27], and ^{157}Ta [28]. The predictions for the spherical orbitals occupied by the odd proton and the corresponding spectroscopic factors are displayed in Table 3. Results of the FRDM calculation [19] have also been included in the comparison. As in the case of the Lu ground-state proton emitters, an excellent agreement between RHB separation energies and experimental data on transition energies for proton emission is observed. In particular, our calculation reproduces the very recent data on proton emission from the closed neutron shell nucleus ^{155}Ta [25]. The significant decrease in proton binding for ^{155}Ta, as compared to ^{157,156}Ta, has been associated with the $N = 82$ closure. In comparison, the FRDM results are found to be in rather poor agreement with experimental data. Except for ^{157}Ta, the spherical orbitals predicted to be occupied by the odd proton agree with the experimental assignments, and the theoretical spectroscopic factor of the $h_{11/2}$ orbital in ^{155}Ta is very close to the experimental value. For ^{157}Ta the RHB model predicts ground-state proton emission from the $h_{11/2}$ orbital. The experimental assignment for the ground-state configuration is $s_{1/2}$, but an alpha decaying state is identified in ^{157}Ta at an excitation energy of only 22(5) keV and assigned to an $h_{11/2}$ isomer [28]. We have also calculated the one-proton separation energy for ^{156}Ta^m: $S_p = -1.250$ MeV, the orbital is $h_{11/2}$ and the spectroscopic factor is 0.79. This is to be compared with the experimental transition energy $E_p = 1.103(12)$ MeV [30], assigned to the $h_{11/2}$ orbital with the experimental spectroscopic factor 0.92(4) [1].

REFERENCES

1. P.J. Woods and C.N. Davids, Annu. Rev. Nucl. Part. Sci. **47**, 541 (1997).
2. S. Aberg, P.B. Semmes, and W. Nazarewicz, Phys. Rev. C **56**, 1762 (1997).
3. C.N. Davids et al, Phys. Rev. Lett. **80**, 1849 (1998).
4. K. Rykaczewski et al, Phys. Rev. C **60**, 011301 (1999).
5. E. Maglione, L.S. Ferreira, and R.J. Liotta, Phys. Rev. Lett. **81**, 538 (1998); Phys. Rev. C **59**, R589 (1999).
6. D. Vretenar, G.A. Lalazissis, and P. Ring, Phys. Rev. C **57**, 3071 (1998).
7. G.A. Lalazissis, D. Vretenar, and P. Ring, Nucl. Phys. **A650**, 133 (1999).
8. D. Vretenar, G.A. Lalazissis, and P. Ring, Phys. Rev. Lett. **82**, 4595 (1999).
9. G.A. Lalazissis, D. Vretenar, and P. Ring, submitted to Phys. Rev. C
10. P. Ring, *Progr. Part. Nucl. Phys.* **37**, 193 (1996).
11. W. Pöschl, D. Vretenar, G.A. Lalazissis, and P. Ring, *Phys. Rev. Lett.* **79**, 3841 (1997).
12. G.A. Lalazissis, D. Vretenar, W. Pöschl, and P. Ring, *Nucl. Phys.* **A632**, 363 (1998).
13. G.A. Lalazissis, D. Vretenar, W. Pöschl, and P. Ring, *Phys. Lett.* **B418**, 7 (1998).
14. G.A. Lalazissis, D. Vretenar, and P. Ring, Phys. Rev. C **57**, 2294 (1998).
15. D. Vretenar, W. Pöschl, G.A. Lalazissis, and P. Ring, *Phys. Rev. C* **57**, R1060 (1998).
16. G.A. Lalazissis, D. Vretenar, P. Ring, M. Stoitsov, and L. Robledo, Phys. Rev. C **60**, 014310 (1999).
17. G.A. Lalazissis, J. König, and P. Ring, *Phys. Rev. C* **55**, 540 (1997).
18. J. F. Berger, M. Girod and D. Gogny, *Nucl. Phys.* **A428**, 32 (1984).
19. P. Möller, J.R. Nix, and K.-L. Kratz, At. Data Nucl. Data Tables **66**, 131 (1997).
20. P. Möller, J.R. Nix, W.D. Myers, and W.J. Swiatecki, At. Data Nucl. Data Tables **59**, 185 (1995).
21. J.C. Batchelder et al., *Phys. Rev. C* **57**, R1042 (1998).
22. K. Livingston et al, Phys. Lett. **B312**, 46 (1993).
23. P.J. Sellin et al., *Phys. Rev. C* **47**, 1933 (1993).
24. R. D. Page et al, Phys. Rev. Lett. **72**, 1798 (1994).
25. J. Uusitalo et al, Phys. Rev. C **59**, R2975 (1999).
26. J. C. Batchelder *et al*, private communication.
27. R.D. Page et al, Phys. Rev. Lett. **68**, 1287 (1992).
28. R.J. Irvine et al, Phys. Rev. C **55**, R1621 (1997).
29. S. Hofman et al, Z. Phys. A **305**, 125 (1982).
30. K. Livingston et al, Phys. Rev. C **48**, R2151 (1993).

Two-Proton Emission in the Hyperharmonics Approach

Ivan G. Mukha

Institut für Kernphysik, Technische Universität, D-64289 Darmstadt, Germany
Associate at Gesellschaft für Schwerionenforschung (GSI), D-64291 Darmstadt, Germany
On leave from Kurchatov Institute, RU-123182 Moscow, Russia

Abstract. Nuclear decays into three-particle channels are considered in a few-body approach of hyperspherical harmonics with emphasis on simultaneous, or direct, emission of two protons. General conditions of direct decays are described and their main features, being experimentally established in decays of light nuclei, are reported.

The analysis method based on an expansion of decay amplitude into a series of hyperspherical harmonics is reviewed. This basis is a generalisation of the spherical function basis in three-body systems. The method is tested on analysis of the direct decay ^{6}Be$\rightarrow \alpha$+p+p where the three-body components in the nuclear structure of ^{6}Be have been studied. In particular, the observed strong proton-proton correlations are treated as a manifestation of a specific three-body quantum effect: the kinematic focusing of fragments over momenta and in space.

The method is applied for the predictions of proton-proton correlations and life-time estimates of the nuclei ^{19}Mg, ^{34}Ca and ^{48}Ni - candidates for two-proton radioactivity. Each direct 2p-decay should result in a set of peaks in the E_{p-p} spectrum whose number and positions depend on the structure of initial nucleus, opposite to the diproton model, predicting the ^{2}He emission with one peak at $E_{p-p} \approx 0$ in all cases.

DIRECT THREE-PARTICLE DECAYS

A simultaneous emission of two protons is a genuine three-particle nuclear decay and is a complementary mode to the known sequential emission of protons via narrow intermediate states. In general, a sequential mechanism of three-particle decay is a chain of two independent binary decays via a narrow intermediate state whose width should be much smaller than the decay energy. The total decay amplitude is then a product of two binary amplitudes. When narrow intermediate states are absent the sequential decay mechanism is not plausible because it contradicts the uncertainty and causality principles. Such non-sequential decays are called direct decays. In sequential decays information about correlations of fragments in the initial nucleus is lost because of strong final state interactions. In direct decays the fragment distributions are not distorted significantly by final state interactions and therefore reflect correlations of the respective clusters in the initial nucleus. In comparison with the two-particle case, three-particle decays are more informative due to additional degrees of freedom where more observables (*eg* energy spectra of fragments, correlations between fragments) are available. Thus, a direct three-particle decay is a promising tool to study nuclear structure.

Main features of direct decays are experimentally established in the two-proton decays of the ^{6}Be ground and first excited states and their analogs in ^{6}He and ^{6}Li*(T=1), (see [1] and references there). Strong nucleon-nucleon correlations are observed which are not connected with the p-p or α-p final state interactions. In particular, the measured energy spectra of α-particles from the mentioned decays display sharp peaks over broad pedestals. These peaks correspond to a strong nucleon-nucleon energy correlation which was first interpreted as emission of ^{2}He, or di-proton, while the pedestals were associated with a sequential proton emission via ^{5}Li [2]. However, such a model fails to explain the angular α-p correlations from the ^{6}Be decay measured in the kinematical complete experiment [3].

HYPERSPHERICAL HARMONICS METHOD

An adequate approach describing few-body nuclear interactions is suggested by L.M. Delves, [4], who introduced the hyperspherical harmonics, or K-harmonics, basis which gives correct angular wavefunctions like the spherical harmonics in the two-particle case. Usage of K-harmonics makes it possible to write the asymptotic of the total wavefunction along three-particle channels in a way which is formally identical with that for two-particle channels. The way to solve the Schrödinger equation for three particles, which decouples the total wavefunction into radial and angular parts is illustrated below and uses the two-particle case for comparison.

The system of three particles i, j, k with total energy E can be characterized by 5 kinematical variables[1]. Instead of the radius in the 2-particle centre of mass system, two Jacobi radii, $\mathbf{x}$ and $\mathbf{y}$, are introduced where one radius is between particles j and k, $\mathbf{x}=\mathbf{r}_j-\mathbf{r}_k$, and the another radius is between the third particle i and c.m. of the selected pair, $\mathbf{y}=(m_j\mathbf{r}_j+m_k\mathbf{r}_k)/(m_j+m_k)-\mathbf{r}_i$. The respective Jacobi momenta $\mathbf{p}_x$, $\mathbf{p}_y$, Jacobi orbital momenta ℓ_x, ℓ_y, and Jacobi energies E_x, E_y are defined by analogous relations (naturally, $E_x+E_y=E$). Then the 5 kinematical variables may be $\Omega_i = (\Omega_x, \Omega_y, x_i)$: the directions Ω_x and Ω_y of the Jacobi momenta and the quantity $x_i=\arctan(\sqrt{E_x/E_y})$ which reflects the energy distribution between the Jacobi subsystems. By introducing the hyperspherical coordinates, hyperradius $\rho^2=r_x^2+r_y^2$ and hyperangle $\theta=\arctan(x/y)$, one may use instead of the orbital operator $\hat{L}$, the grand orbital operator $\hat{K}$, which has eigenfunctions as the functions of hyperangle:

$$\Psi_K^{l_x,l_y} \sim \sin^{l_x}(\theta)\cos^{l_y}(\theta)P_n^{l_x+0.5,l_y+0.5}(\cos 2\theta),$$

where $P_n^{\alpha,\beta}$ are the Jacobi polynomials.

The respective quantum number is called hypermomentum, which minimal value is equal to the sum of the Jacobi orbital momenta, $K=\ell_x+\ell_y+n$ (n=0,1,2,...). This additional quantum number is a three-body analog of the orbital momentum value.

With the $\hat{K}$ eigenfunctions, one may obtain the solution of the three-particle Schrödinger equation $(T+V-E)\Psi_{JM}^T=0$, with the sum of binary potentials $V=V_{ij}+V_{jk}+V_{ki}$, in a form of a hyperradial part of a total wavefunction coupled with the functions of a hyperangle:

---

[1] All particles are assumed to be spinless, the general case including spins is considered in [4].

$$\psi_{JM} = \sum R_{KL}^{l_x l_y}(\kappa\rho) Y_{KLM}^{l_x l_y}(\Omega_i),$$

here the hyperspherical harmonics (HH) functions, or K-harmonics, are

$$Y_{KLM}^{l_x l_y}(\Omega_i) = \Psi_K^{l_x l_y}(\theta) \left[Y_{l_x}(\Omega_x) \otimes Y_{l_y}(\Omega_y) \right]^{LM},$$

$[...]^{JM}$ denotes the **L+S** vector addition to form **J**. The K-harmonics constitute an orthonormal basis like the spherical harmonics, $Y_{lm}(\Omega)$, - in the two-particle case.

After a separation of the hyperangular part in the total wave function one may obtain equations which are equivalent to the Schrödinger equation of a motion of single particle in external field.

In this approach, the centrifugal barrier of a three-body system is proportional to the factor (K+3/2)·(K+5/2) corresponding to the $\ell(\ell+1)$ factor for a two-particle barrier. Usually it is much larger than the two-particle barrier and is not equal to zero even for K=0. The derived hyperradial wavefunctions display formally the same asymptotic behaviour as in the 2-body case [4].

Expansion of a decay amplitude into a series of K-harmonics

The hyperspherical harmonics method is applied for direct three-particle decays by B.V. Danilin et al., [5], where the decay amplitude is suggested to be expanded in a series in the K-harmonics basis, in analogy with the partial wave analysis in a 2-particle case. The general formulae for this approach can be found in [6].

I will highlight the HH method for a description of energy and angular correlations of decay fragments. At low energies, the direct decays should be determined by few components in the amplitude expansion which corresponds to a minimal value of hypermomentum because of a three-particle centrifugal barrier which grows with an increase of K. Thus, in the first approximation one may describe direct decays considering only few fit components with the lowest value of hypermomentum. For example, the energy spectra of fragments can be fitted by a superposition of few components with definite values of quantum numbers and with a specific energy dependence each. A weight of each component gives an information about the norm of the respective configuration in the initial nucleus[2].

The additional notations are given below. The decay to the three particles i, j and k with decay energy Q may be characterized by 5 kinematical variables $\Omega_i = (\Omega_{j-k}, \Omega_{i-jk}, x_i)$ where the quantity $x_i = E_i M Q^{-1}(m_j+m_k)^{-1} = E_i/E_i^{max}$, E_i is the energy of particle i in the c.m. frame of the decaying nucleus, and $m_{i,j,k}$ are the fragment masses, $M=m_i+m_j+m_k$. The energy E_i is related to the energies of the relative motion of particles j and k (E_{j-k}, or E_x), and of their centre of mass and particle i (E_{i-jk}, or E_y) as: $E_i = E_y(m_j+m_k)M^{-1}$, and $E_x + E_y = Q$. The state with spin J decays into three particles with spins s_i, s_j and s_k. To describe the final state the following quantum numbers are used: The *hypermomentum*

---

[2] The long-range Coulomb interaction may influence the fragment distributions and this can be taken into account in a conventional way as a final state interaction.

$K = l_x + l_y + 2n$ (n=0,1,2,...); the orbital angular momenta $\mathbf{l}_x$ and $\mathbf{l}_y$ conjugated to the Jacobi momenta $\mathbf{p}_x$ and $\mathbf{p}_y$ and satisfying conservation of parity π of the decaying state: $\pi = (-1)^{l_x+l_y}$; the total orbital angular momentum $\hat{\mathbf{L}} = \hat{\mathbf{l}}_x + \hat{\mathbf{l}}_y$ satisfying the conservation law of the total angular momentum $\hat{\mathbf{J}} = \hat{\mathbf{L}} + \hat{\mathbf{S}}$; the total spin of all products $\hat{\mathbf{S}} = \hat{\mathbf{s}}_i + \hat{\mathbf{s}}_j + \hat{\mathbf{s}}_k$. The total spin of any pair of particles is $\mathbf{S}_{i-j} = \mathbf{s}_i + \mathbf{s}_j$. The decay amplitude of a state with spin J and its projection M can be expanded in a series in an orthonormal hyperspherical harmonics basis:

$$F_M = \sum_{KLl_xl_y} B_{KLS}^{l_xl_y} \cdot Y_{KLM}^{l_xl_y}(\mathbf{p}_x, \mathbf{p}_y) \cdot C(S_{j-k}, T_{j-k}) \quad (1)$$

The expansion coefficients $B_{KLS}^{l_xl_y}$ corresponding to the hyperspherical functions $Y_{KLM}^{l_xl_y}$ are determined by the decay dynamics and give, when squared, the probabilities for the observed decay modes, classified according to the sets of values of $l_x, l_y, K, L, S, S_{j-k}$. The spin-isospin weight factors $C(S_{j-k}, T_{j-k})$ can be found in [6]. Expression (1) corresponds to the simple case where the spin projections of particles are not measured[3].

In the case of two identical particles, e.g. protons, the most convenient arrangement of the Jacobi coordinates is (p-p,pp-"core"), while the set (p-"core",p-p"core") is close to coordinates in the c.m. system of a decaying nucleus which is normally used in shell-model or mean-field calculations. Conversion in the representation of (1) from one set of Jacobi coordinates $(j-k, i-jk)$ to another set $(k-j, i-kj)$ is accompanied by the change of the coefficients $B_{KLS}^{l_{j-k}l_{i-jk}}$:

$$B_{KLS}^{l_{k-j}l_{i-kj}} = \sum_{l_{j-k},l_{i-jk}} < l_{j-k}l_{i-jk}|l_{i-k}l_{j-ik} > \cdot B_{KLS}^{l_{j-k}l_{i-jk}}, \quad (2)$$

where $< ...|... >$ are Raynal-Revai coefficients [15].

Analyses of the ^6Be, $^6Li^*$, $^6He^*$ decays into $\alpha+N+N$. The approach is tested in analyses of direct decays of A=6 nuclei. Eg, the energy spectrum of α-particles from ^{6}Be can be fitted by the sum of two components only, with S_{p-p}=0 and S_{p-p}=1, and with K_{min}=2. The studied decays are governed by a single K value in the amplitude expansion. Thus, the hypermomentum is confirmed to be a good quantum number.

This interpretation is not unique in describing of single fragments. For example, the α-spectrum measured in [2] was first fitted using the model of sequential emission of protons via the intermediate nucleus ^{5}Li. However, the two mechanisms predict quite different behaviour of α-proton correlations in a kinematical complete experiment. In particular, the three-body approach predicts the different angular dependences of p-p correlations with total spins S_{p-p}=0 and S_{p-p}=1 while in the sequential model these two modes are indistinguishable. In the decisive kinematical complete experiment [1] where the ^{6}Be decay was measured by detecting both α-particles and protons, the different angular distributions of the S_{p-p}=0 and S_{p-p}=1 modes were observed directly thus confirming the three-body decay mechanism.

---

[3] Under some conditions the decay amplitude may depend on the possible orientation of the spin of the initial state which has decayed, see [3].

Nuclear structure reflected in direct decays. The fragment spectra from studied direct decays reflect the three-body nuclear structure. For example, the α+N+N correlations in A=6 nuclei calculated in [7-9] (the fragment correlations both in space and in involved angular momenta) agree quantitatively with the conclusions of data analysis. In particular, the strong momentum and space correlations between two valence protons are found in ^{6}Be. These are the 'di-proton' and 'cigar' configurations. In the first case, two valent protons forms a relatively compact cluster, in the second they are mainly on opposite sides relative to the α-particle. The observed correlations are induced by the specific three-body quantum effect: kinematic focusing of particles over momenta and in space. This phenomenon generalizes the angular dependance of a two-particle scattering with $\ell \neq 0$ to the three-body case with K$\neq$0 and it has a universal nature.

DIRECT TWO-PROTON EMISSION FROM NUCLEI - CANDIDATES OF TWO-PROTON RADIOACTIVITY

This effect is expected to be essential in other nuclei with three-body structure, *eg* the two-proton emitters ^{19}Mg, ^{34}Ca and ^{48}Ni. Their estimated 2p-decay Q-values are less than 1.5 MeV. Since the FWHM of the p-p scattering spectrum is $\sim$3 MeV, the well-known mechanism of a sequential decay via emission of ^{2}He is not plausible here. Predictions of 2p-decay modes of these nuclei done in the HH approach are presented below. The ^{19}Mg decay is first described in details.

Two-proton emission from ^{19}Mg. According to the ref. [10], the main uncertainty in predictions of ^{19}Mg properties is given by poorly known ground-state masses of ^{19}Mg and ^{18}Na: 0.9(3) MeV and 1.5(7) MeV above the ^{17}Ne+p+p threshold, respectively. Due to the large errors in mass estimates, two opposite situations are possible: i) the ^{18}Na+p threshold is below the ^{19}Mg ground state which decays by a sequential emission of protons via ^{18}Na; ii) the ^{18}Na+p threshold is well above ^{19}Mg and a direct two-proton emission into the ^{17}Ne+2p channel dominates. For the last case, I would like to consider a direct two-proton emission in the three-body approach which was first applied for the ^{6}Be data in the ref. [5].

To describe the decay of the ^{19}Mg ground state which has an unknown spin-parity, one should assume some structure of ^{19}Mg. Since ^{19}Mg has 12 protons, two "valent" protons are likely in the $d_{5/2}$ shell and should have the total spin S_{p-p}=0. Thus, the spin-parity of ^{19}Mg are probably 1/2$^-$, the same as the ^{17}Ne ground state. If one assumes that the ^{19}Mg$_{g.s.}$ has mainly the ^{17}Ne+p+p cluster structure, it decays directly into the same fragments. As the mentioned transition ends in the ^{17}Ne(1/2$^-$)+2p channel, the lowest value of the hypermomentum K, allowed by momentum and parity conservation, is zero. One may estimate the three-particle decay width of the suggested ^{19}Mg state using the three-body model [11] where the "generalized R-matrix" approach is suggested. The ordinary R-matrix formula for decaying states is there replaced by a similar expression $\Gamma_K(E)$=2P$_{K+3/2}$(E)γ_K^2, where the penetrability for a three-particle decay is practically the same function as in the R-matrix approach: P$_{K+3/2}$(E)=$\frac{\kappa\rho}{F_{K+3/2}^2+G_{K+3/2}^2}$. The physical meaning of the partial reduced width γ_K^2 is the same as in the two-body case, characterizing the spectroscopic factor for the three-particle exit channel with a hyperradius ρ. Functions $F_{K+3/2}$ and $G_{K+3/2}$ are calculated as regular and irregular Coulomb functions. The Wigner limit of a reduced width is assumed. The value of the calculation parameter, the radius of channel, is chosen of 16 f using an extrapolation of the systematic behaviour of a channel radius extracted from the fitted known widths of nuclei with dominated direct three-particle decay channels: ^{6}Be ($\sim$8 f), ^{10}He ($\sim$11 f), ^{16}Ne ($\sim$16 f) etc, according to [12].

TABLE 1. Calculated widths of the ^{19}Mg ground state for the decay energy Q=0.89 MeV. The main component in the ^{19}Mg wave function is assumed with the hypermomentum K=0.

Radius of channel, f	Penetrability	Width (MeV)
14.	$5.40 \cdot 10^{-7}$	$3.13 \cdot 10^{-6}$
16.	$2.10 \cdot 10^{-6}$	$0.139 \cdot 10^{-4}$
18.	$7.27 \cdot 10^{-6}$	$0.542 \cdot 10^{-4}$
20.	$0.226 \cdot 10^{-4}$	0.000187

TABLE 2. The same as in Table 1, except Q-values.

Channel radius, f	Penetrability Q=0.54 MeV	Width (MeV) Q=0.54 MeV	Penetrability Q=1.24 MeV	Width (MeV) Q=1.24 MeV
14.	$3.84 \cdot 10^{-11}$	$1.73 \cdot 10^{-10}$	0.595	0.000499
16.	$1.65 \cdot 10^{-10}$	$8.57 \cdot 10^{-10}$	0.673	0.00198
18.	$6.40 \cdot 10^{-10}$	$3.72 \cdot 10^{-9}$	0.731	0.00679
20.	$2.24 \cdot 10^{-9}$	$1.45 \cdot 10^{-8}$	0.773	0.0203

TABLE 3. The same as in Table 1, except dominating hypermomentum value, K.

Channel radius, f	Penetrability K=2	Width (MeV) K=2	Penetrability K=4	Width (MeV) K=4
14.	$2.35 \cdot 10^{-8}$	$1.36 \cdot 10^{-7}$	$1.95 \cdot 10^{-10}$	$1.13 \cdot 10^{-9}$
16.	$1.18 \cdot 10^{-7}$	$7.87 \cdot 10^{-7}$	$1.42 \cdot 10^{-9}$	$9.45 \cdot 10^{-9}$
18.	$5.13 \cdot 10^{-7}$	$3.82 \cdot 10^{-6}$	$8.45 \cdot 10^{-9}$	$6.30 \cdot 10^{-8}$
20.	$1.94 \cdot 10^{-6}$	$0.161 \cdot 10^{-4}$	$4.23 \cdot 10^{-8}$	$3.50 \cdot 10^{-7}$

In Tables 1–3 the results of ^{19}Mg width calculations are shown for varied values of the ground state position respective to the ^{17}Ne+p+p threshold, the channel radius ρ and the hypermomentum K, [13]. As one may see, the estimated width of the ^{19}Mg ground state is of 10 eV judging purely by assumption that the $K_{min}=\ell_{p-p}+\ell_{Ne-pp}=0$ configuration dominates the ^{19}Mg wave function. However, due to Pauli principle, such a component should be suppressed because the valent protons have to be in the $d_{5/2}$ shell (or $\ell_{p-Ne}=\ell_{p-pNe}=2$). The configuration with next hypermomentum, K=2, (the estimated width $\sim$1 eV) does not match to the assumed $d_{5/2}$ shell structure as well. Finally, the configuration with K=4 is allowed by the Pauli principle (because the assumed $(d_{5/2})^2$ wave in ^{19}Mg overlaps by 90% with the K=4, $\ell_{p-p}=\ell_{Ne-pp}=0$ component, using eq.(2)) and the respective width of the ground state is of 0.01 eV. The last value corresponds to a very long life-time, of 10^{-14} s, and thus the decay could be classified as a radioactivity phenomenon.

On the basis of the assumed structure of ^{19}Mg one may predict properties of its direct three-particle decay (see details in Appendix). If the K=0 and K=2 components in ^{19}Mg are suppressed by the Pauli principle completely, the decay amplitude and spectra of the fragments should only be defined by the K_{min}=4 configuration with $\ell_{p-p} = \ell_{Ne-pp}$=0. For example, the corresponding E_{p-p} spectrum[4] is shown in fig.1, on left, by a solid curve. Dashed and dotted curves are the results of a calculation of the suppressed decay modes with K=2 and 0, respectively.

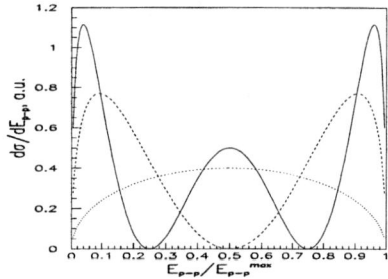
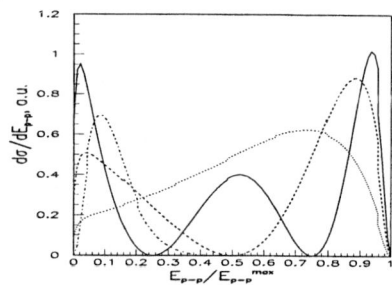

FIGURE 1. On left: possible E_{p-p} spectra from the direct decay ^{19}Mg(1/2$^-$)→^{17}Ne+p+p where $\ell_{p-p}=\ell_{Ne-pp}$=0. The p-p energy is given as a fraction of the maximum E_{p-p} value. The solid, dashed and dotted curves correspond to the decay modes with K=4, 2 and 0, respectively. On right: the same components as shown to the left, and in addition the Coulomb repulsion of the fragments is taken into account. The dot-dashed curve is the result of diproton model.

The long-range Coulomb repulsion must be taken into account when a more realistic approximation is required. In this case the final state interaction model can be used. The full decay amplitude F is then factorized as $|F|^2 = |F_3|^2 \cdot | F_{FSI} |^2$, where $|F_3|$ is the three-particle decay amplitude, and $|F_{FSI}|$ - the final state interaction factor, see Appendix. The calculated E_{p-p} spectra from ^{19}Mg are shown in fig.1, on right: the dotted, dashed, solid curves corresponds to the K=0,2,4 components, respectively. As one can see, the final state interaction influences significantly the E_{p-p} spectrum of the K=0 mode only. The considered configurations in ^{19}Mg have very different probabilities to be observed in its decay. One may quantitatively compare the calculated penetrabilities of the K=0,2,4 modes for the ^{19}Mg direct three-particle decay, which values are $2 \cdot 10^{-6}$, 10^{-7}, 10^{-9}, respectively (see Tables 1–3, the row with the radius of channel of 16 f).

Thus, there could be three exotic modes of the 2p-decay of ^{19}Mg. First, if the K=0 component is not forbidden strictly by the Pauli principle and its admixture to the dominating K=4 configuration is more than 0.1%, the K=0 mode should mainly be observed, with weak p-p correlations as shown in fig.1 by dotted curves. Second, if the admixture of the suppressed K=2 component in ^{19}Mg is more than 1%, the strong p-p correlations (like in the di-proton model) should be detected as well as strong p-p anticorrelations, see dashed curves in fig.1. Third, if the Pauli principle

---

[4] The ^{17}Ne spectrum can easy be obtained from the E_{p-p} spectrum using the formulae $E_{Ne}=\frac{2}{19}E_{Ne-pp}$ and $E_{Ne-pp}+E_{p-p}$=Q.

suppresses the mentioned components strongly, very exotic p-p correlations with three peaks in E_{p-p} spectrum should appear, see solid curves in fig.1. Combinations of these three basic decay modes are possible as well.

One should also consider the decay branch ^{19}Mg($1/2^-$)→^{17}Ne($1/2^-$)+p+p, with L=1. In this decay with K_{min}=2, in order to conserve the momentum and parity, two protons have to be with S_{p-p}=1 and the relative orbital momenta - with $\ell_{p-p}=\ell_{Ne-pp}$=1. This component corresponds with 100% probability to the $\ell_{p-Ne}=\ell_{p-pNe}$=1 configuration, which agrees with two valent protons being in the p-shell of ^{19}Mg and with S_{p-p}=1. This contradicts to the assumed $(d_{5/2})^2$ structure of ^{19}Mg and therefore such a component is unlikely compete with the decay modes considered above.

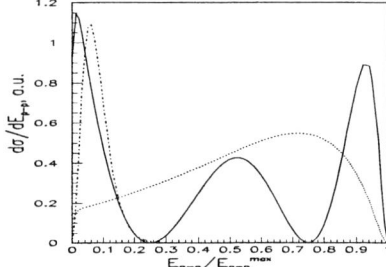 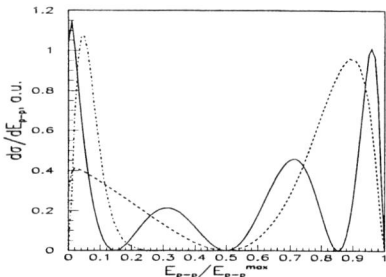

FIGURE 2. The expected E_{p-p} spectra from the direct decays: ^{34}Ca(0^+)→^{32}Ar+p+p with $\ell_{p-p}=\ell_{Ar-pp}$=0, on left; ^{48}Ni(0^+)→^{46}Fe+p+p with $\ell_{p-p}=\ell_{Fe-pp}$=0, on right. The p-p energy is given as a fraction of the maximum E_{p-p} value. The curves are the same as in fig.1, except the solid curve, to the right, which corresponds to the decay mode with K=6. The Coulomb repulsion of fragments is taken into account in all spectra.

Decay modes and life-time estimates of ^{34}Ca and ^{48}Ni. The decay properties of ^{34}Ca are expected to be similar to those of ^{19}Mg. The ^{34}Ca ground state is calculated to be bound respective to the single proton emission (Q_p=−0.9 MeV) and unbound respective to the ^{32}Ar+p+p decay with Q_{2p}=0.75 MeV [16]. According to a conventional shell-model, ^{34}Ca has a complete proton d-shell. Therefore its spin-parity is 0^+ and one may apply all considerations valid for the ^{19}Mg case. The only significant differences are the higher Coulomb barrier and the smaller Q_{2p}-value which result in much larger estimated life-times: $\sim 10^{-10}$ s, 10^{-9} s, $5 \cdot 10^{-7}$ s for the dominating K=0,2,4 components, respectively [13]. The corresponding 2p-decay modes are shown as E_{p-p} spectra in fig.2, on left.

The recently discovered (see J. Giovinazzo et al., Proc. of PROCON99) doublemagic nucleus ^{48}Ni(0^+) with a complete proton $f_{7/2}$-shell is estimated to be bound respective to the 1p-emission (Q_p=−0.46 MeV) and it may decay into the ^{46}Fe+p+p channel with Q_{2p}=1.36 MeV [16]. Since two valent protons are expected to be in the f-shell, the lowest hypermomentum allowed by the Pauli principle is K=6. The corresponding single term in the HH expansion of the decay amplitude is Y_{600}^{00} (see notation in Appendix, eq.(4)) and the respective impressive 2p-decay correlations are shown by solid curve in fig.2, on right. If a small admixture of the $(p_{1/2})^2$-

wave, of 1%, is present in ^{48}Ni, then the K=2 component should dominate in its decay (like in the ^{19}Mg case) with the E_{p-p} spectrum shown by the dashed curve in fig.2, on right. The result of diproton model is shown by dash-dotted curve for a comparison. The high Coulomb barriers cause the very large life-times: $\sim 10^{-6}$ s and $\sim 10^{-4}$ s for the K=2 and K=6 components, respectively [13].

SUMMARY

Two-proton decay is a three-body problem in a case of non-sequential, or direct two-proton emission. As direct decays may be studied by expanding a decay amplitude in a series of hyperspherical harmonics functions, the fragment spectra can be fitted by a few components determined by relative orbital momenta of fragments and a single, minimal value of hypermomentum.

Strong correlations of fragments, observed in direct decays of the A=6 nuclei, reflect exotic three-body configurations in the mother nuclei. These modes are induced by a three-body quantum effect of general nature: kinematic focusing of fragments over momenta and in space due to K$\neq$0.

Because of this three-body phenomenon, strong p-p correlations are expected in direct two-proton emission of other nuclei, *eg* two-proton radioactivity candidates ^{19}Mg, ^{34}Ca, ^{48}Ni. The life-time and decay properties of these nuclei considered in a three-body approach depend strongly on the mass and the structure of ground states. Exotic decay modes, *eg* two-proton radioactivity with strongly oscillating p-p correlations, may appear making these drip-line nuclei attractive objects for experimental studies.

ACKNOWLEDGEMENTS

The author wishes to gratefully acknowledge the collaboration of colleagues. In particular, L. Grigorenko has made several contributions to this work. The support of the Gesellschaft für Schwerionenforshung mbH, the German Federal Minister for Education and Research (BMBF) under Contract 06 DA 820 and the PROCON99 Organizing Committee is acknowledged.

Appendix. Direct 2p-decays of 0$^+$ nuclear states.

Explicit formulae for a simultaneous emission of two protons from ^{19}Mg are presented here, namely, for the transitions with ΔJ^π of 0$^+$ and 1$^+$.

The probability of the ^{19}Mg decay into the ^{17}Ne+2p channel can be derived in analogy with the ^{6}Be 2p-decay, [1]. The ^{17}Ne spectrum differs from the ^{6}Be case by normalization coefficients only and can be fitted by the expression

$$\frac{\partial^2 P}{\partial E_{Ne} \partial \Omega_{Ne}} = \frac{19}{16\pi Q} \sqrt{x_{Ne}(1 - x_{Ne})} \cdot \mid F \mid^2 \cdot \mid F_{FSI} \mid^2, \qquad (3)$$

where $x = E_{Ne}/E_{Ne}^{max}$. For the direct two proton decay with $\Delta J^\pi = 0^+$, the amplitude approximation F has few components, e.g. the expansion with the lowest possible *hypermomentum* values of 0, 2, 4 and 6 is:

$$F(p-p, Ne-pp) \sim B_{000}^{00} Y_{000}^{00} + B_{200}^{00} Y_{200}^{00} + B_{400}^{00} Y_{400}^{00} + B_{600}^{00} Y_{600}^{00}$$
$$= B_{000}^{00} + B_{200}^{00}(2x-1) + B_{400}^{00}(16x^2 - 16x + 3) + B_{600}^{00}(2x-1)(8x(x-1)+1). \quad (4)$$

The decay amplitude (4) expressed via other Jacobi coordinates (p-Ne,p-pNe) is

$$F = B_{000}^{00}[Y_{000}^{00}] + B_{200}^{00}[0.99 Y_{200}^{11} + 0.05 Y_{200}^{00}] + B_{400}^{00}[0.94 Y_{400}^{22} - 0.09 Y_{400}^{11} - 0.33 Y_{400}^{00}]$$
$$+ B_{600}^{00}[0.89 Y_{600}^{33} - 0.04 Y_{600}^{22} - 0.45 Y_{600}^{11} + 0.02 Y_{600}^{00}] \quad (5)$$

where new norms are calculated using the Raynal-Revai transformation (2) of the expansion coefficients B_{K00}^{00} from (4).

As the coordinates (p-Ne,p-pNe) are almost the same as the proton coordinates in the ^{19}Mg c.m. system, the equation (5) can be used for estimates of the single-proton configurations in ^{19}Mg corresponding to the respective p-p modes in the eq.(4). In particular, the first component in the eq.(4), with K=0, $\ell_{p-p}=\ell_{Ne-pp}=0$, with the 100% probability corresponds to the $\ell_{p-Ne}=\ell_{p-pNe}=0$ component in eq.(5) which matches the valent protons being in the s-shell of ^{19}Mg. The second term in (4), when K=2, with the $(0.9985)^2$=0.997 probability coincides with the $\ell_{p-Ne}=\ell_{p-pNe}=1$ component in eq.(5) which represents valent protons in the p-shell. And the last term in (4), with K=4, has the $(0.94)^2$=0.884 probability to overlap with the $\ell_{p-Ne}=\ell_{p-pNe}=2$ component which matches the valent protons in the d-shell. In eq. (3), the final state interaction factor F_{FSI} was used (as in the Migdal-Watson model applied in the ^{6}Be decay case) to take into account the Coulomb repulsion of charged particles and the p-p attraction in a S=ℓ_{pp}=0 wave: $|F_{FSI}|^2 = P_{Ne-p_1} P_{Ne-p_2} \cdot |\Phi_{p-p}|^2$ where $P_{i-j}(E_{i-j}) = (F_0^2 + G_0^2)^{-1}$ is the Coulomb penetration factor for particles i,j and $|\Phi_{p-p}|^2$ is a p-p interaction factor taken in the effective range expansion [14].

REFERENCES

1. O.V. Bochkarev et. al., Nucl. Phys. **A505**, 215-222 (1989).
2. D.F. Geesaman et al., Phys. Rev. **C15**, 1835-1852 (1977).
3. O.V. Bochkarev et al., Sov. J. Nucl. Phys. **55**, 955-969 (1992).
4. L.M. Delves, Nucl. Phys. **20**, 275-308 (1960).
5. B.V. Danilin et. al., Sov.J.Nucl.Phys. **46**, 225 (1987).
6. A.A.Korsheninnikov, Sov.J.Nucl.Phys. **52**, 1304-1315 (1990).
7. V.I. Kukulin et.al., Nucl. Phys., **A453**, 365-382 (1986).
8. B.V. Danilin et.al., Sov.J.Nucl.Phys. **48** (1988) 766; **49** (1988) 217; **53** 71-87 (1991)
9. B.V. Danilin and M.V. Zhukov, Phys. At. Nucl. **56**, 460-475 (1993).
10. D. Beamel et al., "Search for two-proton emission from ^{19}Mg", proposal to the GANIL program committee, 1999.
11. L.V. Grigorenko, "Electromagnetic and weak interactions in light exotic nuclei", Ph.D. thesis, 1997, Chalmers Univ. of Technology, Geteborg, ISBN 91-7197-553-5.
12. L.V. Grigorenko, I.G. Mukha and M.V. Zhukov, "Analysis of a β-delayed multiparticle emission in a few-body approach: the ^{9}Li test", in preparation.
13. L.V. Grigorenko, private communication.
14. R.J.N. Phillips, Nucl. Phys. **A53**, 650-662 (1964).
15. Ya.A. Smorodınskij and V.D. Efros, Sov. J. Nucl. Phys. **17**, 107-123 (1973).
16. B. Alex Brown, Phys. Rev. **C43**, R1513-R1517 (1991).

Exact calculations for deformed proton emitters

E. Maglione[1] and L. S. Ferreira[2]

[1] *Dip. di Fisica "G. Galilei", Via Marzolo 8, I-35131 Padova, Italy*
and INFN, Sezione di Padova, Italy
[2] *Centro de Física das Interacções Fundamentais, and Departamento de Física,*
Instituto Superior Técnico, Av. Rovisco Pais, P-1096 Lisbon Codex, Portugal.

Abstract. Proton decay from deformed nuclei is analysed and the half–lives $T_{1/2}$ are evaluated exactly, assuming that the emitted proton moves in a deformed single particle Nilsson level. The angular momenta and deformations that reproduce the experimental half–lives are determined. All experimental values currently known for decay from the ground state, recently found isomeric states and fine structure, have a perfect and consistent description within our model, with deformations that are soundly established in this mass region.

INTRODUCTION

Nucleon emission from the ground state of spherical, as well as from deformed nuclei, provides information not only on the limits of stability of nuclei, but also gives important spectroscopic information on the nature of the decaying state.

A large variety of proton emitters were observed [1–6] in the region of heavy nuclei with $50 < Z < 82$. These measurements include spherical and deformed nuclei, and define almost completely the borders of proton stability. The Coulomb barrier is sufficiently high in this large Z region, to keep the outgoing proton close to the core nucleus long enough for detection. Outside this charge range, it was also observed recently [7] in $^{58}_{29}$Cu, proton decay from a superdeformed state to a spherical one.

Important aspects of nuclear structure can be learned from proton decay. It is a powerful tool not only to probe small components of the wave functions of the decaying states, but also to determine the deformation of nuclei and the angular momentum of the ground state [8]. This information is more difficult to obtain in stable nuclei. Before the decay, the nucleus goes through a metastable state which has a single particle character, in contrast with resonances in stable nuclei, that lie high in the continuum. Therefore, proton decay will make it possible to follow the behaviour of single-particle resonances in nuclei beyond the drip line, and consequently may guide future experiments in the search for new emitters.

The theoretical description of proton emission from spherical systems was studied by various authors [1,2,9–11] using the semi-classical WKB method or standard reaction theory with the distorted-wave Born approximation. The experimental results were well reproduced within these models.

The first attempts to calculate the half-life for emission from a deformed nuclei, were made by the authors of Ref. [4,12,13].

THE MODEL

An important characteristic of nuclei on the drip line, is that they have a Fermi level very close or even immersed in the continuum. The interpretation of this feature must take into account the unstable property of the states. In principle, this should be done by solving the time-dependent Schrödinger equation and following the evolution of the decaying states. However, one can transform the time dependent process into a stationary problem by imposing to the wave functions outgoing boundary conditions, i.e.,

$$\lim_{r \to \infty} u_{lj}(r) = N_{lj}(G_l(kr) + iF_l(kr)) \tag{1}$$

with k= $\sqrt{2\mu E/\hbar^2}$, and N_{lj} a normalization constant. The functions F and G are the regular and irregular Coulomb functions. This provides all the bound states and states with complex energies that have negative imaginary parts. If the absolute values of those imaginary parts are small, one can associate the corresponding eigenstates with physical resonances such that the real part of the energy represents the position of the resonance and the imaginary part is proportional to the decaying width namely, $\mathcal{E}_n = E_n - i\Gamma_n/2$. For spherical systems this imaginary part defines completely the decaying width if there is only one open channel. For proton emitters, the measured half lives $T_{1/2} = \ln 2\hbar/\Gamma$, are longer than a microsecond and correspond to decay widths smaller than 10^{-16}MeV. To find accurately such small imaginary parts might be technically difficult to achieve. An alternative procedure can be used [8], with the help of the asymptotic wave function. In this fashion, the decay width corresponding to the decay to the channel lj is then given by,

$$\Gamma_{lj}(r) = \frac{\hbar^2 k}{\mu} \frac{|u_{lj}(r)|^2}{|G_l(kr) + iF_l(kr)|^2}. \tag{2}$$

It is immediate to see from the boundary condition, that this expression is independent of r beyond the range of the nuclear potential. A comparison between results obtained from Eq. (2) and the imaginary part of the energy for the nuclei under consideration, shows relative differences smaller than one part in a billion [14], for half–lives longer than 10^{-16} seconds.

Spherical calculations for some proton emitters, were not able to reproduce the experimental results [1,11]. These nuclei, in fact, are expected to be deformed, and a different procedure has to be applied to interpret them.

We take as a basic feature of our model, that the decaying nucleon moves in a single-particle Nilsson level of a deformed potential. The search of complex energy eigenvalues in a nonspherical system was solved only recently [15]. The wave function $\Psi(\vec{r})$, corresponding to this state is obtained by solving exactly [15] the Schrödinger equation for a deformed Saxon–Woods potential with a deformed spin-orbit term, and realistic parameters.

In practice, $\Psi(\vec{r})$ is expanded in spherical waves, i. e.

$$\Psi_m(\vec{r}) = \sum_{j \geq m} u_{ljm}(r) \left[Y_l(\hat{r}) \chi \right]_{jm} \tag{3}$$

where m is a conserved quantity in the axially symmetric case, u and χ are the radial and spin functions, and Y the spherical harmonics. The orbital angular momentum l is determined by the parity of the state. Projecting on the state $[Y_{l'}(\hat{r})\chi]_{j'm}$ one obtains a set of coupled channel equations which have the form,

$$\left(\frac{d^2}{dr^2} - k^2 - \frac{l(l+1)}{r^2} \right) u_{ljm}(r) = \sum_{l'j'} \left(V^m_{1\,ljl'j'} + V^m_{2\,ljl'j'} \frac{d}{dr} \right) u_{l'j'm}(r), \tag{4}$$

where the quantities $V^m_{1\alpha\alpha'}$ and $V^m_{2\alpha\alpha'}$ are the matrix elements of the interaction between the angular and spin parts of the partial waves, and α designates the set of quantum numbers lj. The first derivative of the wave function is coming from the deformed spin-orbit potential.

Equations (4) are solved with the conditions of regularity at the origin and outgoing waves at large distances, for each partial wave. There are technical difficulties to solve such equations, that can be avoided integrating instead the set of first order coupled equations for the logarithmic derivative,

$$\frac{d}{dr}\phi^m_{\alpha\alpha'}(r) - \sum_{\alpha''} V^m_{2\alpha\alpha''}\phi^m_{\alpha''\alpha'}(r) + \sum_{\alpha''} \phi^m_{\alpha\alpha''}(r)\phi^m_{\alpha''\alpha'}(r) -$$
$$\left(k^2 + \frac{l_\alpha(l_\alpha+1)}{r^2} \right)\delta_{\alpha\alpha'} - V^m_{1\alpha\alpha'} = 0 \tag{5}$$

where the matrix $\phi^m(r)$ is defined by $\phi^m(r) = d\mathcal{U}_m(r)/dr \times \mathcal{U}_m^{-1}(r)$ and $\mathcal{U}_m$ is the matrix whose columns are the N unknown linearly independent solutions of Eq. (4). The solution of this problem is obtained from the integration of $N \times N$ coupled equations.

The method provides all Nilsson bound single particle levels of the parent nucleus, and a set of states with complex energies that can be interpreted as physical resonances. The parent nucleus is described by a wave function of a particle–plus–rotor in the strong coupling limit. Using standard notation, the angular momentum projected wavefunction is given by,

$$\Phi_m^{J_i M_i, K_i} = \left(\frac{\hat{J}_i}{16\pi^2}\right)^{1/2} \{\mathcal{D}_{M_i K_i}^{J_i} \Psi_{K_i} + (-1)^{J_i+K_i} \mathcal{D}_{M_i -K_i}^{J_i} \Psi_{\bar{K}_i}\} \tag{6}$$

where $\hat{J}_i = 2J_i + 1$, $\mathcal{D}$ are the rotation matrices and Ψ is the intrinsic single particle wavefunction of Eq. (3).

Since electromagnetic transitions are usually faster than the decay by proton emission, the decay is more probable when the nucleus is in the lowest energy state with $J_i = K_i$.

The exit channel wavefunction will be the tensorial product of the internal wavefunctions of the daughter nucleus Ψ_d, times the wavefunction corresponding to the relative motion of the proton with respect to the mother nucleus. The daughter wavefunction is

$$\Psi_d^{J_d M_d K_d} = \left(\frac{\hat{J}_d}{8\pi^2}\right)^{1/2} \mathcal{D}_{M_d K_d}^{J_d}. \tag{7}$$

Taking the overlap between the initial and final states, the partial decay width becomes, in analogy with Eq. (2)

$$\Gamma_{l_p j_p}^{J_d}(r) = \frac{\hbar^2 k}{\mu} \frac{2\hat{J}_d <J_d, 0, j_p, K_i | K_i, K_i>^2}{\hat{K}_i} \frac{|u_{l_p j_p}(r)|^2}{|G_{l_p}(r) + iF_{l_p}(r)|^2} u_{K_i}^2, \tag{8}$$

where $|J_d - K_i| \geq j_p \geq K_i$ and $u_{K_i}^2$ is the probability that the single particle level in the daughter nucleus is empty, evaluated in the BCS approach. From energy considerations, the most probable decay is usually the one that leaves the daughter nucleus in the ground state. In this way the proton has the largest possible energy. Therefore the relation $J_d = M_d = K_d = 0$ holds, and angular momentum conservation implies that $j_p = J_i = K_i$.

The Coulomb potential of a deformed nucleus has nonspherical components. Therefore, Eq. (8) is independent of r only at larger distances.

In all these derivations it was assumed that the deformation of the daughter nucleus is the same as the mother one. For more details see Ref. [8].

RESULTS AND DISCUSSION

In the present work, we apply this exact formalism to analyse available experimental data on odd deformed decaying nuclei. We will consider as deformed, those systems where spherical calculations provide unreasonable spectroscopic factors. They are the proton emitters $^{109}_{53}$I, $^{113}_{55}$Cs, $^{117}_{57}$La, $^{131}_{63}$Eu, $^{141}_{67}$Ho, and $^{151}_{71}$Lu from the ground state, the decay from the isomeric state of $^{141}_{67}$Ho and $^{151}_{71}$Lu and decay of $^{131}_{63}$Eu to the first 2^+ excited state of $^{130}_{62}$Sm.

The Nilsson levels for these nuclei were calculated from a deformed Saxon–Woods potential with the "universal" parameters of Ref. [17]. The results are compiled in

TABLE 1. Total angular momentum and deformation that reproduce the experimental half–lives for the measured deformed odd proton emitters compared with the predictions of Ref. [21,22].

	Proton decay		Möller–Nix	
	J	β	J	β
^{109}I	1/2+	0.14	1/2+	0.16
^{113}Cs	3/2+	0.15 ÷ 0.20	3/2+	0.21
^{117}La	3/2+	0.20 ÷ 0.30	3/2+	0.29
^{131}Eu	3/2+	0.27 ÷ 0.34	3/2+	0.33
^{141}Ho	7/2−	0.30 ÷ 0.40	7/2−	0.29
^{141m}Ho	1/2+	0.30 ÷ 0.40		
^{151}Lu	5/2−	−0.18 ÷ −0.14	5/2−	−0.16
^{151m}Lu	3/2+	−0.18 ÷ −0.14		

Table 1, and have been presented in Ref. [8,18–20]. A remarkable agreement can be observed in all nuclei between the total angular momentum and deformation that reproduce the experimental half–lives, and the values predicted by the mass formula of Möller and Nix [21,22]. A quite important feature is the description of the half–life of the isomeric states, with the same deformation used for the ground state. This shows the consistency of the model. We will limit our discussion here, to a few cases only.

$^{151}_{71}$LU

This nucleus has always been considered as spherical since the decay of the ground state could be reproduced with a $h_{11/2}$ level. However, it was recently found [23] an isomeric state at 77 keV, that cannot be interpreted with realistic spectroscopic factors neither as a $s_{1/2}$ nor a $d_{3/2}$. The "experimental" spectroscopic factors are in this case 2-3 times larger than the "theoretical" ones. This discrepancy was observed in other nuclei, and attributed to deformation [1,11]. It is therefore quite interesting to analyse ^{151}Lu in this perspective and see if a deformed calculation can give a more consistent interpretation than the spherical one.

The Nilsson levels for the nucleus ^{151}Lu are presented in Fig. 1 as a function of the deformation parameter β. Since the proton and neutron Fermi levels are closer to the magic number 82, one might expect [21,22,25–27] an oblate $\beta < 0$ deformation. In the oblate case, the Fermi surface is always at the $K = 5/2^-$ level coming from the spherical $h_{11/2}$ state, while levels very close to it are the $K = 1/2^+$ and $K = 3/2^+$ coming from the $s_{1/2}$ and $d_{3/2}$ shells respectively. These are the candidates for the ground and isomeric states.

The half–lives for proton emission from these three Nilsson levels are shown in

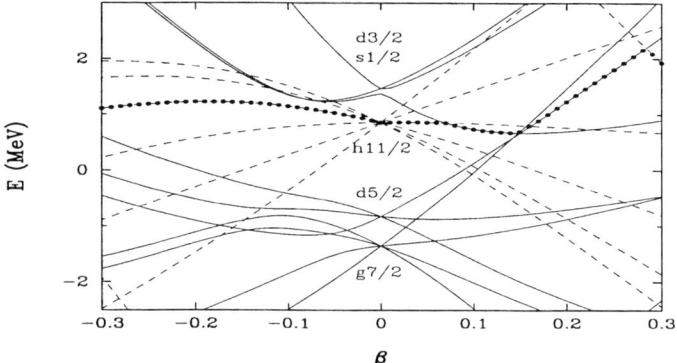

FIGURE 1. Proton Nilsson levels corresponding to $^{151}_{71}$Lu. The dotted line indicates the Fermi level.

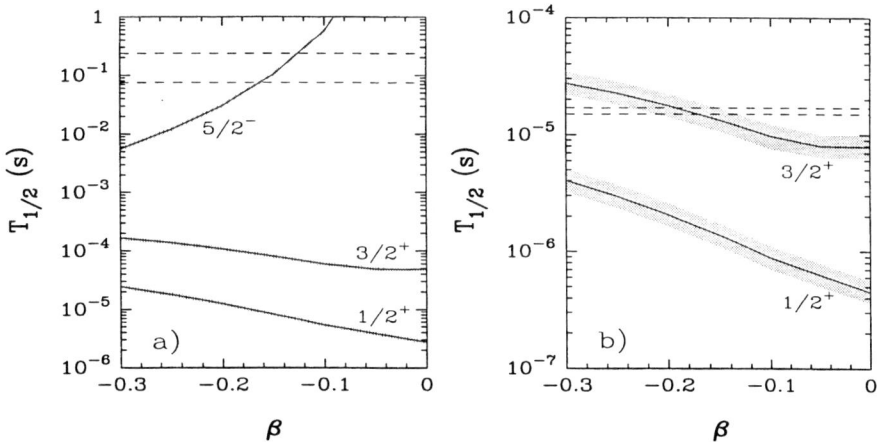

FIGURE 2. Half–life of the states around the Fermi surface at negative deformation in $^{151}_{71}$Lu, for the ground state (a) and isomeric state (b). The shadowed area represents the uncertainty due to the experimental error bars in the energy. The experimental values (dashed lines) are from Ref. [23]. In the case of the ground state the half–life for proton decay was deduced from the measured one using a proton branch $b_p = 70 \pm 35\%$ [24].

Fig. 2 as a function of oblate deformation. In our calculation, the depth of the single particle potential is readjusted to give independently the correct experimental energies for the ground and isomeric states. Since these experimental energies differ by only 77 keV, the half–life of both states should change with energy in accordance with the relation $T_{1/2} \propto \exp a/\sqrt{E}$. In fact, Fig. 2 shows how this behaviour is accurately satisfied with the same value a for all states. This provides a good check on the accuracy of our calculation.

From Fig. 2, it is apparent that the ground state decay can be reproduced only as a decay from the $K = 5/2^-$ level. Taking into account the spectroscopic factor $u^2 \sim 0.5$, which is reasonable for levels around the Fermi surface, the comparison with the experimental value for the half–life, indicates a deformation of $-0.18 < \beta < -0.14$. This value is astonishingly close to the ones predicted by other authors, namely $\beta = -0.16$ for the $K = 5/2^-$ ground state [21,22], $\beta = -0.15$ [25] and $\beta \approx -0.2$ of Ref. [26]. Another important consistency condition, is the possibility of explaining the recently measured isomeric state with the same value of β. In this region of deformation, the states closer to the Fermi surface, are the $K = 3/2^+$ and $K = 1/2^+$. The last one has a very short half–life, while the $K = 3/2^+$ state reproduces very well the experimental half–life of the isomeric state, with the same deformation as the ground state. These results fully support our hypothesis of an oblate deformed ^{151}Lu.

$^{131}_{63}\text{EU}$

Due to energy considerations, one expects proton decay to proceed mainly to the ground state of the daughter nucleus. However, in rotational nuclei the first excited state could be very low in energy, therefore, a sizeable branching ratio could be expected, known as fine structure. In fact, this was measured for the first time [28] in the radioactive decay from the ground state of ^{131}Eu. In a previous experiment [4], only the ground state line was observed at 950(8)keV. The recent experiment reports an updated value for this energy at 932(7) keV, and a second proton peak with energy 811(7)keV. Since the half lives of the two states were almost the same, this line was interpreted as proton decay from the ground state to the first excited 2^+ state of the daughter nucleus ^{130}Sm.

The Nilsson single particle levels for ^{131}Eu are shown in Fig. 3(a), as function of deformation. The half–life for proton emission for all levels that lie or are close to the Fermi surface, up to a deformation $\beta = 0.4$, are shown in Fig. 3(b). This figure is Fig. 2c of Ref. [18] scaled by a factor of ≈ 1.85, since the new experimental value for the energy of the outgoing proton was used.

Using the spectroscopic factor $u^2 \approx 0.5$ there are three states $K = 5/2^+_a$, $K = 3/2^-$ and $K = 3/2^+$ that could reproduce the half–life for the decay. The first one with β in the range of 0.15 – 0.28, the second around 0.2 – 0.24 and the last one between 0.24 and 0.34. The decay to the excited state can be discussed in an analogous manner. The total width for the decay is a sum of partial widths

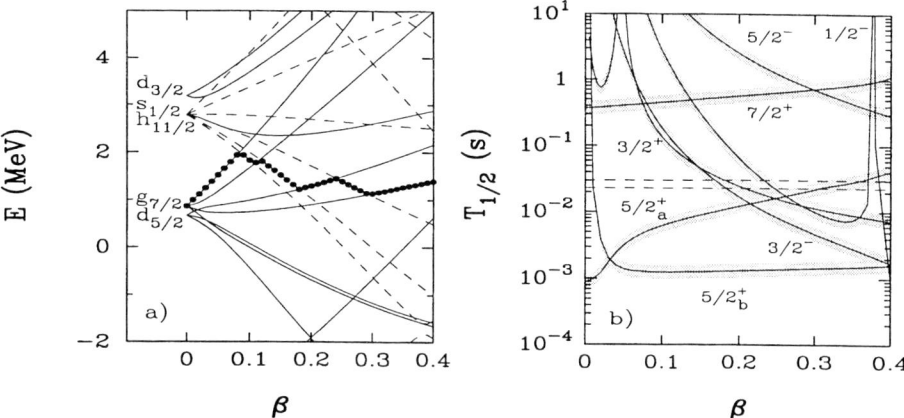

FIGURE 3. (a) As in Fig. 1 for $^{131}_{63}$Eu. (b) As in Fig. 2 for $^{131}_{63}$Eu. The curve labelled $5/2^+_a$ is coming from the spherical shell $d_{5/2}$; $7/2^+$, $5/2^+_b$ and $3/2^+$ are all coming from the $g_{7/2}$; finally, $5/2^-$, $3/2^-$ and $1/2^-$ are all coming from the $h_{11/2}$. The experimental values (dashed lines) are from Ref. [28].

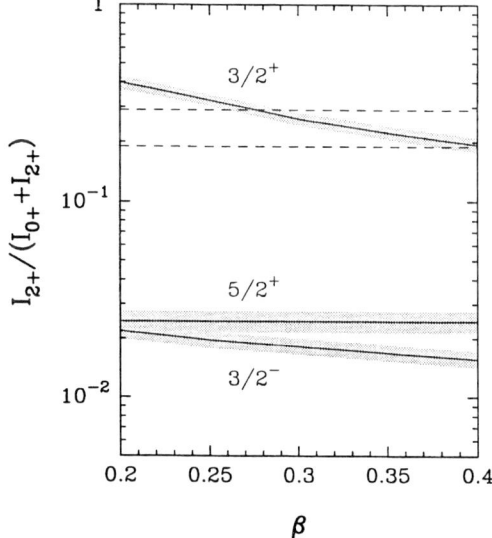

FIGURE 4. Branching ratio for the decay to the 2^+ state of ^{130}Sm as a function of deformation.

for all possible values of the angular momentum j_p of the emitted proton. From total angular momentum conservation, $J_i + 2 \geq j_p \geq \max(|J_i - 2|, J_i)$. The partial widths that give the main contributions correspond to situations where N_{l_p,j_p} is large or the centrifugal barrier low, i.e. l_p small, the latter being the most important condition. If $J_i = l_p + 1/2$, usually only one term survives in the sum, the one that has $j_p = J_i$. In this case, the ratio of the decay width to the 2^+ with respect to the one to the 0^+ depends practically only on the energy of the 2^+ and the l_p, j_p comes only in a Clebsch–Gordan. The dependence on deformation is small, since the N_{l_p,j_p} cancels out in the ratio. A strong dependence on deformation can be observed only if N_{l_i,K_i} is much smaller that the other components, or if $J_i = l_p - 1/2$ when two channels with the same centrifugal barrier contribute.

The branching ratio for the decay to the 2^+ is shown in Fig. 4. The spectroscopic factor u^2, present in the numerator and denominator of this ratio, is the same and cancels out in the final result. As can be seen in the figure, the branching ratio for the $K = 5/2+$ state is practically constant, since the $j_p = d5/2$ dominates the decay due to the higher centrifugal barrier of the $j_p = g7/2$ and $g9/2$ components. This is not true for the $K = 3/2+$ state where the dominating component is the $d5/2$ that has the same centrifugal barrier, but a larger N.

From Fig. 4, it is clear that the levels $K = 5/2_a^+$ and $K = 3/2^-$ can be disregarded, since they do not fit the experimental value. The only state capable of reproducing the data is the $K = 3/2^+$, with deformation larger than 0.27. Our result and assignment are in perfect agreement with the findings of Ref. [28].

Another nucleus where the fine structure could be observed is ^{141}Ho. In this case there is no doubt on the $K = 7/2-$ assignment of the state, as explained in Ref. [18].

Using a deformation of $\beta = 0.29$ [21,22] the energy of the 2^+ should be [29] at 140 keV, and the branching ratio obtained is 5%, a value probably not reachable with present experimental apparatus.

CONCLUSIONS

In conclusion, we have shown in this paper that our model to calculate exactly the decay from single particle Nilsson levels is able to describe quite accurately and consistently, all the available experimental data on odd deformed proton emitters from the ground and isomeric states. The available data on fine structure are perfectly reproduced. The calculation provides information on the deformation and angular momentum J of the decaying nuclei, which would be difficult to extract using other probes in this unstable mass region. Our results support the nuclear structure predictions of Refs. [21,22] for nuclei at the proton drip line.

ACKNOWLEDGEMENTS

We would like to thank the organizers for the invitation to partecipate in this symposium.

REFERENCES

1. P. J. Woods and C. N. Davids, Annu. Rev. Nucl. Part. Sci. **47**, 541 (1997), and references therein.
2. C. N. Davids et al., Phys. Rev. **C55**, 2255 (1997).
3. J. C. Batchelder, et al., Phys. Rev. **C57**, R1042 (1998).
4. C. N. Davids, et al., Phys. Rev. Lett. **80**, 1849 (1998).
5. K. Rykaczewski, et al., Phys. Rev. **C60**, 011301(R) (1999).
6. F. Soramel, et al., contribution to these proceedings.
7. D. Rudolph, et al., Phys. Rev. Lett. **80**, 3018 (1998).
8. E. Maglione, L. S. Ferreira and R. J. Liotta, Phys. Rev. Lett. **81**, 538 (1998).
9. W. F. Feix and E. R. Hilf, Phys. Lett. **120B**, 14 (1983).
10. A. Gillitzer, T. Faestermann, K. Hartel, P. Kienle and E. Nolte, Z. Phys. **A326**, 107 (1987).
11. S. Åberg, P. B. Semmes and W. Nazarewicz, Phys. Rev. **C56**, 1762 (1997).
12. V. P. Bugrov and S. G. Kadmensky, Sov. J. Nucl. Phys. **49**, 967 (1989).
13. S. G. Kadmensky and V. P. Bugrov, Phys. of Atom. Nucl. **59**, 399 (1996).
14. L. S. Ferreira and E. Maglione, to be published.
15. L. S. Ferreira, E. Maglione and R. J. Liotta, Phys. Rev. Lett. **78**, 1640 (1997).
16. R. G. Thomas, Prog. Theor. Phys. **12**, 253 (1954).
17. S. Cwiok, J. Dudek, W. Nazarewicz, J. Skalski and T. Werner, Comp. Phys. Comm. **46**, 379 (1987).
18. E. Maglione, L. S. Ferreira and R. J. Liotta, Phys. Rev. **C59**, R589 (1999).
19. L. S. Ferreira and E. Maglione, Phys. Rev. C, in press.
20. E. Maglione and L. S. Ferreira, submitted to Phys. Rev. C.
21. P. Möller, J. R. Nix, W. D. Myers and W. J. Swiatecki, At. Data Nucl. Data Tables **59**, 185 (1995).
22. P. Möller, R. J. Nix and K. L. Kratz, At. Data Nucl. Data Tables **66**, 131 (1997).
23. C. R. Bingham, et al., Phys. Rev. **C59**, R2984 (1999).
24. P. J. Sellin, et al., Phys. Rev. **C47** (1993) 1933.
25. G. A. Lalazissis, D. Vretenar and P. Ring, nucl-th/9907038 preprint, Tech Uni. München, July 1999.
26. Y. Aboussir, J. M. Pearson, A. K. Dutta and F. Tondeur, At. Data Nucl. Data Tables **61**, 127 (1995).
27. R. Bengtsson, P. Möller, J. R. Nix and Jing-ye Zhang, Physica Scripta **29**, 402 (1984).
28. A. A. Sonzogni, et al., Phys. Rev. Lett. **83**, 1116 (1999).
29. L. Grodzins, Phys. Lett. **2**, 88 (1962).

Theoretical Predictions for Beta-Delayed One- and Two-Proton Emission

T. Siiskonen and P. O. Lipas

Department of Physics, University of Jyväskylä, P.O. Box 35 (Y5), FIN-40351 Jyväskylä, Finland

Abstract. Theoretical estimates for isospin-forbidden beta-delayed proton and two-proton emission are calculated in a shell-model framework. The parent nuclei studied are ^{23}Al and ^{31}Ar. The results are compared to new experimental data. The reliability of the shell-model estimates is discussed.

INTRODUCTION

The processes of beta-delayed proton and two-proton emission have proven to be excellent spectroscopic tools; see, e.g., Honkanen et al. [1] and Axelsson et al. [2]. In these processes, effects of isospin mixing are clearly seen in those channels which proceed via the isobaric analog state (IAS) in the beta-decay daughter nucleus. By definition, the IAS has the isospin of the ground state of the parent nucleus. Thus proton and two-proton emissions from the IAS to low-lying states in the particle-emission daughter nucleus are isospin forbidden, since they require an isospin change of $\Delta T = 3/2$ and 1, respectively. Isospin impurities are therefore needed in the IAS wave function.

The parent nuclei considered in the present study are ^{23}Al and ^{31}Ar. Their basic (Hartree-Fock) configurations, as well as those of the daughter nuclei, lie in the interior of the sd shell. Therefore we feel justified in choosing the complete sd shell as the valence space for our shell-model (SM) calculations. Any positive-parity intruder states lie so high in energy that beta feeding to them is negligible, and negative-parity states are not fed by allowed beta decay. The isospin-mixed wave functions were obtained with the matrix elements of Ormand and Brown [3] and the SM code OXBASH [4].

Earlier SM studies of these nuclei include Refs. [5] (^{23}Al) and [2,6] (^{31}Ar). For ^{23}Al [5], isospin mixing and proton emission from the IAS were examined. We give here a more detailed description of the proton decay channels, together with new experimental data [7]. For ^{31}Ar, the proton [2] and two-proton [2,6] decay channels were not examined in the SM calculations, while we do so here and make comparisons with new experimental data.

THEORETICAL BACKGROUND

The aim of our SM calculations with isospin mixing is to study the distribution of Fermi strength, especially near the IAS, and the subsequent particle emission.

If isospin were a strictly conserved quantity, the Fermi matrix element for β^+ decay would be given by the familiar model-independent expression

$$B(F^+)_T = \langle \mathcal{M}_{F+} \rangle_T^2 = T(T+1) - M_{Ti}(M_{Ti}+1), \tag{1}$$

where M_{Ti} is the isospin component of the initial state and the subscript T refers to good isospin. Use of an isospin-nonconserving (INC) Hamiltonian modifies this expression, as discussed in Refs. [2,8]. In the most compact form this can be expressed as $B(F^+) = a^2 B(F^+)_T$, where $a^2 \leq 1$ is a measure of the T-symmetry breaking.

In first-order perturbation theory the mixing amplitude between states $|\phi_1\rangle$ and $|\phi_2\rangle$ is given by

$$\delta = \frac{\langle \phi_1 | H_{\text{INC}} | \phi_2 \rangle}{E_2 - E_1}, \tag{2}$$

where the states $|\phi_i\rangle$ and energies E_i are obtained with an isospin-conserving interaction. We note that δ is sensitive to the level spacing $\Delta E = E_2 - E_1$, which is not given reliably by the SM, as discussed in Ref. [9]. This uncertainty makes the estimation of δ, and consequently a, rather qualitative. In Ref. [9] a mean value, obtained by varying the energy of the IAS, was used. We have applied a different scheme where the mixing is obtained directly from the SM observables [2,8]. Since the uncertainties in the calculated quantities are large, the resulting isospin admixtures and particle-emission widths should be considered as estimates. Comparison with experiment can shed light on the reliability of the SM wave functions.

If we write the Fermi matrix element $\langle \mathcal{M}_{F+} \rangle = \langle \alpha' J M | t_+ | \alpha J M \rangle$ in occupation-number representation and apply the Wigner-Eckart theorem, we obtain [2]

$$\langle \mathcal{M}_{F+} \rangle = \frac{1}{\sqrt{2J+1}} \sum_j \sqrt{2j+1}\, \Omega_j (\alpha' J \| [a_{jn}^\dagger \tilde{a}_{jp}]_0 \| \alpha J), \tag{3}$$

where Ω_j is the radial overlap integral

$$\Omega_j = \int_0^\infty dr\, r^2 R_j^n(r) R_j^p(r) \leq 1. \tag{4}$$

The reduced matrix element, with the factor $(2J+1)^{-1/2}$, in Eq. (3) is the relevant one-body transition density, calculated with the SM code OXBASH [4]. The radial overlap integrals were calculated with Woods-Saxon single-particle wave functions, with parameters from Ref. [10]. In the cases we have considered, deviations of the Ω_j from unity are of the order 10^{-3}, and we have simply put $\Omega_j = 1$.

The proton and two-proton emission widths were calculated with spectroscopic factors θ_l^2 obtained from the INC shell-model calculation. The spectroscopic amplitude for the emission of a cluster of orbital angular momentum l and k particles is defined as [11]

$$\theta_l \equiv \theta_l(k;\lambda) = \left(\frac{A}{A-k}\right)^{\lambda/2} G \frac{(\alpha J\|A_l^\dagger(k)\|\alpha'J')}{\sqrt{2J+1}}. \tag{5}$$

The quantity λ is defined, in terms of harmonic-oscillator states, as $\lambda = \sum_i N_i = \sum_i (2n_i + l_i)$. The factor G is an easily calculated statistical quantity which depends on how the cluster particles are distributed on the SM orbitals, and $A_l^\dagger(k)$ is the cluster creation operator. The necessary barrier penetrabilities P_l, which depend sensitively on the decay Q value, were calculated by the code COCAG [12] using regular and irregular Coulomb wave functions. The expression for the single-particle width is taken from Ref. [13]. The resulting expression for the cluster-emission width is

$$\Gamma_l = 2\theta_l^2 P_l(Q) \frac{3(\hbar c)^2}{2\mu c^2 R_0^2}, \tag{6}$$

which contains the reduced mass μ of the daughter-ejectile system and the standard nuclear radius R_0. The total width for a given initial and final state is $\Gamma = \sum_l \Gamma_l$. For proton emission, the value of the orbital angular momentum is given directly by the l of the single-particle orbital involved. For two-proton emission, l is obtained from the angular-momentum and parity selection rules (we assume that the two protons are emitted in a relative s wave).

RESULTS AND DISCUSSION

Parent Nucleus ^{23}Al

Experimentally, beta-delayed proton emission from ^{23}Al has been studied by several groups (e.g. Refs. [5,7]). The observed decays proceed via a few intermediate states in the beta-plus daughter ^{23}Mg. Among these states is the IAS, at an excitation energy of 7.795 MeV [14]. In the works cited, it was not possible to measure any low-energy proton spectrum corresponding to initial states below the IAS. However, at least two strong proton branches were observed starting from states above the IAS. In Ref. [5] the experimental data were collected by the helium-jet technique. This may explain the differences from the data of Ref. [7] discussed below. We also note differences in the results of the basically similar SM calculations in Ref. [5] and the present work.

In the SM spectrum for ^{23}Mg obtained with the INC Hamiltonian, the energy of the IAS is 7.817 MeV ($J^\pi = 5/2^+$ and dominant $T = 3/2$). Candidates for producing appreciable isospin mixing in the IAS are the $5/2^+$ states, with dominant

$T = 1/2$, that are calculated 30 keV below and 188 keV above the IAS, at $E_x = 7.787$ and 8.005 MeV respectively. These three states carry 99.5% of the total Fermi strength (with $\Omega_j = 1$), the main mixing component being the state 30 keV below, as one would expect. Experimentally, there is a positive-parity state of $J \geq 3/2$ at 7.780 MeV [14], i.e., 15 keV below the IAS (7.795 MeV). The first identified $5/2^+$ state below the IAS is at 7.582 MeV, i.e., 213 keV away, so its mixing with the IAS should be small. We assume that the experimental 7.780 MeV state corresponds to the 7.787 MeV SM state. The energy levels are given in Table 1.

Above the IAS, the first observed state that might produce mixing lies at 7.852 MeV [14] and has unknown spin and parity. We propose to identify it with the $7/2^+$ SM state at 8.067 MeV, so this state will not contribute to the isospin impurity of the IAS. The first experimentally identified $5/2^+$ state above the IAS is 360 keV away, at 8.155 MeV; we consider this state as the experimental counterpart of the SM state at 8.005 MeV.

We have noted above that the penetrability $P_l(Q)$ has a very strong dependence on the Q value. On the other hand, the inaccuracy of SM energy predictions is at least of the order of 100 keV. It follows that, at small Q values, the proton-emission widths Γ_l calculated from Eq. (6) can vary by several orders of magnitude due to these inaccuracies. Our goal in the present work is to make meaningful comparisons between calculated and experimental proton intensities. Therefore we elect to use the experimental Q values rather than the SM ones as input in Eq. (6). This choice is analogous to the use of reduced gamma-decay probabilities, rather than lifetimes with their strong energy dependence, in comparing nuclear-structure calculations with experiment.

The proton intensity I for a state i is calculated using the gamma and proton-emission widths as

$$I_i = \beta_i^+ \frac{\Gamma_i^p}{\Gamma_i^\gamma + \Gamma_i^p}, \qquad (7)$$

where β_i^+ is the beta-plus feeding. The proton branch from the 7.780 MeV state is cut off by the small penetrability. On the other hand, at higher excitation energies beta feeding is so small that only weak proton branches appear above $E_p = 1$ MeV.

TABLE 1. Experimental [14] and calculated states in ^{23}Mg. See text and Fig. 1 for more details.

Experiment		Shell model		
E_x (MeV)	J^π	E_x (MeV)	J^π	Number
7.582	$5/2^+$			
7.780	$\geq 3/2^+$	7.787	$5/2^+$	
7.795	$5/2^+$	7.817	$5/2^+$	1[a]
7.852		8.067	$7/2^+$	2
8.155	$5/2^+$	8.005	$5/2^+$	3
8.453	$\geq 3/2^+$	8.493	$5/2^+$	4

[a] Isobaric analog state.

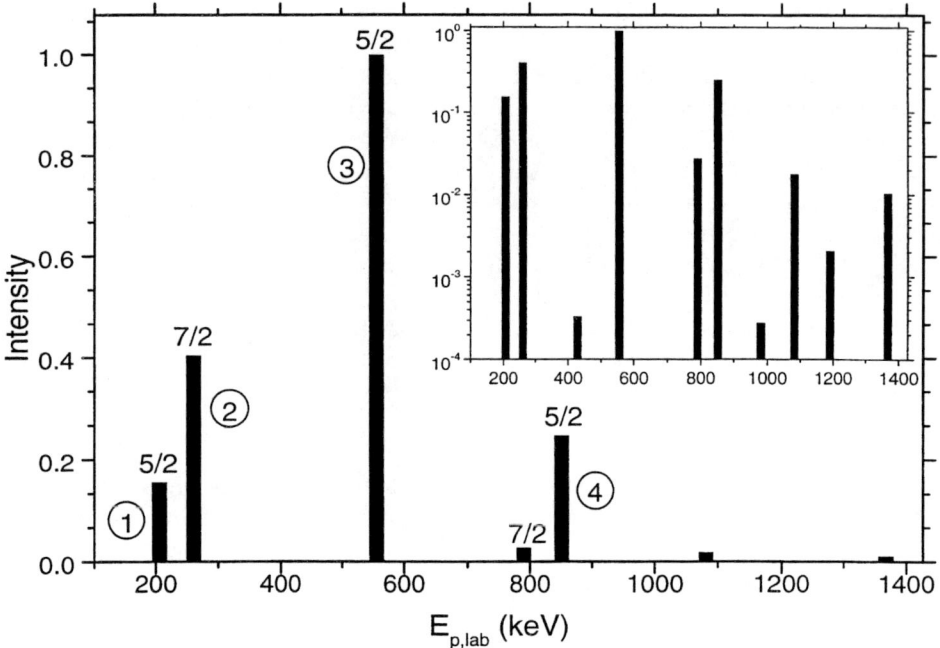

FIGURE 1. Shell-model proton spectrum for the decay $^{23}\text{Mg}^* \to {}^{22}\text{Na} + p$. Peak number 1 corresponds to protons from the IAS. The calculated spins are shown.

In Fig. 1 the SM proton spectrum is shown on a linear and semilogarithmic scale, with the Q values taken from experiment as discussed above. The lowest visible peak (number 1) in the spectrum corresponds to the transition from the IAS. We see from comparison with Fig. 2 that the relative intensities of peaks 1, 2, and 3 are nicely reproduced. The SM initial state for peak 3 is 97% $T = 1/2$, so proton emission from it is essentially completely allowed; the missing 3% is admixed in the IAS. However, the main $T = 1/2$ contribution (22%, see discussion below) to the IAS comes from the near $5/2^+$ state below. Peak 4 has too little intensity in the SM spectrum. In this proton-energy range the proton intensity, Eq. (7), is completely determined by the beta feeding β^+, and our SM value for β^+ is clearly too small.

Although the statistics in the 250 keV region leave some doubt as to whether the first peak in the spectrum of Fig. 2 is truly double, with a separation of 57 keV, we have adopted this experimental interpretation [7]. The measured proton width for the IAS is 1.9 ± 0.8 meV. However, our SM prediction, 17.5 meV, is an order of magnitude larger. This is solely due to the fact that the SM gives a mixing state very close to the IAS, and hence an excessive admixture of $T = 1/2$ in the IAS. The smallness of the experimental proton width indicates that the experimental 7.780 MeV state does not have $J^\pi = 5/2^+$.

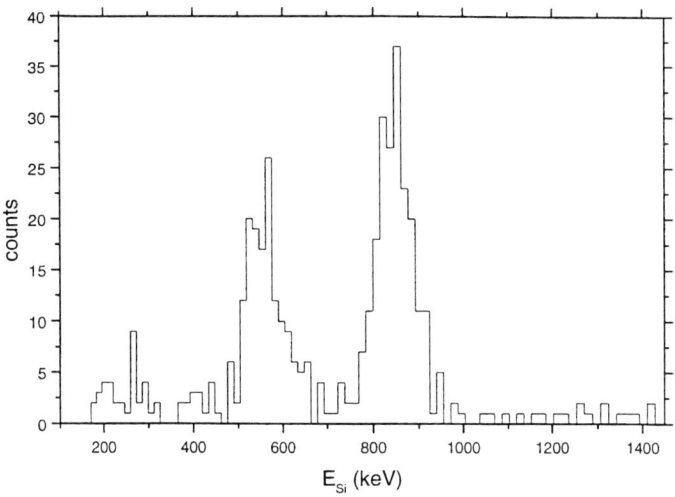

FIGURE 2. Experimental proton spectrum for the decay ^{23}Mg$^* \to {}^{22}$Na$+p$ [7].

Parent Nucleus ^{31}Ar

Beta-delayed proton and two-proton emission from ^{31}Ar has been studied very carefully in recent experiments [2,15,16]. The main task in these experiments has been the examination of two-proton decay channels. The primary question is whether the decay is simultaneous (^{2}He or uncorrelated protons, mode I) or sequential (mode II).

Our SM calculations give information on the decay mode. For sequential emission, we have calculated the first-proton widths. Furthermore, nearly all the intermediate states decay by a second proton emission, from ^{30}S* to ^{29}P. Also simultaneous emission can be handled easily in our framework (see, e.g., Refs. [11,17]). The cluster wave functions are generated with an interaction [4] which conserves the assumed SU(3) symmetry of the states. However, the SM results do not yield information on angular correlations, which means that the nature of mode I cannot be studied. This would require knowledge of the final-state interaction between the protons. The wave functions of the nuclei were again calculated with the INC interaction of Ref. [3].

There are many possible initial states in ^{31}Cl. To keep the calculations tractable we have considered only the decays from the IAS. Then the only quantity which determines the relative branching between modes I and II is the emissison width obtained from Eq. (6).

The mode-I sum spectrum calculated by the SM is shown in Fig. 3. The experimental spectrum has peaks at 7.6, 6.2, 5.7 and 5.2 MeV corresponding to two-proton transitions from the IAS to the first four states in ^{29}P [18]. In the SM spectrum the corresponding mode-I transitions from the IAS are shown together

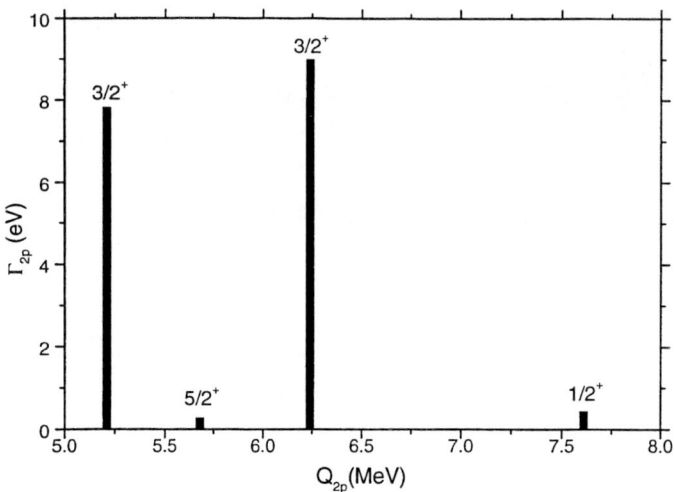

FIGURE 3. Shell-model sum spectrum for the decay ^{31}Cl(IAS) $\rightarrow$ ^{29}P+2p (simultaneous emission). The calculated spins of the final states are shown.

with the final-state spins in ^{29}P. The only $l = 0$ transition, to the 5/2$^+$ state in ^{29}P, is suppressed by a small spectroscopic factor. This also applies to the ground-state (1/2$^+$) transition with $l = 2$.

Emission of the first proton from the IAS was analyzed in the same way as in the previous section. In addition we have assumed that the gamma widths of the intermediate states in ^{30}S are neglible. The first-proton SM spectrum is shown in Fig. 4. Such states above 5 MeV in ^{30}S were taken into account which correspond to proton laboratory energies below 6.7 MeV. At higher proton energies the two-proton transition intensities drop quickly as the penetrability for the second proton decreases. The energy release Q_{2p} for the transition ^{31}Cl(IAS) $\rightarrow$ ^{29}P(g.s.) is 7.61 MeV, and the minimum excitation energy in ^{30}S is 4.31 MeV for proton emission.

Above 8.5 MeV excitation energy in ^{30}S the density of the accessible SM intermediate states increases rapidly. Therefore a detailed peak-by-peak calculation does not give very much useful information, and a statistical approach would be more appropriate.

Comparison of the widths in Figs. 3 and 4 shows clearly that the sequential decay mode is favored over simultaneous emission. The sequential widths are two orders of magnitude larger, and the weak mode-I transitions vanish in the background. This conclusion was also reached in the analysis of the experimental data [18].

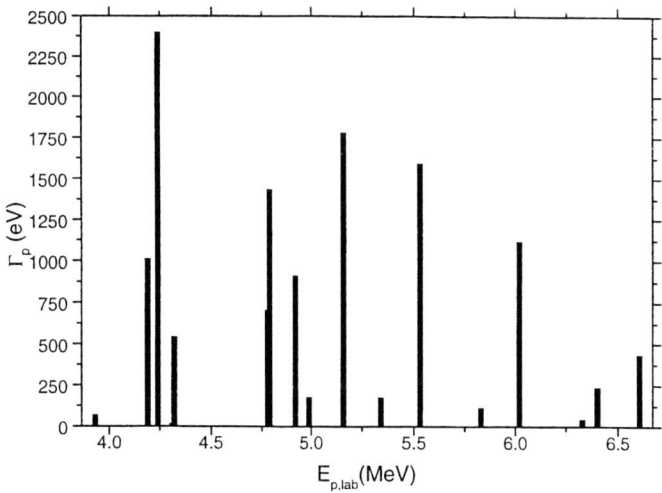

FIGURE 4. Shell-model first-proton spectrum from the IAS in ^{31}Cl.

CONCLUSIONS

Properties of beta-delayed proton and two-proton emission were calculated in ^{23}Al and ^{31}Ar. On the whole, the theoretical results agree well with experiment in both cases.

For ^{23}Al, new experimental data helped to resolve earlier discrepancies between calculations and experiment, and allowed a clear identification of proton-emission channels from ^{23}Mg.

For ^{31}Ar, the competition between sequential and simultaneous two-proton emission from the IAS in ^{31}Cl was examined. The shell-model results give a clear indication that the dominant decay mode is sequential, proceeding via many intermediate states in ^{30}S.

More detailed information on the calculations will be given elsewhere [19], including also beta-delayed proton and two-proton emission in ^{35}Ca.

REFERENCES

1. Honkanen, A., Dendooven, P., Huhta, M., Lhersonneau, G., Lipas, P. O., Oinonen, M., Parmonen, J.-M., Penttilä, H., Peräjärvi, K., Siiskonen, T., and Äystö, J., Nucl. Phys. **A621**, 689-705 (1997).
2. Axelsson, L., et al., Nucl. Phys. **A634**, 475-496 (1998); **A641**, 529(E) (1998).
3. Ormand, W. E., and Brown, B. A., Nucl. Phys. **A491**, 1-23 (1989).
4. Brown, B. A., Etchegoyen, A., and Rae, W. D. M., OXBASH, the Oxford University-Buenos Aires-MSU shell-model code, Michigan State University Cyclotron Laboratory Report No. 524, 1988.

5. Tighe, R. J., Batchelder, J. C., Moltz, D. M., Ognibene, T. J., Rowe, M. W., Cerny, J., and Brown, B. A., Phys. Rev. C **52**, R2298-R2301 (1995).
6. Bazin, D., *et al.*, Phys. Rev. C **45**, 69-79 (1992).
7. Peräjärvi, K., *et al.* (unpublished).
8. Ormand, W. E., and Brown, B. A., Nucl. Phys. **A440**, 274-300 (1985).
9. Ormand, W. E., and Brown, B. A., Phys. Lett. B **174**, 128-132 (1986).
10. Bohr, A., Mottelson, B. R., *Nuclear Structure, Vol. I*, Benjamin, New York, 1969, pp. 238-240.
11. Anyas-Weiss, N., Cornell, J. C., Fisher, P. S., Hudson, P. N., Menchaca-Rocha, A., Millener, D. J., Panagiotou, A. D., Scott, D. K., Strottman, D., Brink, D. M., Buck, B., Ellis, P. J., and Engeland, T., Phys. Rep. **12**, 201-272 (1974).
12. Sextro, R. G., Ph.D. thesis, Lawrence Berkeley Laboratory Report LBL-2360 (1973).
13. Macfarlane, M. H., and French, J. B., Rev. Mod. Phys. **23**, 567-691 (1960).
14. Endt, P. M., Nucl. Phys. **A521**, 1 (1990).
15. Mukha, I., *et al.*, Nucl. Phys. **A630**, 394c (1998).
16. Fynbo, H. O. U., *et al.*, Phys. Rev. C **59**, 2275-2277 (1999).
17. Brown, B. A., Phys. Rev. Lett. **65**, 2753-2756 (1990).
18. Fynbo, H. O. U., *et al.* (unpublished).
19. Siiskonen, T., and Lipas, P. O., (unpublished).

Resonances in deformed nuclei: R-matrix theory and oscillator expansion

A.T. Kruppa[1,2], W. Nazarewicz[3-5] and P.B. Semmes[6]

[1] *Institute of Nuclear Research of the Hungarian Academy of Sciences,*
H-4001 Debrecen, P.O. Box 51, Debrecen, Hungary
[2] *Joint Institute for Heavy Ion Research, Oak Ridge, Tennessee 37831*
[3] *Department of Physics and Astronomy, University of Tennessee, Knoxville, Tennessee 37996*
[4] *Physics Division, Oak Ridge National Laboratory, Oak Ridge, Tennessee 37831*
[5] *Institute of Theoretical Physics, Warsaw University, ul. Hoża 69, PL-00681, Warsaw, Poland*
[6] *Physics Department, Tennessee Technological University, Cookeville, Tennessee 38505*

Abstract. Single-particle resonances in deformed nuclei are considered using the coupled-channel Schrödinger equation method with the outgoing boundary conditions. Two variants of this method are investigated: the non-adiabatic one (based on the weak-coupling scheme) and the adiabatic one (based on the strong coupling). The R-matrix theory and the Gamow state approach are discussed and compared with each other. It is shown that the widths of very narrow proton resonances can be calculated by combining the harmonic oscillator expansion method with the R-matrix approach.

I INTRODUCTION

The Gamow (resonant) state approach is among the best time-independent methods for discretizing the single-particle continuum. The Schrödinger equation for deformed proton emitters leads to an eigenvalue problem with coupled-channel differential equations. Solving these equations accurately and reliably requires special care. Calculating the parameters of the resonances can be carried out directly by employing the pure outgoing wave boundary condition of a Gamow state or indirectly by using the R-matrix theory of Wigner and Eisenbud.

In order to avoid the cumbersome numerical solution of the coupled-channels eigenvalue problem, the simple harmonic oscillator expansion method, which leads to a matrix eigenvalue problem, can be applied to get the energy of a Gamow state. We show that the combination of the R-matrix theory and the harmonic oscillator expansion method is able to reproduce the widths of very narrow proton resonances.

In Sec. II the coupled-channel equations with and without Coriolis coupling are given. In Sec. III we discuss the different boundary conditions leading to the R-matrix and Gamow formalism. The exact R-matrix calculations are discussed in

Sec. IV. Finally, Sec. V shows an example of the application of the harmonic oscillator expansion method to the proton decay of ^{131}Eu.

II COUPLED EQUATIONS FOR PROTON EMISSION

We describe the scattering of an inert, spin one-half projectile by a nucleus. The Hamiltonian is

$$H = H_0(\xi) + T + \sum_\lambda V_\lambda(r)(Q_\lambda(\xi) \cdot Y_\lambda(\hat{r})), \tag{1}$$

where $H_0(\xi)$ is the internal Hamiltonian of the target, ξ denotes the internal coordinates of the target, and T is the kinetic energy of the relative motion. The target-projectile relative coordinate is $\vec{r}$. The projectile is considered to be inert so it is not necessary to specify its Hamiltonian in Eq. (1). The third term in Eq. (1) is the target-projectile interaction. It is given by an appropriately chosen inner product of tensor operators.

A Non-adiabatic method: weak coupling

In the non-adiabatic method, the wave function of the parent nucleus (i.e., particle emitter) can be written in the weak-coupling form:

$$\Psi^{JM}(\vec{r}, \xi) = \frac{1}{r} \sum_{Ijl} u_{Ijl}(r) \Phi^{JM}_{Ijl}(\hat{r}, \xi) \tag{2}$$

where the channel function is

$$\Phi^{JM}_{Ijl}(\hat{r}, \xi) = \sum_{m,\mu} \langle jmI\mu|JM\rangle \mathcal{Y}_{jlm}(\hat{r}, m_s) \psi_{I,\mu}(\xi). \tag{3}$$

In Eq. (3)

$$\mathcal{Y}_{jlm}(\hat{r}, m_s) = \sum_{m_l, m_s} \langle lm_l \frac{1}{2} m_s | jm \rangle \, i^l Y_{lm_l}(\hat{r}) \chi_{m_s}, \tag{4}$$

with χ_{m_s} being the spin function of the projectile (in our case: emitted nucleon). The states $\psi_{I,\mu}(\xi)$ are eigenstates of the target (i.e., daughter nucleus) Hamiltonian $H_0(\xi)$:

$$H_0(\xi) \psi_{I,\mu}(\xi) = \epsilon_I \psi_{I,\mu}(\xi). \tag{5}$$

The radial functions $u^J_{Ijl}(r)$ are the solutions of the coupled differential equations

$$\left[-\frac{\hbar^2}{2\mu}\left(\frac{d^2}{dr^2} - \frac{l(l+1)}{r^2}\right) + \epsilon_I - E\right] u_{Ijl}^J(r) + \sum_{\lambda I'j'l'} V_\lambda(r)\, \mathcal{V}_{Ijl,I'j'l'}^J(\lambda)\, u_{I'j'l'}^J(r) = 0, \tag{6}$$

where the coupling matrix element is defined by

$$\mathcal{V}_{Ijl,I'j'l'}^J(\lambda) = (4\pi)^{-1/2}(-1)^{J-1/2-I'+j+j'+\frac{1}{2}(l'-l)}\hat{j}\hat{j}'\hat{I}'$$
$$\times \langle j - \tfrac{1}{2} j' \tfrac{1}{2} | \lambda 0 \rangle\, \langle I'0\lambda 0|I0\rangle\, W(jIj'I';J\lambda). \tag{7}$$

The coupling matrix (7) is derived under the assumption of the rigid rotational motion of the daughter nucleus. The detailed form of the form factors and the derivation of (7) can be found in Ref. [1].

B Adiabatic approximation: strong coupling

In the adiabatic approximation, one assumes the complete degeneracy of the rotational states in the daughter nucleus: $\epsilon_I = \epsilon_0 = 0$. Using a Racah identity, one can prove

$$\sum_I (-1)^{J+K+j+1/2} \langle jKJ-K|I0\rangle \mathcal{V}_{Ijl,I'j'l'}^J(\lambda) =$$
$$\mathcal{V}_{jl,j'l'}^K(\lambda)(-1)^{J+K+j'+1/2}\langle j'KJ-K|I'0\rangle, \tag{8}$$

where

$$\mathcal{V}_{jl,j'l'}^K(\lambda) = (-1)^{K+j+j'+1/2}\hat{j}\hat{j}'\hat{\lambda}^{-1}\langle j\tfrac{1}{2}j'-\tfrac{1}{2}|\lambda 0\rangle\langle jKj'-K|\lambda 0\rangle. \tag{9}$$

By introducing the functions

$$g_{jl}^{JK}(r) = \sqrt{2}\sum_I (-1)^{J+K+j+1/2}\, \langle jKJ-K|I0\rangle u_{Ijl}^J(r), \tag{10}$$

the set (6) reduces to the following coupled equations:

$$\left[-\frac{\hbar^2}{2\mu}\left(\frac{d^2}{dr^2} - \frac{l(l+1)}{r^2}\right) - E\right]g_{jl}^{JK}(r) + \sum_{\lambda j'l'} V_\lambda(r)\mathcal{V}_{jl,j'l'}^K(\lambda) g_{j'l'}^{JK}(r) = 0. \tag{11}$$

Since the coupling potential in (11) is independent of J, from now on this index is dropped from $g_{jl}^{JK}(r)$ in the adiabatic method.

One can easily show that Eq. (11) describes the scattering of the projectile by a deformed potential. The adiabatic single-particle wave function

$$\Psi^K(\vec{r}) = \sum_{jl} g_{jl}^K(r)\mathcal{Y}_{jlK}(\hat{r},m_s) \tag{12}$$

is equivalent to a deformed Nilsson orbit. The fact that the eigenstates of (11) do not depend on J implies that in the adiabatic limit the rotational band built upon the Nilsson orbital (12) is degenerate (i.e., its moment of inertia is infinite).

III R-MATRIX AND GAMOW THEORY

By specifying the boundary conditions, the coupled differential equations (6) and (11) correspond to an eigenvalue problem. It is always assumed that the solutions are regular at the origin, i.e., $u^J_{Ijl}(0) = g^K_{jl}(0) = 0$. In the following, we shall use the shorthand notation $u_c(r)$ either for $u^J_{Ijl}(r)$ or for $g^K_{jl}(r)$, and l_c will denote the single-nucleon orbital angular momentum in the channel c.

The R-matrix theory, developed by Wigner, Eisenbud [2], and later by Lane and Thomas [3], is intended to give the parameterization of the scattering S-matrix on the real energy axis and subsequently the parameterization of the scattering cross section. The R-matrix boundary condition is

$$a\frac{u'_c(a)}{u_c(a)} = B_c, \qquad (13)$$

where the boundary condition parameters, B_c, are arbitrary real numbers. It is assumed that the short-range interaction between the projectile and target can be neglected beyond the (large) channel radius a. The boundary condition (13) defines a discrete complete set of functions $u_c^\lambda(r)$ corresponding to the real eigenvalues E_λ. They are normalized to one inside the channel surface, $\sum_c \int_0^a |u_c(r)|^2 dr = 1$. Written in terms of the real reduced width amplitudes,

$$\gamma_{\lambda c} = \left(\frac{\hbar^2}{2m_c a}\right)^{1/2} u_c^\lambda(a), \qquad (14)$$

the R-matrix can be written as

$$R_{cc'} = \sum_\lambda \frac{\gamma_{\lambda c}\gamma_{\lambda c}}{E_\lambda - E}. \qquad (15)$$

(In Eq. (14), m_c is the reduced mass.) The R-matrix is related to the scattering S-matrix in a complicated way [3]. It can be demonstrated that if all the eigenstates are taken into account in Eq. (15), then the calculated S-matrix does not depend on the boundary condition parameters or on the exact choice of the channel radius.

The Gamow (or Siegert) states of Eqs. (6) and (11) are defined by the following boundary condition

$$\frac{u'_c(a)}{u_c(a)} = \frac{k_c O'_{l_c}(k_c a)}{O_{l_c}(k_c a)}, \qquad (16)$$

where O_{l_c} is the outgoing wave (e.g., Coulomb function for protons). A solution with complex wave number $k_c = \sqrt{2m_c(E - \epsilon_c)}/\hbar$ leads to a Gamow state with complex energy $E_r - i\Gamma_r/2$. At that energy the scattering S-matrix has a pole [4] in the complex energy plane. Lane and Thomas in Ref. [3] call that pole of the S-matrix a "radioactive state". The real-energy states defined by the R-matrix theory of Wigner and Eisenbud should not be confused with the Gamow states.

The advantage of the Gamow state is that the particle-decay half-life can be readily obtained from the width Γ_r, $T_{1/2} = \hbar \ln 2/\Gamma_r$. Since the R-matrix theory gives the scattering S-matrix only on the real energy axis, further considerations are needed to determine the parameters of a resonance if the R-matrix boundary condition is selected for the solution of the coupled equations. Assuming that in a given energy region only one term $\lambda = \lambda_0$ dominates in Eq. (15), Thomas showed [5] how to obtain the Gamow resonance energy E_r and its width Γ_r within the R-matrix theory. Specifically, if the R-matrix boundary condition parameters are set so that

$$B_c = S_{l_c}(E_\lambda), \qquad (17)$$

the complex-energy pole of the S-matrix, $\mathcal{E}$, satisfies the equation

$$E_{\lambda_0} - \mathcal{E} + (\mathcal{E} - E_{\lambda_0})\dot{\Delta}_{\lambda_0}(E_{\lambda_0}) - \frac{1}{2}i\Gamma_{\lambda_0}(E_{\lambda_0}) = 0, \qquad (18)$$

where

$$\frac{1}{2}\Gamma_\lambda(E) \equiv \sum_c P_{l_c}(E)\gamma_{\lambda c}^2 \qquad (19)$$

and

$$\dot{\Delta}_\lambda(E_\lambda) \equiv -\sum_c \dot{S}_{l_c}(E_\lambda)\gamma_{\lambda c}^2. \qquad (20)$$

The functions $P_{l_c}(E)$ and $S_{l_c}(E)$ are referred to as the penetration and shift functions, respectively. (They are expressed by the Coulomb F_{l_c} and G_{l_c} functions, see, e.g., Ref. [3].) The dot in Eqs. (18) and (20) denotes the derivative with respect to energy. Assuming that (20) is negligible, one obtains $E_r = E_\lambda$ and the resonance width is given by the frequently quoted expression $\Gamma_r = \Gamma_\lambda(E_\lambda)$.

IV EXACT R-MATRIX CALCULATIONS

The phrase "exact R-matrix calculation" means that the coupled differential equations are solved numerically. The solution with the R-matrix boundary condition is generated by a modified version of the code CCGAMOW [6] which is based on the piecewise perturbation technique.

Since we are interested in describing decaying systems, our main objective is to calculate the position and width of Gamow states. Hence the result of the R-matrix calculation will be compared with the calculation using the Gamow state boundary condition; the latter will be referred to as the exact one. The Gamow states are generated by the code CCGAMOW using extended precision. All calculations based on the R-matrix theory are done in double precision. It is not necessary to use extended precision in R-matrix theory but it is unavoidable to apply extended

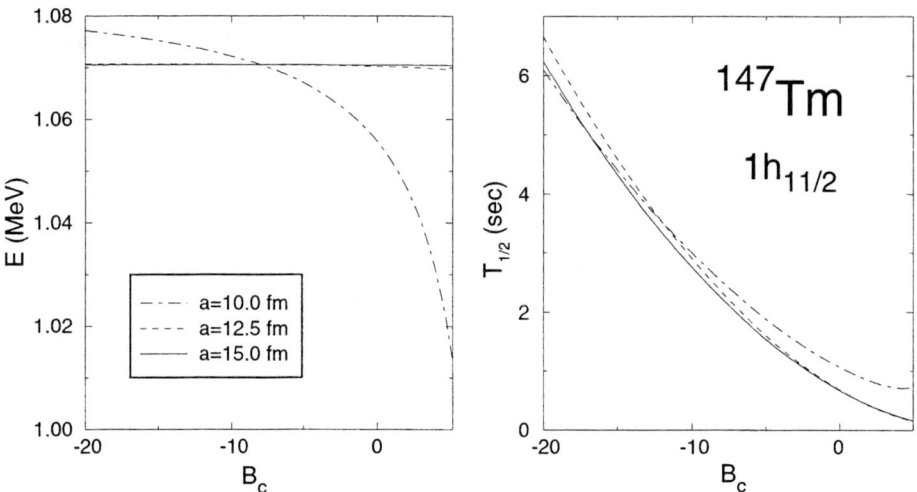

FIGURE 1. Energy and half-life of the spherical $1h_{11/2}$ proton orbit in ^{147}Tm calculated in the R-matrix theory as a function of the boundary condition parameter B_c for different values for the channel radius.

precision in Gamow state calculations if the width of the resonance is very small ($\Gamma_r < 10^{-16}$ MeV).

Consider first a narrow spherical proton emitter, e.g., the $1h_{11/2}$ orbit of ^{147}Tm discussed in Ref. [8]. The R-matrix calculations have been performed for three different values of the channel radius. Figure 1 shows the calculated eigenvalue and the half-life as functions of the boundary condition parameter. The eigenvalue is quite stable if the channel radius is large and it is very close to the exact position of the resonance (1.70562 MeV). The calculated width however varies greatly. In this form, the R-matrix theory is unable to give a reliable prediction for the resonance's half-life.

Fortunately, the "natural boundary condition" assumption of Eq. (17) turns out to work well. In the iterative R-matrix technique, first we take a boundary condition parameter, calculate the eigenvalue E_λ, and check whether the condition (17) is satisfied or not. If this condition is violated, the boundary condition parameter is modified and the whole procedure is repeated until the correct solution is found. The result of this type of calculation is shown in Fig. 2. Here the half-life is shown as a function of the channel radius. At each channel radius, the optimal boundary condition is determined and then the half-life is calculated. The exact result is reproduced extremely well. Note the difference of the scales on Fig. 1 and Fig. 2. The small deviation between the Gamow and R-matrix results is probably related

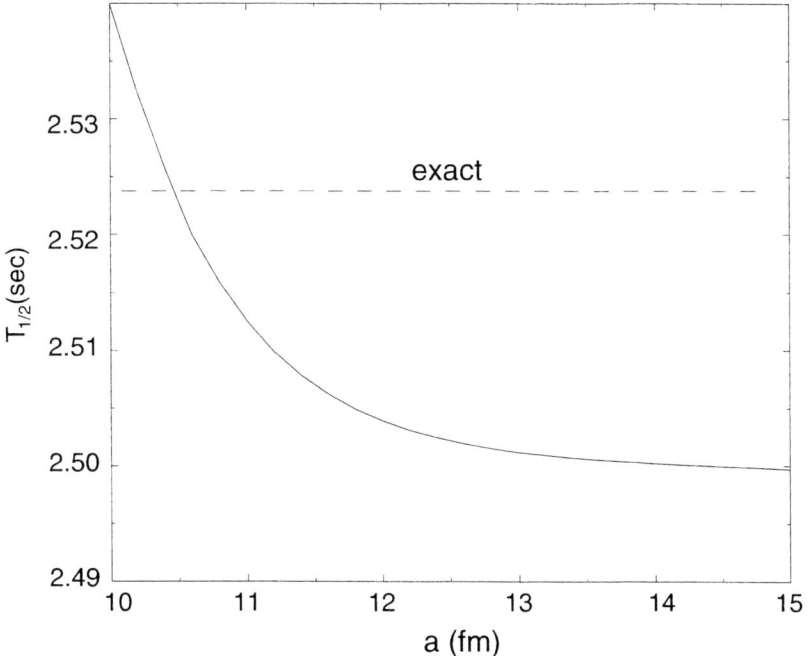

FIGURE 2. Half-life of the spherical $1h_{11/2}$ orbit in ^{147}Tm calculated in the R-matrix theory as a function of the channel radius using the "natural boundary condition".

to the fact that only the one-channel R-matrix expression is used. The iterative R-matrix technique, "the natural boundary condition" method, can be applied also in the case of a deformed mean field. In the many-channel case the accuracy of the iterative R-matrix approach is similar to the one-channel calculation.

V OSCILLATOR EXPANSION

The eigenfunctions of an axially deformed average nuclear field, the Nilsson orbits, can be expanded in a complete set of functions. In this way one can avoid the numerical solution of the eigenvalue problem of coupled differential equations. The wave function (12) may be approximated [7] by

$$\Psi^K(\vec{r}) \approx \sum_{n_\rho, n_z, m_s} C(K, n_\rho, n_z, m_s) |n_\rho, n_z, K, m_s\rangle, \qquad (21)$$

where $|n_\rho, n_z, K, m_s\rangle$ is the eigenfunction of the axially deformed harmonic oscillator in the cylindrical basis. The coefficients $C(K, n_\rho, n_z, m_s)$ are determined by

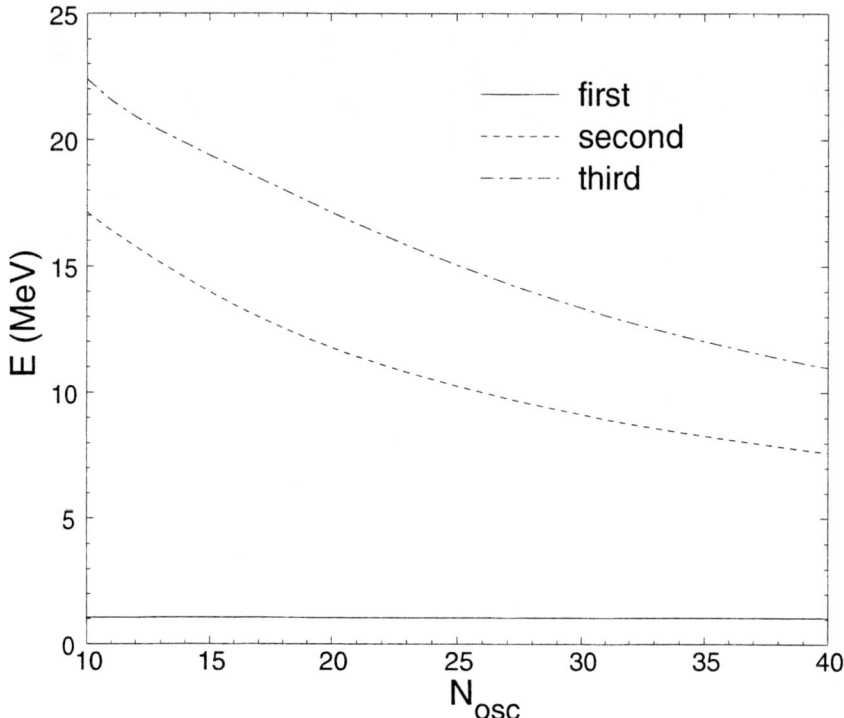

FIGURE 3. The energies of the three lowest spherical proton $h_{11/2}$ orbitals of ^{147}Tm in the spherical Woods-Saxon model as a function of the size of harmonic oscillator basis $N_{\rm osc}$. The length parameter of the spherical basis is 2 fm.

matrix diagonalization using the code SWBETA [7]. After the the eigenvalue problem of the Hamiltonian matrix has been solved, the wave function (21) can be transformed into a similar form as in Eq. (12) with

$$g_{jl}^K(r) \approx \sum_n A(K,n,j,l) R_{nl}(r), \qquad (22)$$

where $R_{nl}(r)$ is the radial function of the spherical harmonic oscillator. (We carried out the transformation from cylindrical variables ρ, z, ψ to polar variables r, Θ, ϕ because it is easier to formulate the R-matrix theory in these variables.)

It was recognized long ago that by using the harmonic oscillator expansion (or any expansion in a square integrable basis) not only can the bound states be determined, but also the positions of narrow resonances. If M basis functions are used in the expansion, then M eigenvalues are obtained from the matrix diagonalization. When the size of the basis is increased, the eigenvalues of all the positive energy solutions

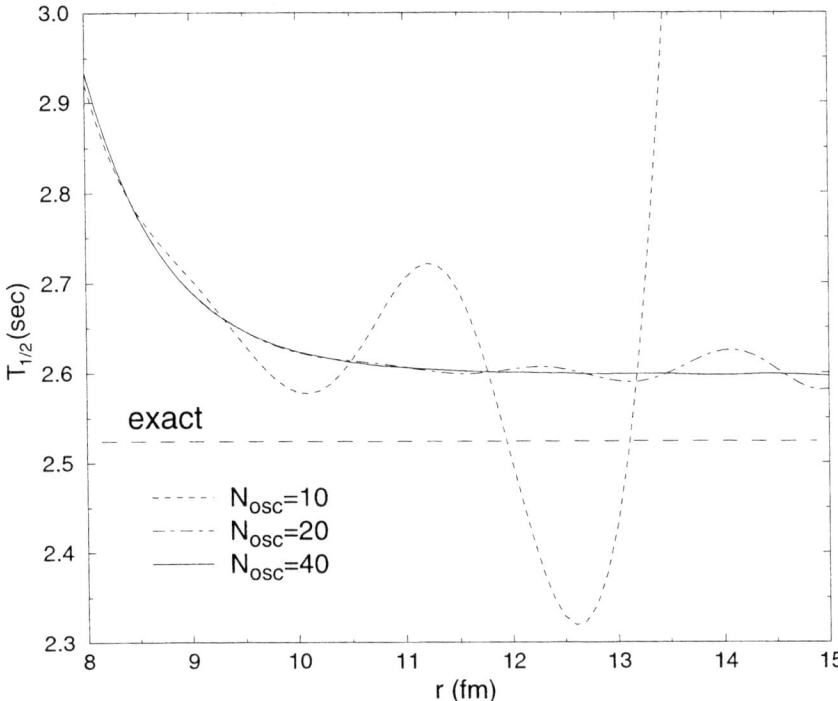

FIGURE 4. Half-life of the spherical $1h_{11/2}$ orbit of ^{147}Tm as a function of the channel radius calculated in the R-matrix theory based on the harmonic oscillator expansion. Three different basis sizes are used in the calculation.

tend toward zero. The sign of a resonance is that specific positive energy solutions are relatively stable with respect to increasing the size of the basis. This is shown in Fig. 3 for the three lowest spherical $h_{11/2}$ proton orbits of ^{147}Tm. The lowest $h_{11/2}$ state obtained from the diagonalization is a good approximation to the resonance; its energy is very stable with respect to the size of the oscillator basis used, at least in the range of N_{osc} considered. (In the calculations we take all the deformed oscillator states with principal quantum number $N \leq N_{osc}$.) The higher-lying states cannot be represented by the expansion procedure; they correspond to the high-energy $h_{11/2}$ continuum.

Thus the position of the resonance can be found, but how can its width be determined? Several proposals exist in the literature. They are dubbed "L^2 stabilization methods" (since only square integrable functions are used in the expansion). Here we combine the oscillator expansion method and the R-matrix formalism.

In the R-matrix theory, the coupled equations are solved with the boundary

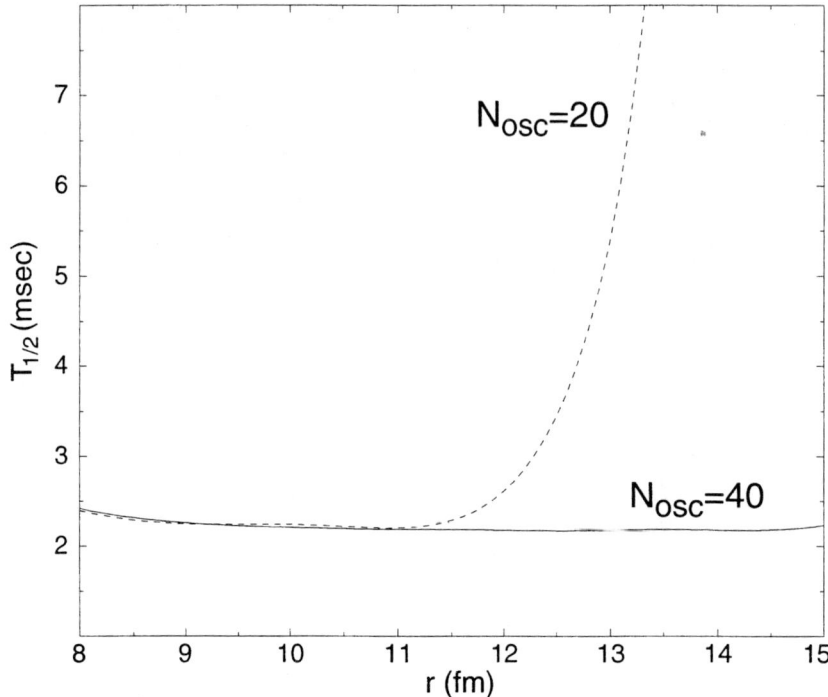

FIGURE 5. Half-life of the $K^\pi = 3/2^+$ deformed Nilsson resonance resonance in ^{131}Eu calculated in the R-matrix theory based on the harmonic oscillator expansion as a function of the channel radius. Two different basis sizes are used in the calculation.

conditions given in advance. This cannot be achieved in the framework presented here, but the procedure can be reversed: Starting from the approximate solution (22), the corresponding boundary condition parameter can be derived at each r, and then the machinery of the R-matrix theory can be applied. It is to be noted, however, that although the boundary condition is independent of the wave function normalization, the reduced width amplitude depends critically on it. Thus, at each r the approximate solution (22) must be normalized to one inside the channel surface.

Figure 4 shows the half-life of the lowest spherical $h_{11/2}$ orbit of ^{147}Tm as a function of the channel radius using three different basis sizes. A perfect stabilization of the result is obtained at large $N_{\rm osc}$. The quality of the oscillator-expansion method is similar to that of the R-matrix result discussed in Fig. 2. The deformed nucleus ^{131}Eu with deformation $\beta_2=0.32$ is considered in Fig. 5. The Nilsson orbit is characterized by the quantum numbers $K^\pi = 5/2^+$. The calculations use the

Becchetti-Greenlees potential parameters, with the depth of the potential adjusted to the position of the resonance at 0.950 MeV. The calculations are carried out using N_{osc}=20 and 40 oscillator quanta. Again, at large N_{osc} the result is very stable. In summary, the results presented in Figs. 4 and 5 show that the harmonic oscillator expansion can be successfully used for the determination of widths of very narrow resonances if the size of the basis is large enough.

Acknowledgments

This work was supported in part by the Hungarian NRF (OTKA T026244 and T029003) and by the U.S. Department of Energy under Contract Nos. DE-FG02-96ER40963 (University of Tennessee), DE-FG05-87ER40361 (Joint Institute for Heavy Ion Research), DE-AC05-96OR22464 with Lockheed Martin Energy Research Corp. (Oak Ridge National Laboratory), and DE-FG02-92ER40694 (Tennessee Technological University).

REFERENCES

1. Tamura T., *Rev. Mod. Phys.* **37**, 679-708 (1965).
2. Wigner E.P. and Eisenbud L., *Phys. Rev.* **72**, 29-41 (1947).
3. Lane A. M. and Thomas R.G., *Rev. Mod. Phys.* **30**, 257-353 (1958).
4. Humblet J. and Rosenfeld L., *Nucl. Phys.* **26**, 529-578 (1961).
5. Thomas R. G., *Prog. Theor. Phys.* **12**, 253-264 (1954).
6. T. Vertse, A.T. Kruppa, L.Gr. Ixaru, and M. Rizea, to be submitted to *Comput. Phys. Commun.*
7. Cwiok S., Dudek J., Nazarewicz W., Skalski J., and Werner T., *Comput. Phys. Commun.* **46**, 379-399, (1987).
8. Åberg S., Semmes P.B., and Nazarewicz W., *Phys. Rev. C* **57**, 1762-1773 (1997).

Proton emission from Gamow resonance

T. Vertse[1,2], A. T. Kruppa[1,2], B. Barmore[2-4], W. Nazarewicz[3-5], L. Gr. Ixaru[6] and M. Rizea[6]

[1] *Institute of Nuclear Research of the Hungarian Academy of Sciences, H-4001 Debrecen, P.O. Box 51, Debrecen, Hungary*
[2] *Joint Institute for Heavy Ion Research, Oak Ridge, Tennessee 37831*
[3] *Department of Physics and Astronomy University of Tennessee, Knoxville, Tennessee 37996*
[4] *Physics Division, Oak Ridge National Laboratory, Oak Ridge, Tennessee 37831*
[5] *Institute of Theoretical Physics, Warsaw University, ul. Hoża 69, PL-00681, Warsaw, Poland*
[6] *Institute of Physics and Nuclear Engineering "Horia Hulubei", P.O. Box MG-6, Bucharest, Romania*

Abstract. We developed two computer codes: CCGAMOW and NONADI for calculating the complex energy eigenvalues and eigenfunctions of deformed Gamow resonances with high accuracy by using the piecewise perturbation method. The code CCGAMOW calculates resonant Nilsson orbitals using the adiabatic approximation in which the energies of the ground and excited rotational states of the daughter nucleus are degenerate. In the code NONADI this approximation is lifted and the rotational degree of freedom of the core and the Coriolis coupling in the parent nucleus are taken into account. The difference between adiabatic and non-adiabatic approaches is found to be non-negligible for the proton emission from the ground state of ^{141}Ho.

I INTRODUCTION

Nuclear states decaying predominantly by proton emission are often described by using single-particle (s.p.) proton states with complex energy eigenvalues, i.e. by Gamow resonances. Gamow states were introduced by Gamow [1] in order to describe α-decay. Since they describe a time-dependent process within a stationary picture, Gamow states are not wave functions in the normal quantum mechanical sense. They have complex energy eigenvalues which correspond to poles of the S-matrix extended to complex energy, $\mathcal{E}$ and wave number, k. Nevertheless, they are extremely useful mathematical tools to describe narrow resonances or long-lived quasistationary states [2].

In this paper we use Gamow states to describe proton emission. We shall focus attention on the mathematical and numerical approaches needed for calculating narrow Gamow resonances in spherically symmetric and axially deformed nuclei.

We compare the adiabatic and non-adiabatic approaches to proton emission from the deformed ^{141}Ho nucleus and demonstrate the importance of the non-adiabatic description.

II DEFINITION OF GAMOW STATES

Let us first consider the simplest case in which the proton moves in a spherically symmetric, finite potential $v(r)$, which is a sum of the central nuclear potential, v_N, the nuclear spin-orbit term, v_{so}, and the Coulomb potential, v_C:

$$v(\mathbf{r}) = v(r) = v_N(r) + v_{so}(r) + v_C(r). \tag{1}$$

The Gamow state, ψ_n, is the eigenvector of the single-particle Hamiltonian $\hat{h} = \hat{t} + v(r)$, where the kinetic energy operator is $\hat{t} = -\frac{\hbar^2}{2\mu}\Delta$ (μ is the reduced mass):

$$\hat{h}\psi_n = \mathcal{E}_n\psi_n . \tag{2}$$

The eigenfunction ψ_n is characterized by the angular momentum quantum numbers (l, j, m):

$$\psi_n = \psi_{l,j,m}(\mathbf{r}, k) = \frac{u_{l,j}(r,k)}{r}[Y_l(\hat{r})\chi_{1/2}]_{j,m}. \tag{3}$$

Let us abbreviate l and j as a single subscript $i = \{l, j\}$ and introduce the complex wave number, k ($k^2 = \frac{2\mu}{\hbar^2}\mathcal{E}$). We can write the radial equation as

$$u_i''(r,k) = \left[\frac{l(l+1)}{r^2} + V(r) - k^2\right]u_i(r,k), \tag{4}$$

where $V(r) = \frac{2\mu}{\hbar^2}v(r)$ is the potential in units of k^2. The Gamow solution should be regular at the origin,

$$u_i(0, k) = 0, \tag{5}$$

and asymptotically, where only v_C is present, it should join to an outgoing Coulomb wave, $O_l = G_l + iF_l$. Therefore, at $r = r_{as}$ (where r_{as} is the asymptotic radius, which is much larger than the range of the average potential) the logarithmic derivative of the solution should be

$$D(r_{as}, k) = u_i'(r_{as}, k)/u_i(r_{as}, k) = O_l'(\eta, kr_{as})/O_l(\eta, kr_{as}), \tag{6}$$

where $\eta = \frac{Ze^2\mu}{\hbar^2 k}$ is the Sommerfeld parameter. The solutions defined by Eqs. (4-6) are either bound states, $\mathcal{E}_n = E_b < 0$, with negative real energies and imaginary wave numbers $k_n = i\gamma_n$ ($\gamma_n > 0$), or Gamow states, $\mathcal{E}_n = E_r - i\frac{\Gamma}{2}$, with a nonzero imaginary part $\Gamma \neq 0$, and $k_n = \kappa_n - i\gamma_n$.

The asymptotic behavior of these solutions is determined by k_n; at a very large distance the outgoing solution (for $\eta = 0$) is proportional to $e^{ik_n r}$. The resonance is called a decaying Gamow resonance if $\Gamma > 0$ or a capturing resonance if $\Gamma < 0$. For a real potential v, the pair of resonances lies symmetrically with respect to the imaginary k-axis; hence decaying Gamow states have $k_d = \kappa_n - i\gamma_n$ and capturing ones have $k_c = -\kappa_n - i\gamma_n$, with $\kappa_n > 0$ and $\gamma_n > 0$. The radial wave function of the capturing Gamow state is the complex conjugate of that of the decaying one, $u_c(r) = u_d(r)^*$. Both u_c and u_d oscillate with increasing amplitude as a function of r.

Berggren proposed a new completeness relation, which includes Gamow states [3], by generalizing the scalar product. He introduced a bilinear basis set and a regularization procedure ($\mathcal{R}eg$). With this generalization, the norm is

$$\langle u_c | u_d \rangle = \mathcal{R}eg \int_0^\infty u_d^2(r) dr = 1 \; . \tag{7}$$

A convenient method for regularization is to rotate r to the first quadrant of the complex r-plane beyond a certain distance r_{max}. This is often referred to as the exterior complex scaling method.

The total width of the Gamow resonance with complex energy $\mathcal{E}_d$ is given by

$$\Gamma = -\frac{1}{2}\mathcal{I}m(\mathcal{E}_d). \tag{8}$$

The half-life of the state is $T_{1/2} = \hbar \ln 2 / \Gamma$.

III NUMERICAL CALCULATION OF SPHERICAL GAMOW STATES

For realistic potentials, the radial equation has to be solved by means of numerical integration. In Ref. [4] the code GAMOW was introduced, which uses the Fox-Goodwin method for solving the radial equation. A more powerful method, the piecewise perturbation, is used for the same purpose in Ref. [5] (code ANTI). The main features are similar in the two codes. The total r domain of Eq. (4) consists of a real r domain, $[0, r_{max}]$, and a complex ray $I_3 = [r_{max}, r_{as}]$, where r_{as} is complex. The real domain is further divided into two intervals, $I_1 = [0, r_{min}]$ and $I_2 = [r_{min}, r_{max}]$. For some approximate value of k_n, one finds the "left" solution of Eq. (4), $u^L(r,k)$, that satisfies the boundary condition at the origin, and it is calculated by integrating numerically from $r = 0$ to r_m. The "right" solution of Eq. (4), $u^R(r, k_n)$, satisfies the asymptotic boundary condition at $r = r_{as}$; the numerical integration proceeds on I_3 inward from the point r_{as} to the real axis at r_{max}, and then continues along the real r-axis in region I_2 until reaching the matching point, r_m. The complex eigenvalue, k_n, can be found by finding the zero of the transcendental function

$$\Phi(k) = D^L(r_m, k) - D^R(r_m, k) = 0, \tag{9}$$

where the logarithmic derivatives of u_L and u_R at the matching distance are

$$D^L(r_m, k) = \frac{u'_L(r_m, k)}{u_L(r_m, k)}, \quad D^R(r_m, k) = \frac{u'_R(r_m, k)}{u_R(r_m, k)}. \tag{10}$$

The root finding can be done by means of the Newton-Raphson technique. Note that in order to get the contribution of the asymptotic region to the norm in Eq. (7), r must be complex in region I_3. The corresponding rotation angle (which for large values of r_{as} coincides with $\arg(r)$) should satisfy the condition

$$\pi - \arg(k) > \arg(r) > -\arg(k) \tag{11}$$

so that the magnitude of the solution converges to zero as $r_{as} \to \infty$ along the complex ray. The contribution to the norm beyond r_{as} is neglected, and at r_{as} we use the asymptotic series of the outgoing Coulomb function O_l (and its derivatives)

$$O_l(\eta, \rho) = \exp\left[i\left(\rho - \eta \ln 2\rho - l\frac{\pi}{2} + \sigma_l\right)\right] \left\{1 + \sum_{n \geq 1} \prod_{j=1}^{n} \frac{(i\eta + l + j)(i\eta - l + j - 1)}{j(2i\rho)}\right\} \tag{12}$$

with $\rho = kr_{as}$. We use the convergence acceleration procedures of Wynn [6] and Levin [7] to speed the convergence of the summation.

For the known proton emitters, the width Γ is so small that extremely high numerical accuracy is required for $\mathcal{E}_d$. The width can also be calculated from the outgoing probability current [9]

$$\Gamma(r) = i\frac{\hbar^2}{2\mu} \frac{u_d'^*(r, k_n) u_d(r, k_n) - u_d'(r, k_n) u_d^*(r, k_n)}{\int_0^r |u_d(x, k_n)|^2 dx}, \tag{13}$$

which, by construction, does not depend on r. If we use extended precision arithmetic, the width calculated using Eq. (13) is indeed r-independent, and it agrees well with the value obtained from Eq. (8). Another way to estimate the width is to use the R-matrix expression of Thomas [10],

$$\Gamma = \frac{\hbar^2 \kappa_n}{\mu} \frac{\mathcal{R}e(u_d(a, k_n))^2}{|O_l(\mathcal{R}e(\eta), \kappa_n r_{as})|^2}, \tag{14}$$

where we approximate the (real) R-matrix resonant wave function with the real part of the normalized Gamow resonance. This approximation works fairly well for the narrow Gamow resonances corresponding to the known proton emitters [11]. For large values of r_{as}, expression (14) is generally within 5% of the values calculated from Eqs. (8) and (13).

IV GAMOW STATES IN THE DEFORMED POTENTIAL

A Adiabatic approach

The generalization of the s.p. Hamiltonian to an axially symmetric, deformed potential, $v(\mathbf{r})=v(r,\theta)$, leads to a system of n coupled, differential equations. In the *intrinsic* frame of reference, defined by the principal axis of the deformed average potential, the proton moves in an orbit with good quantum numbers π (parity) and Ω (projection of the total s.p. angular momentum $\mathbf{j}$ onto the symmetry axis). The s.p. wave function can be expanded in spherical partial waves

$$\psi^{\Omega,\pi}(\mathbf{r},k) = \sum_{l,j}^{l_{max},j_{max}} \frac{u_{l,j}(r,k)}{r}[Y_l(\hat{r})\chi_{1/2}]_{j,\Omega}, \qquad (15)$$

in which the radial wave functions are the solutions of a set of coupled differential equations

$$u_i''(r,k) = \sum_{i'} \left[\frac{l_i(l_i+1)}{r^2}\delta_{i,i'} + V_{i,i'}(r) - k^2\delta_{i,i'}\right] u_{i'}(r,k). \qquad (16)$$

Here $i = 1, 2, \ldots, n$ runs over all partial waves which can be coupled to the given Ω and π, and $V_{i,i'}(r) = \langle lj\Omega|V(\mathbf{r})|l'j'\Omega\rangle$ are the matrix elements of the deformed potential. The system of coupled equations can be written in matrix form as

$$\mathbf{u}''(r,k) = \left(\frac{\underline{L}}{r^2} + \underline{V}(r) - k^2\underline{1}\right)\mathbf{u}(r,k), \qquad (17)$$

where the underlined quantities denote $n \times n$ matrices. In Eq. (17) $\underline{V}$ is the potential matrix, $\underline{L}$ is the diagonal matrix $l_i(l_i+1)\delta_{ij}$, $\underline{1}$ is the identity matrix, and

$$\mathbf{u}(r,k) \equiv [u_1(r,k), u_2(r,k), \ldots, u_n(r,k)]^T \qquad (18)$$

is an n dimensional column vector. This eigenvalue problem is solved with boundary conditions given by Eqs. (5) and (6). First, the radial wave functions are regular at the origin, $u_i(0,k)=0$. At large values of r ($r>r_{as}$), all the off-diagonal coupling terms vanish and (17) reduces to a decoupled set of n differential equations. Therefore, an adiabatic Gamow state should satisfy Eq. (6) in every channel ($i = 1, \ldots, n$) with the same value of k.

The problem of determining complex eigenvalues and eigenfunctions can be reduced to a set of initial value problems for the system of coupled equations (17). As in the spherically symmetric case, one calculates "left" and "right" solutions which are then matched at r_m.

For solving the initial value problems with high accuracy, a package of subroutines based on the piecewise perturbation methods are used (see Ch. 3 of Ref. [12] and

Ref. [5]). This package is aimed at solving initial value problems for systems of ordinary differential equations of the form

$$\boldsymbol{u}''(r,k) = \left(\frac{\underline{L}}{r^2} + \frac{\underline{S}(r)}{r} + \underline{P}(r) - k^2\underline{E}\right)\boldsymbol{u}(r,k), \quad a < r < b, \tag{19}$$

along a straight-line segment $s = [a,b]$ in the complex plane. The matrices $\underline{S}(r)$ and $\underline{P}(r)$ are symmetric and their elements are complex functions of (complex) r. Furthermore, it is assumed that each matrix element is well approximated by a polynomial of second degree inside any reasonably large subinterval of the segment.

The package consists of two sets of subroutines. One set is designed for the vicinity of the origin $r \in [0, r_{min}] = I_1$, where the centrifugal term has the largest importance. It produces the regular solution inside the I_1-interval by a perturbative technique in which the centrifugal term is taken as the reference potential and the sum of the other three terms is taken as a perturbation. The other set of routines is designed for the remaining part of the r-domain, i.e., for I_2 and I_3. Here the integration is performed on a lattice of non-equidistant mesh-points which is determined by the variation of the potential and the accuracy required. On each subinterval, matrix elements of the sum of the three potentials are first approximated by their average values. The deviations from the second degree polynomial are considered to be perturbations which are then taken up to the second order. In both regions, the respective packages produce the vector $\boldsymbol{u}$ and its derivatives. A detailed description of this package will be given elsewhere [13].

The code which calculates the Gamow states in a deformed potential using the adiabatic approximation is called CCGAMOW. Besides the energy eigenvalue and the normalized wave function, it computes the partial widths using the current expression [9],

$$\Gamma_i(r) = i\frac{\hbar^2}{2\mu}\frac{u_i'^*(r,k_n)u_i(r,k_n) - u_i'(r,k_n)u_i^*(r,k_n)}{\sum_{i'}^n \int_0^r |u_{i'}(r',k_n)|^2 dr'}, \tag{20}$$

where the sum of the partial widths,

$$\Gamma(r) = \sum_i^n \Gamma_i(r), \tag{21}$$

gives the total decay width as a function of r. A serious check of our calculation is that for large r-values the condition $\mathcal{I}m(\mathcal{E}) = -1/2\,\Gamma(r)$ is satisfied if we use extended precision arithmetic. (In principle, $\Gamma(r)$ should be independent of r at any r.) The partial widths of Eq. (20) are in reasonably good agreement with those calculated by using the Thomas formula of Eq. (14) at large values of r_{as}. (This R-matrix expression was used recently in Refs. [14,15] dealing with deformed proton emitters.)

In our calculations, we assume that the spin-orbit term v_{so} is spherical; i.e., it does not contribute to the off-diagonal couplings of Eq. (17). As discussed by

Nilsson [17], the impact of the deformed component of the spin-orbit term, δv_{so}, on the Nilsson orbitals is weak. In addition, there is some arbitrariness in defining the average spin-orbit interaction, and the influence of δv_{so} on the final result is well below this uncertainty.

B Non-adiabatic approach

The deformed Gamow state can be associated with a deformed, resonant Nilsson orbital in a finite potential. Since it describes the s.p. motion in the intrinsic frame, it breaks angular momentum conservation. In order to restore rotational invariance, one can adopt the strategy of the particle-plus-rotor model and couple the intrinsic, deformed state to the deformed core. This is the strong-coupling scheme of Ref. [16].

Another strategy, adopted in this work, is the weak-coupling approach in which the wave function of the parent nucleus is obtained by coupling the spherical single-proton wave functions to the deformed states of the daughter nucleus. In this scheme the intrinsic wave function is not introduced, and the parent state preserves the total angular momentum J, its projection M, and the parity π. The Hamiltonian of the deformed core-plus-particle system can be written in the laboratory frame as

$$\hat{H} = \hat{H}_0(\xi) + \hat{t} + \sum_\lambda v_\lambda(r)(Q_\lambda(\xi) \cdot Y_\lambda(\hat{r})), \tag{22}$$

where $\hat{H}_0(\xi)$ is the internal Hamiltonian of the core (daughter nucleus), with internal coordinates ξ. The eigenstates $\psi_{I,\mu}(\xi)$ of $\hat{H}_0(\xi)$ are that of the symmetric top, and the corresponding eigenvalues, ϵ_I, can be either taken from experiment or modeled according to the rotational expression $\epsilon_I = \kappa I(I+1)$. In Eq. (22) $\hat{t}$ is the kinetic energy of the particle and the third term is a multipole expansion of the core-particle interaction. (A similar Hamiltonian was introduced in Ref. [18] to describe alpha emission from deformed nuclei.)

The parent wave function,

$$\Psi^{JM}(\mathbf{r},\xi) = \sum_{Ijl} \frac{u_{Ijl}(r)}{r} \Phi^{JM}_{Ijl}(\hat{r},\xi), \tag{23}$$

is composed of the radial function and the channel function,

$$\Phi^{JM}_{Ijl}(\hat{r},\xi) = \sum_{m,\mu} \langle jm I\mu | JM \rangle \mathcal{Y}_{jlm}(\hat{r},m_s) \psi_{I,\mu}(\xi), \tag{24}$$

in which $\mathcal{Y}_{jlm}(\hat{r},m_s)$ is the spin-angular part of the single-proton wave function. The radial functions, $u^J_{Ijl}(r)$, are the solutions of the coupled, differential equations

$$\left[-\frac{\hbar^2}{2\mu}\left(\frac{d^2}{dr^2} - \frac{l(l+1)}{r^2}\right) + \epsilon_I - E\right] u^J_{Ijl}(r) + \sum_{\lambda I'j'l'} v_\lambda(r)\, \mathcal{V}^J_{Ijl,I'j'l'}(\lambda)\, u^J_{I'j'l'}(r) = 0, \qquad (25)$$

where the matrix element $\mathcal{V}^J_{Ijl,I'j'l'}(\lambda)$ corresponds to the rotational coupling (for details, see Refs. [19–21]) and the channels are characterized by quantum numbers $i = \{Ijl\}$. In order to solve the system of coupled equations (25), the program CCGAMOW had to be extended. Since the resulting coupled-channel code, NONADI, does not employ the adiabatic approximation, the k values in the different Ijl channels,

$$k_I^2 = \frac{2\mu}{\hbar^2}(E - \epsilon_I), \qquad (26)$$

and the Sommerfeld parameter, $\eta_I = \frac{Ze^2\mu}{\hbar^2 k_I}$, both depend on the excitation energy ϵ_I. Consequently, the boundary condition at large distance, Eq. (6), has to be modified as:

$$D_i(r_{as}, k_I) = u'_i(r_{as})/u_i(r_{as}) = O'_l(\eta_I, k_I r_{as})/O_l(\eta_I, k_I r_{as}). \qquad (27)$$

The eigenvector of Eq. (25), with proper boundary conditions, represents the Gamow states in the laboratory system of reference. The normalization of the solution is done exactly in the same way as in CCGAMOW. The method of solution also follows the adiabatic case, but the degeneracy in I is lifted. For more details, see Refs. [22,23].

V RESULTS

We have analyzed proton emission from deformed nuclei using both adiabatic (CCGAMOW) and non-adiabatic (NONADI) approaches. Results of the analysis are presented elsewhere [22,23]. Here we only present an illustration that illuminates the differences between the adiabatic and non-adiabatic methods. For our example, we consider proton emission from the ground state of ^{141}Ho, which according to the adiabatic calculations of Refs. [14,15,24,25] is the $7/2^-[523]$ deformed Nilsson orbit. The parameters of the s.p. potential are those of Chepurnov [26] save the strength, V_0, which has been fixed by the Q-value of the proton decay to ensure the correct barrier penetrability. We assume that the proton emission feeds the members of the ground state rotational band in ^{140}Dy having a constant moment of inertia. The excitation energy of the 2^+ state is unknown; hence, it has been taken as $\epsilon_2 = 0.16$ MeV based on systematics. As one can see in Table 1, the parent wave function is dominated by the $h_{11/2}$ spherical proton component. On the other hand, the partial width to the 0^+ ground state, Γ_0, is primarily determined by the $f_{7/2}$ component in the wave function. Though the total summed weights, $|c_{lj}|^2$, of

TABLE 1. Weights $|c_{Ilj}|^2$ of the main configurations in the ground state $J^\pi = 7/2^-$ wave function in ^{141}Ho (β_2=0.27, β_4=-0.06) calculated in non-adiabatic and adiabatic approaches. The results of the adiabatic calculations are shown in the third and fourth columns where the 7/2$^-$[523] deformed Nilsson (resonant) orbit is calculated both by CCGAMOW and by NONADI (with $\epsilon_I = 0$).

channel I l j	NONADI $\epsilon_2 = 0.16 MeV$ $\|c_{Ilj}\|^2$	NONADI $\epsilon_2 = 0.0$ $\|c_{Ilj}\|^2$	CCGAMOW $\|c_{lj}\|^2$
2 5 11/2	0.028	0.074	
4 5 11/2	0.289	0.462	
6 5 11/2	0.377	0.260	
8 5 11/2	0.115	0.020	
$\sum_I \|c_{Ilj}\|^2$	0.809	0.816	0.817
0 3 7/2	0.011	0.027	
2 3 7/2	0.046	0.063	
4 3 7/2	0.046	0.017	
6 3 7/2	0.018	0.001	
$\sum_I \|c_{Ilj}\|^2$	0.121	0.108	0.108

the different partial waves in the parent wave function are influenced very little by the removal of the degeneracy of the daughter states, their distribution among the members of the rotational band, $|c_{Ilj}|^2$, are changed considerably due to Coriolis coupling. The removal of the degeneracy reduces the ground-state component in the wave function (0.011) by a factor of 2.5 with respect to the adiabatic case (0.027). This in turn reduces $\Gamma_0^{NA} = 2.6 \times 10^{-20}$ MeV to one-third of the adiabatic value of $\Gamma_0^A = 8.3 \times 10^{-20}$ MeV. In the adiabatic approach, the total width was approximated by $\Gamma^A = \Gamma_0^A + \Gamma_2^A = 8.6 \times 10^{-20}$ MeV. (Γ_2^A was estimated by repeating the adiabatic calculation at the modified Q-value corresponding to the transition to 2^+ state. This gives $\Gamma_2^A = 2.3 \times 10^{-21}$ MeV.) The total width of $\Gamma^{NA} = 2.8 \times 10^{-20}$ MeV turned out to be only one-third of the adiabatic value. This example shows that the effect of the Coriolis coupling might be important in certain cases and, in general, cannot be neglected.

Acknowledgments

This work was supported in part by the Hungarian NRF (OTKA T026244 and T029003) and by the U.S. Department of Energy under Contract Nos. DE-FG02-96ER40963 (University of Tennessee), DE-FG05-87ER40361 (Joint Institute for Heavy Ion Research), and DE-AC05-96OR22464 with Lockheed Martin Energy Research Corp. (Oak Ridge National Laboratory).

REFERENCES

1. Gamow G., *Z. Phys.* **51**, 204 (1928).
2. Baz A.I, Zel'dovich Ya.B., and Perelomov A.M., *Scattering Reactions and Decay in Nonrelativistic Quantum Mechanics* (Israel Program for Scientific Translations, Jerusalem, 1969).
3. Berggren T., *Nucl. Phys.* **A109**, 265 (1968).
4. Vertse T., Pál K.F., and Balogh Z., *Comput. Phys. Commun.* **27**, 309 (1982).
5. Ixaru L.Gr., Rizea M., and Vertse T., *Comput. Phys. Commun.* **85**, 217 (1995).
6. Wynn P., *MTAC* **10**, 91 (1956).
7. Levin D., *Int. J. Comput. Math.* **B3**, 371 (1973).
8. Thompson I.J. and Barnett A.R., *Comput. Phys. Commun.* **36**, 363 (1985).
9. Humblet J. and Rosenfeld L., *Nucl. Phys.* **26**, 529 (1961).
10. Thomas R.G., *Prog. Theor. Phys.* **12**, 253 (1954).
11. Arima A. and Yoshida S., *Nucl. Phys.* **A219**, 475 (1974).
12. Ixaru L.Gr., *Numerical Methods for Differential Equations* (Reidel, Dordrecht 1984).
13. Vertse T., Kruppa A.T., Ixaru L.Gr., and Rizea M., to be submitted to Comput. Phys. Commun.
14. Maglione E., Ferreira L.S., and Liotta R.J., *Phys. Rev. Lett.* **81**, 538 (1998).
15. Maglione E., Ferreira L.S., and Liotta R.J., *Phys. Rev.* C **59**, R589 (1999).
16. Bohr A. and Mottelson B.R., *Nuclear Structure*, vol. 2 (W.A. Benjamin, New York, 1975).
17. Nilsson S.G., *Mat. Fys. Medd. Dan. Vid. Selsk.* **29**, No. 16 (1955).
18. Berggren T. and Olanders P., *Nucl. Phys.* **A473**, 189 (1987).
19. Tamura T., *Rev. Mod. Phys.* **37**, 679 (1965).
20. Bugrov V.P. and Kadmenskiĭ S.G., *Sov. J. Nucl. Phys.* **49**, 967 (1989).
21. Kruppa A.T., Nazarewicz W., and Semmes P.B., Proceedings of the *International Symposium on Proton-Emitting Nuclei*, Oct. 7-9, 1999, Oak Ridge, Tennessee, USA
22. Kruppa A.T., Barmore B., Nazarewicz W., and Vertse T., in preparation.
23. Barmore B., Kruppa A.T., Nazarewicz W., and Vertse T., to be published.
24. Davids C.N. et al. *Phys. Rev. Lett.* **80**, 1849 (1998).
25. Rykaczewski K. et al. *Phys. Rev.* **C60**, R011301 (1999).
26. Chepurnov V.A., *Yad. Fiz.* **6**, 955 (1967)

DYNAMICAL CALCULATION OF PROTON EMISSION FROM A DEFORMED TO A SPHERICAL NUCLEUS

Patrick Talou

Theoretical Division, Los Alamos National Laboratory, Los Alamos, New Mexico 87545

Abstract. The proton decay of a deformed nucleus which undergoes shape modifications during the emission process is investigated through the use of the TDSE approach developed recently. Qualitative and illustrative preliminary calculations are presented for high-energy excited states in ^{58}Cu. Tunneling probabilities, decay rates and angular distributions of proton emission are calculated and their dependency upon the (time-dependent) deformation of the potential is investigated.

INTRODUCTION

If necessary, this international symposium has proved the richness of the study of proton emitters and of the diversity of experimental as well as theoretical ways to tackle it [1,2]. New interesting and unexpected features are now revealed [3,4]. Among them, prompt particle emission from high-spin states in the deformed second well to spherical states in the daughter nucleus has been recently observed in several nuclei around $N = Z = 28$ [5]. This emission process during which the nucleus undergoes significant shape modifications is an important conceptual problem for the stationary schemes commonly used for the particles decay interpretation. On the contrary, the approach developed by Carjan *et al.* a few years ago for α-emission and fission [6] and based on the numerical solution of the time-dependent Schrödinger equation for initial quasi-stationary states can hopefully solve this issue.

Here, we report the first attempt to treat fully the proton emission from a deformed to a spherical nucleus, i.e. with a time-dependent potential. This talk is by no way intended to encompass the full problem nor to directly interpret the recent experimental data by D. Rudolph *et al.*, but instead to show the feasibility of such new calculations.

MODEL AND FORMALISM

As a simple and commonly used model, we consider a single-proton tunneling through an average deformed potential barrier created by its interaction with the nucleons of the daughter nucleus. The shape of the nuclear surface is parametrized with the help of

Cassinian ovals [7]. For axially symmetric nuclei, only one parameter, denoted by ϵ, is required to describe the nuclear shapes. For small values of ϵ (≤ 0.4), one can easily relate this quantity to the quadrupole β_2 parameter. As an example, $\epsilon = 0.2$ corresponds to $\beta_2 \simeq 0.23$ ($\epsilon = 0$ corresponds to a spherical nucleus).

The single-particle potential V_{sp} is the sum of a generalized deformed Woods-Saxon nuclear term and a deformed Coulomb part with a uniform charge distribution inside the nucleus. In order to simplify these first illustrative calculations, the spin-orbit term has been removed.

The time-dependent Schrödinger equation (TDSE) describing the time evolution of the proton in the potential V_{sp} reads (in cylindrical coordinates)

$$i\hbar \frac{\partial}{\partial t} \psi_p(\rho, z, t) = \left[-\frac{\hbar^2}{2\mu} \left(\frac{1}{\rho} \frac{\partial}{\partial \rho} + \frac{\partial^2}{\partial \rho^2} + \frac{\partial^2}{\partial z^2} - \frac{\Lambda^2}{\rho^2} \right) + V_{sp}(\rho, z, t) \right] \psi_p(\rho, z, t). \quad (1)$$

We solved this equation using a time propagator method called MSD2 [8] on a discretized spatial grid. The initial wave function $\psi_p(t = 0)$ has been obtained via the prescription of Gurvitz [9]. From the solution $\psi_p(\rho, z, t)$ in time, one can infer several important physical quantities as detailed in ref. [10]. Among them are:

- the tunneling probability, $P_{tun}(t)$, i.e., the probability that the proton has escaped from the nucleus at time t;

- the (time-dependent) decay rate $\lambda(t)$;

- the tunneling angular distribution, $P_{tun}(t, \theta)$ estimated in spherical coordinates and which gives the angular distribution of proton emission with respect to the nuclear axis of symmetry (z-axis).

The time-dependence of the nuclear deformation $\epsilon(t)$, hence of the single-particle potential $V_{sp}(\rho, z, \epsilon(t))$, has been arbitrarily chosen to follow a linear behaviour

$$\epsilon(t) = \begin{cases} \epsilon_i \left(1 - \frac{t}{T_d}\right) & \text{for } t \leq T_d \\ 0 & \text{for } t > T_d \end{cases} \quad (2)$$

where T_d is the duration of the nuclear deformation, i.e., the time it takes for the nucleus to go from its initial configuration (here, $\epsilon_i = 0.2$) to its final one (here, $\epsilon_f = 0$). The other important time scale of the problem is the half-life $T^p_{1/2}$ of the proton decay of the spherical nucleus ($\epsilon = 0$). Within this model, two natural limits appear:

- $T^p_{1/2} \ll T_d$: the deformation of the nucleus in its final configuration is equal to its initial one: $\epsilon_f = \epsilon_i = const$. The proton wave function already escaped the daughter nucleus before any significant modification of the interacting potential could happen;

- $T^p_{1/2} \gg T_d$: the deformation of the nucleus "jumps" to its final configuration as early as the first times of the decay. In other words, $\epsilon(t = 0^+) = \epsilon_f = 0$. Hence the proton interacts as early as $t = 0^+$ with a constant potential characterized by ϵ_f. We will call this situation the "sudden approximation".

In both cases, the potential V_{sp} is constant, deformed ($\epsilon = 0.2$) and spherical ($\epsilon = 0$) respectively. Between these two limiting situations, the relation (2) applies.

FIRST RESULTS

In these very preliminary calculations, several strong hypotheses have been made. As mentioned earlier, no spin-orbit term has been taken into account, and the time-dependence of the nuclear surface deformation has been chosen (quite arbitrarily) to follow a simple linear behaviour in time (cf. Eq. 2). In addition, we chose to study high-energy excited decaying states (by modifying artificially the depth V_0 of the potential) in order to cut drastically the computation time required. Hopefully, these concessions would be lifted in a near future.

We applied the above formalism to the decay of single-proton quasi-stationary states in ^{58}Cu. The depth of the nuclear Woods-Saxon potential is $V_0 = 60$ MeV. In the following, we will focus only on "$1g$" states ($l = 4$), $\Lambda = 0$ to 4, which lie at the energy $E \simeq 8.4$ MeV for $\epsilon = 0$. (Λ is the projection of the angular momentum on the z-axis of symmetry and is a good quantum number.)

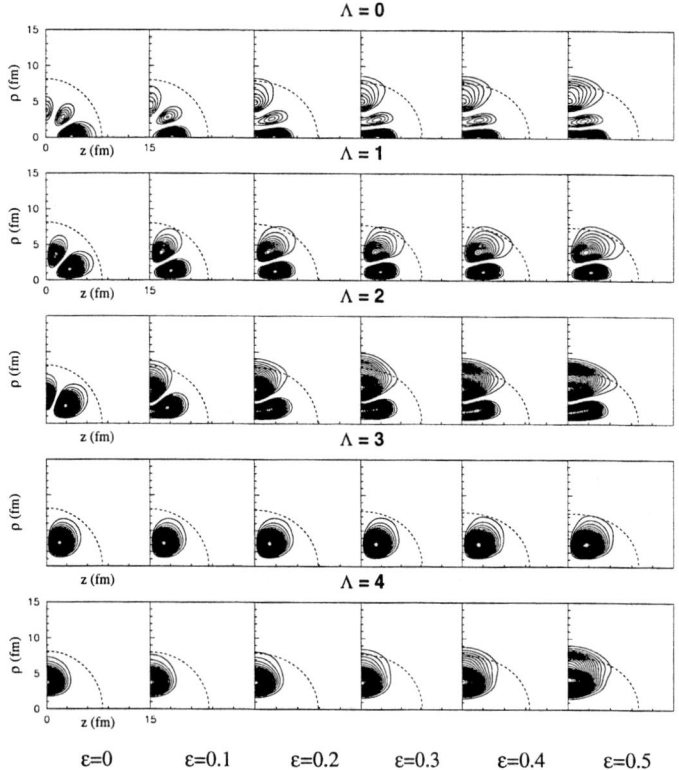

FIGURE 1. $|\psi_p(\rho, z, t = 0)|^2$ of "$1g$" states ($\Lambda = 0..4$) used as initial quasi-stationary wave functions in the TDSE calculations.

The square moduli of these wave functions are represented in the cylindrical half-plane in Fig.1 at several nuclear deformations, from $\epsilon = 0$ to 0.5. An important feature emerges from this figure: the structure (topology) of high-Λ states is not strongly modified as the deformation increases. On the other hand, low-Λ states can be dramatically influenced as one turns on the deformation. In fact, the rule $\Lambda \leq l$ implies that high-Λ states are unlikely to be distributed over a large number of spherical basis states, hence are more robust to deformation than their low-Λ partners. For our present purpose, this means that the influence of a time-dependent nuclear deformation will be generally more important for low-Λ than for high-Λ quasi-stationary states.

In the following, we will present the results obtained for the $\Lambda = 1$ state only. On the left side of Fig.2 are represented the tunneling probabilities $P_{tun}(t)$, up to $T_{max} = 4 \times 10^{-21}$ sec, for different values of the nuclear deformation duration T_d.

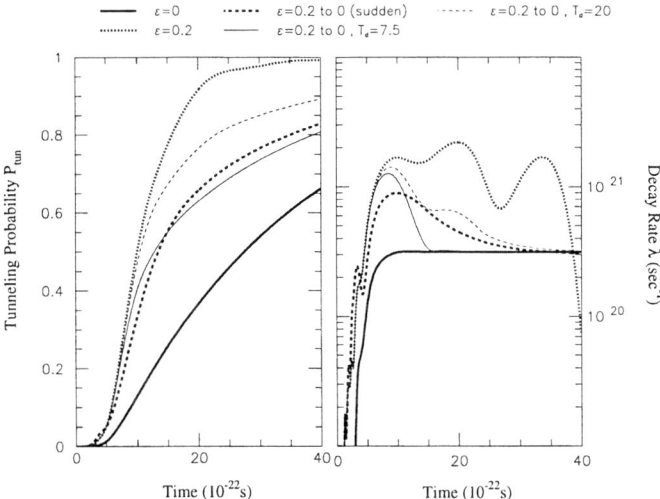

FIGURE 2. Tunneling probabilities $P_{tun}(t)$ and decay rates $\lambda(t)$ for different time-dependencies of $V_{sp}(\epsilon(t))$.

The spherical case ($\epsilon = 0$) is plotted for reference. The "$\epsilon = 0.2$" and "$\epsilon = 0.2 \to 0$ (sudden)" cases correspond to the situations $T^p_{1/2} \ll T_d$ and $T^p_{1/2} \gg T_d$, respectively. In the first case, the proton wave function leaks out from the nuclear surface very quickly, and the tunneling probability is already greater than 0.9 at $t = 2 \times 10^{-21}$ sec. In fact, this time should be considered as the end of the decay, since subsequently, contributions from other lower states become non-negligible.

An interesting result is the important difference observed between the decay rates at $\epsilon = 0$ and in the sudden approximation. In both cases, the potential is stationary and spherical. Hence, the only difference comes from the topological difference between the two initial wave functions. As noted above, such a "dramatic" behaviour is typical of low-Λ states.

An important related quantity is the (time-dependent) decay rate $\lambda(t)$ shown on the right side of Fig. 2. (Again, the behaviour of $\lambda(t)$ for $\epsilon = 0.2$ should be observed only up to $t = 2 \times 10^{-21}$ sec.) The spherical case shows the typical behaviour of the decay of a quasi-stationary state through an *isotropic* potential barrier, i.e., a first non-exponential transient time followed by a stationary value corresponding to an exponential decay [12]. As for $\epsilon = 0.2$, the decay rate strongly oscillates because of the mixing of angular momenta due to the anisotropy of the barrier [11].

Interestingly, the decay rate $\lambda(t)$ in the sudden case seems to tend at large times towards the 'asymptotic' decay rate $\lambda_{\epsilon=0}(\infty)$. In fact, although not obvious from the figure presented here, $\lambda(t)$ keeps evolving below $\lambda_{\epsilon=0}(\infty)$. In this sudden case, the wave function does not have time to adapt to its new potential and some of its components get trapped at time $t = 0^+$.

Another very interesting feature appearing in these calculations is the existence of an 'optimal' time T_d^{opt} (here, $\simeq 12 \times 10^{-22}$sec) for which $\lambda(t)$ reaches $\lambda_{\epsilon=0}(\infty)$ the fastest. At each time, the wave function follows *adiabatically* the deformed potential and is a quasi-stationary state of it. At T_d^{opt}, the state is exactly the spherical quasi-stationary wave function. For $T_d < T_d^{opt}$, the wave function does not have time enough to adapt adiabatically to the ever changing potential. On the contrary, for $T_d > T_d^{opt}$, the mixing of angular momenta during tunneling prevents the decay rate from reaching $\lambda_{\epsilon=0}(\infty)$.

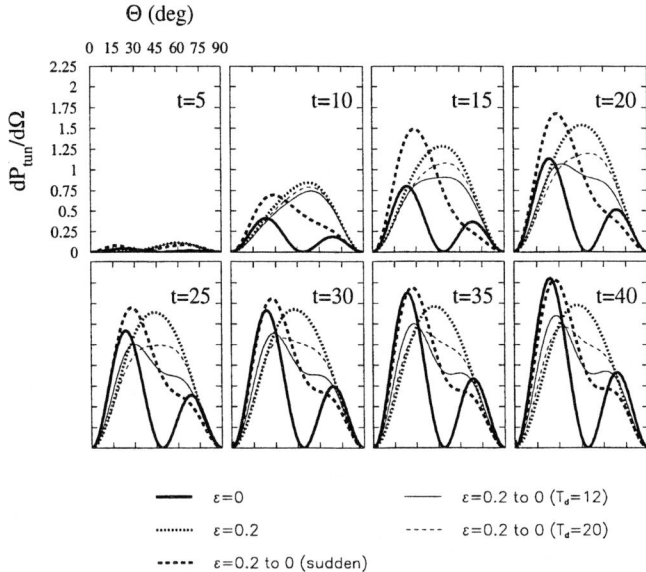

FIGURE 3. Tunneling angular distributions (with respect to the z-axis of symmetry) for several time-dependencies of $V_{sp}(\epsilon(t))$.

Finally, we calculated the tunneling angular distributions $P_{tun}(t,\theta)$ in each case studied above. As expected, the time-dependence of the potential has a strong influence on this quantity (see Fig.3) as can be inferred from the limiting cases $\epsilon = 0$, $\epsilon = 0.2$ and $\epsilon = 0.2 \to 0$ (sudden). The spherical calculation simply exhibits two peaks corresponding to the spherical harmonic $Y_4^1(\theta,\phi)$, while the deformed $\epsilon = 0.2$ pattern comes from an interplay between the initial wave function structure and the anisotropic potential barrier [13].

PERSPECTIVES

In these preliminary results, we have shown the feasibility of calculations accounting for the modification of the shape of a nucleus while decaying by proton emission. This has been done through the use of the TDSE approach which allows a full dynamical treatment of the multi-dimensional tunneling process involved. Several physical quantities are indeed significantly affected by the modifications of the parent-nucleus shape during its decay. Future and more realistic calculations will be performed in order to interpret the recent experimental data from D. Rudolph et al. [4], and to predict possible experimental signatures of such a phenomenon (like during a single-particle levels quasi-crossing [14]).

REFERENCES

1. P.J. Woods, this conference.
2. W. Nazarewicz, this conference.
3. A.A. Sonzogni, this conference.
4. D. Rudolph, this conference and references therein.
5. D. Rudolph et al., Phys. Rev. Lett. **80**, 3018 (1998).
6. O. Serot, N. Carjan and D. Strottman, Nucl. Phys. **A569**, 562 (1994); N. Carjan, O. Serot and D. Strottman, Zeit. Phys. A **349**, 353 (1994).
7. V.V. Pashkevich, Nucl. Phys. **A169**, 275 (1971).
8. T. Iitaka, Phys. Rev. E **49**, 4684 (1994).
9. S.A. Gurvitz and G. Kalbermann, Phys. Rev. Lett. **59**, 262 (1987).
10. N. Carjan, this conference and references therein.
11. P. Talou, N. Carjan and D. Strottman, Nucl. Phys. **A647**, 21 (1999).
12. P. Talou, D. Strottman and N. Carjan, Phys. Rev. C **60**, 054318 (1999).
13. D. Strottman, *Achievements and Perspectives in Nuclear Structure*, Crete, Greece, Ed. S. Åberg, Physica Scripta, Proceedings, 1999; N. Carjan, *The Nucleus: New Physics for the New Millenium*, Faure, South Africa, Ed. J. Sharpey-Schafer, Plenum Press, Proceedings, 1999.
14. P. Talou, N. Carjan and D. Strottman, Phys. Rev. C **58**, 3280 (1998).

Decay Rates for Spherical and Deformed Proton Emitters[1]

Cary N. Davids and Henning Esbensen

Physics Division, Argonne National Laboratory, Argonne, IL 60439

Abstract. Using Green's function techniques, we derive expressions for the width of a proton decaying state in spherical and deformed nuclei. We show that the proton decay widths calculated by the "exact" expressions of Maglione et al. are equivalent to the distorted wave expressions of Bugrov et al., and that of Åberg et al. in the spherical case.

INTRODUCTION

The calculation of the decay rates for proton emitters is of current interest. Several authors have presented expressions for the proton decay width of spherical nuclei [1–3], and deformed nuclei [1,4,5]. The proton-emitting states are extremely narrow, with observable widths not exceeding 10^{-10} eV. In this work we treat the decaying states as stationary states. The decay of the states is imposed by imposing an outgoing wave Green's function to solve the Schroedinger equation.

Two apparently different methods have been used for deriving the decay rates for proton emitters. In both cases one first determines the wave function for the relative motion of the proton and the daughter nucleus in the resonant state. We shall consider the spherical and deformed cases separately.

SPHERICAL NUCLEI

Maglione, Ferreira, and Liotta [1], using what we will call the direct (Dir) method, describe the parent nucleus as a single nucleon moving in the potential set up by the daughter nucleus. They give the single particle radial wavefunction at a large distance R outside the nucleus to be an outgoing Coulomb wave:

$$\psi_{\ell j}^{out}(r) = \frac{u_{\ell j}(r)}{r} = \frac{N_{\ell j} O_\ell(kr)}{r} \text{ at } r = R, \tag{1}$$

---

[1] This work was supported by the U.S. Department of Energy, Nuclear Physics Division, under Contract W-31-109-ENG-38.

where $O_\ell(kr) = [G_\ell(kr) + iF_\ell(kr)]$, $N_{\ell j}$ is a normalization constant, k is the wave number, and F_ℓ and G_ℓ are the regular and irregular Coulomb functions, respectively. They calculate the probability flux penetrating a sphere at large distances to obtain an expression for the mean lifetime τ:

$$\frac{1}{\tau} = |N_{\ell j}|^2 v,$$

where v is the velocity. One may then obtain the decay width

$$\Gamma_{\ell j}^{\text{Dir}} = \frac{\hbar^2 k}{\mu} |N_{\ell j}|^2 \tag{2}$$

where μ is the reduced mass.

Åberg et al. [3] derive an expression for the width using what they refer to as the DWBA method. We prefer to call it the Distorted Wave (DW) method, since it uses distorted waves but not the Born approximation. The width is given by

$$\Gamma_{\ell j}^{\text{DW}} = \frac{4\mu}{\hbar^2 k} \left| \int_0^\infty F_\ell(kr)[V(r) - V_C^0(r)] u_{\ell j}(r) dr \right|^2$$

where $V(r)$ is the total potential between the outgoing proton and the daughter nucleus, and $V_C^0(r)$ is the point source Coulomb potential, and $u_{\ell j}(r)/r$ is again the radial wavefunction obtained by numerically integrating the Schroedinger equation with a one-body potential.

To compare the two widths $\Gamma_{\ell j}^{\text{Dir}}$ and $\Gamma_{\ell j}^{\text{DW}}$, we note that the wavefunctions $F_\ell(kr)$ and $u_{\ell j}(r)$ are both solutions of the radial Schroedinger equation:

$$\frac{d^2 u_{\ell j}(r)}{dr^2} + [k^2 - \frac{\ell(\ell+1)}{r^2} - \frac{2\mu}{\hbar^2} V(r)] u_{\ell j}(r) = 0 \tag{3}$$

$$\frac{d^2 F_\ell(kr)}{dr^2} + [k^2 - \frac{\ell(\ell+1)}{r^2} - \frac{2\mu}{\hbar^2} V_C^0(r)] F_\ell(kr) = 0. \tag{4}$$

Multiply equation (3) by $-F_\ell(kr)$ and equation (4) by $u_{\ell j}(r)$ and add the two equations:

$$u_{\ell j}(r) \frac{d^2 F_\ell(kr)}{dr^2} - F_\ell(kr) \frac{d^2 u_{\ell j}(r)}{dr^2} + \frac{2\mu}{\hbar^2} F_\ell(kr)[V(r) - V_C^0(r)] u_{\ell j}(r) = 0.$$

Integrate over r from 0 to a large radius R, well outside the nucleus.

$$\frac{2\mu}{\hbar^2} \int_0^R F_\ell(kr)[V(r) - V_C^0(r)] u_{\ell j}(r) dr = \int_0^R \left[F_\ell(kr) \frac{d^2 u_{\ell j}(r)}{dr^2} - u_{\ell j}(r) \frac{d^2 F_\ell(kr)}{dr^2} \right] dr$$

$$= \int_0^R \frac{d}{dr} \left[F_\ell(kr) \frac{du_{\ell j}(r)}{dr} - u_{\ell j}(r) \frac{dF_\ell(kr)}{dr} \right] dr$$

$$= \left[F_\ell(kr) \frac{du_{\ell j}(r)}{dr} - u_{\ell j}(r) \frac{dF_\ell(kr)}{dr} \right]_0^R$$

At the lower limit $r = 0$, both $F_\ell(kr)$ and $u_{\ell j}(r)$ vanish. For R well outside the nucleus, from equation (1), substitute $u_{\ell j}(R) = N_{\ell j}[G_\ell(kR) + iF_\ell(kR)]$. Therefore

$$\frac{2\mu}{\hbar^2}\int_0^R F_\ell(kr)[V(r) - V_C^0(r)]u_{\ell j}(r)dr = N_{\ell j}\left[F_\ell(kR)\frac{dG_\ell(kR)}{dr} - G_\ell(kR)\frac{dF_\ell(kR)}{dr}\right].$$

The square bracket on the right hand side of the equation equals $-k$ times the Wronskian of the Coulomb functions, which has the value 1. In addition, we may safely extend the upper bound of the integral on the left hand side from R to ∞, since the right hand side is radius-independent for large R. So

$$\left|\int_0^\infty F_\ell(kr)[V(r) - V_C^0(r)]u_{\ell j}(r)dr\right|^2 = \left|\frac{-\hbar^2 k N_{\ell j}}{2\mu}\right|^2$$

and

$$\Gamma_{\ell j}^{DW} = \frac{4\mu}{\hbar^2 k} \times \frac{\hbar^4 k^2 |N_{\ell j}|^2}{4\mu^2} = \frac{\hbar^2 k}{\mu}|N_{\ell j}|^2 = \Gamma_{\ell j}^{Dir}.$$

It is of interest to compare the numerical results for the half-lives ($t_{1/2} = \hbar\ln(2)/\Gamma$) obtained with the direct method and the distorted wave method for spherical nuclei. We show in Table 1 the calculated half-lives for three spherical decaying states having orbital angular momentum $\ell = 0$, 2, and 5, respectively. The radial wavefunctions $u_{\ell j}(r)$ were calculated by integrating the radial Schroedinger equation, using for the proton-daughter nucleus potential the real part of the Becchetti-Greenlees optical model potential [7]. The potential depth was adjusted to match the energy eigenvalue to the proton decay Q-value, corrected for recoil and atomic screening:

$$Q_{p,n} = E_p\frac{m_p}{\mu} + E_{sc},$$

where m_p is the proton mass and E_{sc} is the atomic screening correction.

TABLE 1. Comparison of the proton half-lives of three spherical proton radioactivities calculated with the direct (Dir) and distorted wave (DW) methods.

Nucleus	E_p(keV)	j_p	ℓ_p	$t_{1/2,p}^{Dir}$	$t_{1/2,p}^{DW}$
^{167}Ir^g	1064(6)	$\frac{1}{2}^+$	0	35.78 ms	35.77 ms
^{147}Tm^m	1119(5)	$\frac{3}{2}^+$	2	171.4 μs	171.3 μs
^{167}Ir^m	1238(7)	$\frac{11}{2}^-$	5	1.9937 s	1.9935 s

It is seen that the calculated half-lives agree to better than 0.05% between the two methods. Since the two methods should give identical results, the difference must reflect the accuracy of the numerical techniques that have been used.

DEFORMED NUCLEI

We consider a deformed odd-A nucleus, consisting of a single particle that is strongly coupled to an axially-symmetric even-even core. To calculate the outgoing proton wavefunction we use the exact Gell-Mann-Goldberger transformation and the distorted-wave Green's function with outgoing Coulomb wave boundary conditions. At large distances,

$$\Psi_{KIM}^{(+)}(\mathbf{r}) = -\frac{2\mu}{\hbar^2 k} \sum_{\ell j R} \frac{O_\ell(kr)}{r} |(\ell j R) IM\rangle \left\langle (\ell j R) IM | \frac{F_\ell(kr')}{r'} \left| V(\mathbf{r}') - V_C^0(r') \right| \Psi_{KIM} \right\rangle. \tag{5}$$

Here $V(\mathbf{r}')$ is the total deformed potential acting between the proton and the core nucleus. The angular momentum part of the Green's function includes the rotational states $|RM_R\rangle$ ($R = 0, 2 \ldots$) of the core and the single-particle state $|\ell j m\rangle$, which are coupled to the total spin (IM):

$$|(\ell j R) IM\rangle = \sum_{mM_R} \langle jmRM_R | IM \rangle |\ell j m\rangle |RM_R\rangle.$$

The total wavefunction of the initial state is of the form [8]

$$\Psi_{KIM} = \sqrt{\frac{\hat{I}}{16\pi^2}} [D_{MK}^I(\omega')\phi_K + (-1)^{I+K} D_{M-K}^I(\omega')\phi_{\overline{K}}], \tag{6}$$

where $\hat{I} = 2I + 1$. The single-particle wavefunction ϕ_K is described (as in the Nilsson model) in terms of the intrinsic (body-fixed) coordinates of the daughter nucleus. It can be expanded in spherical components

$$\phi_K(\mathbf{r}') = \sum_{\ell j} \phi_{\ell j}^{(i)}(r') |\ell j K\rangle_0, \tag{7}$$

where the sum is over $j \geq |K|$ and the subscript "0" denotes a state in the intrinsic frame. For the final state, the rotational state wavefunction of the daughter is

$$|RM_R\rangle = \sqrt{\frac{\hat{R}}{8\pi^2}} D_{M_R 0}^R(\omega').$$

We pick out a particular outgoing channel, in which the proton carries off angular momentum $\ell_p j_p$, with projection m_p, leaving the daughter nucleus with angular momentum R and projection M_R. For the evaluation of the matrix element in equation (5), the final state $|(\ell_p j_p R) IM\rangle$ must be expressed in terms of the single-particle wavefunction in the intrinsic system:

$$|\ell_p j_p m_p\rangle = \sum_{K'} D_{m_p K'}^{j_p}(\omega') |\ell_p j_p K'\rangle_0.$$

Combining these, we obtain

$$|(\ell_p j_p R)IM\rangle = \sqrt{\frac{\hat{R}}{8\pi^2}} \sum_{K' m_p M_R} \langle j_p m_p R M_R | IM \rangle D^R_{M_R 0}(\omega') D^{j_p}_{m_p K'}(\omega') |\ell_p j_p K'\rangle_0$$

$$= \sqrt{\frac{\hat{R}}{8\pi^2}} \sum_{K'} \langle j_p K' R 0 | I K' \rangle D^I_{MK'}(\omega') |\ell_p j_p K'\rangle_0, \quad (8)$$

where we have used a well-known relation involving a sum of D-functions [9].

Evaluation of the Matrix Element

We can now calculate the matrix element found in equation (5). Integrating over the orientation of the daughter nucleus ω' produces an expression that is diagonal in the quantum numbers IMK. This is evident from the first part of (6). The second term will select the value $-K$ from the sum (8) over final state K'-values. In fact, the two terms are of equal magnitude, so we obtain

$$\left\langle ((\ell_p j_p R)IM | \frac{F_{\ell_p}(kr')}{r'} \left| V(\mathbf{r}') - V^0_C(r') \right| \Psi_{KIM} \right\rangle = \sqrt{\frac{2\hat{R}}{\hat{I}}} \langle j_p K R 0 | I K \rangle \mathcal{M}_{\ell_p j_p K}, \quad (9)$$

where

$$\mathcal{M}_{\ell_p j_p K} = \left\langle \ell_p j_p K \left| \frac{F_{\ell_p}(kr')}{r'} [V(\mathbf{r}') - V^0_C(r')] \right| \phi_K \right\rangle_0 \quad (10)$$

is evaluated in the intrinsic frame. It is noted that the matrix element (9) is independent of the M-quantum number.

Partial Decay Width

We can now write the outgoing wavefunction (5) for a specific channel $(\ell_p j_p RIK)$ as

$$\Psi^{(+)}_{\ell_p j_p R, KIM}(\mathbf{r}) = -\frac{2\mu}{\hbar^2 k} \sqrt{\frac{2\hat{R}}{\hat{I}}} \langle j_p K R 0 | I K \rangle \frac{O_{\ell_p}(kr)}{r} |(\ell_p j_p R)IM\rangle \mathcal{M}_{\ell_p j_p K}.$$

The deformed decay width follows as

$$\Gamma^{DW}_{\ell_p j_p RIK} = \frac{4\mu}{\hbar^2 k} \frac{2\hat{R}}{\hat{I}} \langle j_p K R 0 | I K \rangle^2 \left| \mathcal{M}_{\ell_p j_p K} \right|^2. \quad (11)$$

Apart from a pairing term, this expression is identical to that obtained by Kadmensky and Bugrov [4,5]. This can be shown by inserting the expansion (7) of the initial state into the matrix element (10) and using the expression

$$|\ell j m\rangle = \sum_{m_\ell m_s} \langle \ell m_\ell \tfrac{1}{2} m_s | j m \rangle Y_\ell^{m_\ell}(\hat{\mathbf{r}}') \chi(m_s)$$

for the single-particle states. Since the interaction does not change the proton spin, the single-particle matrix element will be diagonal in m_s. It is also diagonal in K. Thus one obtains

$$\mathcal{M}_{\ell_p j_p K} = \sum_{\ell j m_s} \langle \ell_p m_{\ell_p} \tfrac{1}{2} m_s | j_p K \rangle \langle \ell m_\ell \tfrac{1}{2} m_s | j K \rangle$$

$$\times \left\langle Y_{\ell_p}^{m_{\ell_p}}(\hat{\mathbf{r}}') \frac{F_{\ell_p}(kr')}{r'} \left| V(\mathbf{r}') - V_C^0(r') \right| \phi_{\ell j}^{(i)}(r') Y_\ell^{m_\ell}(\hat{\mathbf{r}}') \right\rangle.$$

This expression has been simplified by noting that $m_{\ell_p} = m_\ell = K - m_s$, thus eliminating sums over those variables. The $Y_2^0(\hat{\mathbf{r}}')$ term in the deformed potential $V(\mathbf{r}')$ apparently allows for an angular momentum exchange at the nuclear surface between the outgoing proton and the daughter nucleus, leading to non-diagonal terms in the matrix element.

Direct Method

We can also use the direct method to determine the decay width of a deformed proton emitter. To do this we expand the wavefunction $\Psi_{KIM}(\mathbf{r})$ of the initial state on the complete set of angular momentum basis states:

$$\Psi_{KIM}(\mathbf{r}) = \sum_{\ell j R} |(\ell j R) I M \rangle \langle (\ell j R) I M | \Psi_{KIM} \rangle.$$

Using equations (9) and (10) with $F_{\ell_p}(kr)/r[V(\mathbf{r}) - V_C^0(r)]$ replaced by 1 we can immediately write the overlap matrix element as:

$$\langle (\ell j R) I M | \Psi_{KIM} \rangle = \sqrt{\frac{2\hat{R}}{\hat{I}}} \langle j K R 0 | I K \rangle \langle \ell j K | \phi_K \rangle.$$

Inserting the expansion (7) for the outgoing channel $\ell_p j_p$ we obtain

$$\langle \ell_p j_p K | \phi_K \rangle = \phi_{\ell_p j_p}^{(i)}(r) \to A_{\ell_p j_p} \frac{O_{\ell_p}(kr)}{r}, \quad \text{for } r \to \infty, \tag{12}$$

assuming that the intrinsic states are matched to outgoing Coulomb waves as in equation (1). The outgoing wave is therefore

$$\Psi_{\ell_p j_p R, KIM}^{(+)}(\mathbf{r}) = N_{\ell_p j_p RIK}^{\text{Dir}} \frac{O_{\ell_p}(kr)}{r} |(\ell_p j_p R) I M \rangle,$$

where

$$N^{\text{Dir}}_{\ell_p j_p RIK} = \sqrt{\frac{2\hat{R}}{\hat{I}}} \langle j_p KR0|IK\rangle A_{\ell_p j_p}.$$

This gives the decay width, according to equation (2), as

$$\Gamma^{\text{Dir}}_{\ell_p j_p RIK} = \frac{\hbar^2 k}{\mu} \frac{2\hat{R}}{\hat{I}} \langle j_p KR0|IK\rangle^2 |A_{\ell_p j_p}|^2. \qquad (13)$$

This expression is consistent with the result given in equation (8) of ref. [1]. In addition, it allows the calculation of partial decay widths to excited states of the daughter nucleus.

It can also be demonstrated explicitly that the distorted wave method and the direct method give identical results for the decay width of a deformed proton emitter. This can be seen by replacing the interactions in the matrix element (10) by the associated single-particle Hamiltonian minus the kinetic energy operators,

$$V(\mathbf{r}) - V_C^0(r) \to \left[H + \hbar^2 \nabla^2/2\mu\right]_{right} - \left[H_0 + \hbar^2 \nabla^2/2\mu\right]_{left}. \qquad (14)$$

The subscripts 'right' and 'left' indicate that the operators must act to the right and to the left, respectively, when inserted in the matrix element of Eq. (10).

Making the substitution (14) in equation (10), the contributions from the two single-particle Hamiltonians must cancel because the two wave functions have the same energy. Thus the matrix element (10) can be expressed as

$$\mathcal{M}_{\ell_p j_p K} = \frac{\hbar^2}{2\mu} \left\langle \ell_p j_p K \left| \frac{F_{\ell_p}(kr)}{r} (\nabla^2_{right} - \nabla^2_{left}) \right| \phi_K \right\rangle.$$

Using Green's theorem and the expansion (7) for the initial state one obtains

$$\mathcal{M}_{\ell_p j_p K} = \frac{\hbar^2}{2\mu} \left(F_{\ell_p}(kr) \frac{d[r\phi^{(i)}_{\ell_p j_p}(r)]}{dr} - [r\phi^{(i)}_{\ell_p j_p}(r)] \frac{dF_{\ell_p}(kr)}{dr} \right)_{r\to\infty}$$

$$= -\frac{\hbar^2 k}{2\mu} A_{\ell_p j_p},$$

where we again have used the asymptotic form (12) of $\phi^{(i)}_{\ell_p j_p}(r)$. Inserting this into the DW decay width expression (11) we see that it becomes identical to the Dir width, equation (13).

For the deformed case we show in Table 2 the calculated half-lives for three decaying states ($\beta_2 = 0.3$) and total angular momentum $j = 3/2^+, 5/2^+$, and $7/2^-$, respectively.

TABLE 2. Comparison of the proton half-lives of deformed proton radioactivities calculated with the direct (Dir) and distorted wave (DW) methods.

Nucleus	E_p(keV)	j_p	ℓ_p	β_2	$C_{\ell j}$ [a]	$t^{Dir}_{1/2,p}$	$t^{DW}_{1/2,p}$
^{131}Eu	932(7)	$\frac{3}{2}^+$	2	0.3	-0.208	27.92 ms	24.09 ms
^{131}Eu	932(7)	$\frac{5}{2}^+$	2	0.3	-0.0999	176.2 ms	214.8 ms
^{141}Ho^g	1169(8)	$\frac{7}{2}^-$	3	0.3	0.240	4.087 ms	3.266 ms

[a] A. A. Sonzogni, private communication (1999)

Here the calculated half-lives only agree to within 20% between the two methods. The discrepancy is probably due to the truncation in the eigenfunction space, such that only the nearest spherical states were included. The initial state is therefore not the exact or complete solution to the deformed Hamiltonian, and the Gell-Mann-Goldberger transformation method will therefore not provide exactly the same result as the direct method. The comparison of the results of the two methods is therefore a test of how close the truncated solution comes to being correct. Further investigation in this area is needed. Perhaps a coupled-channels approach, such as that developed by [10], will offer closer agreement between the direct and distorted wave methods.

CONCLUSIONS

We have shown that the distorted wave method and the direct method of calculating the width of spherical and deformed proton emitter are equivalent. In the spherical case numerical agreement is demonstrated to better than 0.05%, while for the deformed case the agreement is only within about 20%. Improved methods of calculating the wavefunctions should reduce this discrepancy. We recommend using either of these methods in place of the WKB method, which has certain problems related to the frequency factor (see ref. [3] for a discussion of this point). For the cases where the radial wavefunction is known over the $0 < r < 25$ fm range, the direct method is preferred for its calculational simplicity. However, if the radial wavefunction is known reliably only in the region of the nuclear surface, the distorted wave method is to be preferred.

ACKNOWLEDGMENTS

The authors wish to thank A. A. Sonzogni for providing the results of his calculations for deformed nuclei using the distorted wave method.

REFERENCES

1. Maglione, E., Ferreira, L. S., and Liotta, R. J., Phys. Rev. Lett. **81**, 538 (1998).
2. Bugrov, V. P., Kadmensky, S. G., Furman, V. I., and Khlebostroev, V. G., Sov. J. Nucl. Phys. **41**, 717 (1985).
3. Åberg, S., Semmes, P. B., and Nazarewicz, W., Phys. Rev. C **56**, 1762 (1997); Phys. Rev. C **58**, 3011 (1998).
4. Bugrov, V. P. and Kadmensky, S. G., Sov. J. Nucl. Phys. **49**, 967 (1989).
5. Kadmensky, S. G. and Bugrov, V. P., Phys. of Atomic Nuclei **59**, 424 (1996).
6. Glendenning, N. K., *Direct Nuclear Reactions*, Academic Press, New York, 1983, p. 47.
7. Becchetti, F. D. and Greenlees, G. W., Phys. Rev. **182**, 1190 (1969).
8. Bohr, A. and Mottelson, B. R., *Nuclear Structure*, W. A. Benjamin, Reading, 1975, Vol. II, Equation 4-19.
9. Bohr, A. and Mottelson, B. R., *Nuclear Structure*, W. A. Benjamin, New York, 1969, Vol. I, Equation 1A-43.
10. Ferreira, L. S., Maglione, E., and Liotta, R. J., Phys. Rev. Lett. **78**, 1640 (1997).

Theoretical Approaches and Experiments on Proton Decay

Stanislav G. Kadmensky

Department of Nuclear Physics, Voronezh State University, Russia, 394000.

Abstract. It is shown that the multiparticle theory of proton radioactivity (MTPR), based on the integral formulae for proton widths, has sufficiently high accuracy and totality for description of deep subbarrier proton decay of nuclei. The theoretical scheme of calculation for proton widths of odd-odd deformed nuclei is created. The connection of fine proton spectrum structure with types of proton orbit in 141Ho ^{141}Ho is analysed. It is shown that nuclear deformation parameters, found for investigation of proton decay of deformed odd-even and odd-odd nuclei and predicted by some systematics are the same.

INTRODUCTION

The comparison of experimental proton widths [1] with corresponding theoretical widths is the only possible way to get some information about a structure and shape of nuclei located near the proton drip-line, determining the limit of nucleus existence in nature. The comparison of this information with the predictions of different systematics based on the investigation of nuclei located near betastability band gives the possibility to verificate the accuracy of principal representations of nuclear physics.

A good base for the theoretical calculations is the multiparticle theory of proton radioactivity (MTPR), based on the integral formula for proton widths and developed at first for the case of proton decays of spherical nuclei [2-3] and then generalised for the case of proton decays of deformed nuclei [4-5]. In the frame of MTPR the classification of proton transitions was built and the proton spectroscopic factors taking into account the influence of manybody nuclear effects on the proton decay probabilities were introduced.

On the base of MTPR the widths for proton transitions from not only ground states of spherical nuclei [2,1,6], but also from multiquasiparticle isomeric states [3] was described. For the case of deformed nuclei in contrast to R-matrix theory [7] MTPR presents no problems with a choice of angular dependent channel radius and therefore gives the possibility to calculate successfully [4-5,8-9] the proton widths of some odd-even nuclei.

Recently the new methods [6,10-11] of calculation of single particle characteristics of proton decays were represented. The aim of this talk is to compare the possibilities of different theoretical approaches to proton decay description and to continue the investigations of proton decays of odd-even and odd-odd deformed nuclei and the fine structure of proton spectra on MTPR base.

Multiparticle Theory of Proton Radioactivity

Let us consider the main characteristics of MTPR, using the methods and designations of articles [2,4]. The state of proton radioactive parent nucleus can be described by the stationary wave function $(\Psi_{\sigma_i}^{J_iM_i\pi_i})_0$ with the energy E_i^0, received from the Gamov quasistationary wave function $\Psi_{\sigma_i}^{J_iM_i\pi_i}$, if the deep subbarrier proton decay condition $G_{l_p}(k_cR_1) \gg F_{l_p}(k_cR_1)$ takes place in the region $R_0 \le r \le R_1$, where radius R_1 lies on the right from the proton channel radius R_0. The partial width Γ_{ipc} of proton decay to channel c ($c \equiv J_f\pi_f\sigma_f j_p l_p$) for the parent nucleus is determined by the integral formula:

$$\Gamma_{ipc} = 2\pi|B_{ipc}|^2 =$$
$$2\pi \left| \int \hat{A} \left[\frac{F_{l_p}(k_cr)U_c}{r}(V_{pA-1} - \frac{(Z-1)e^2}{r}) \right] (\Psi_{\sigma_i}^{J_iM_i\pi_i})_0 d\tau \right|^2, \qquad (1)$$

where V_{pA-1} and $Q_c = \frac{\hbar^2 k_c^2}{2m} = E_i^0 - E_f^0$ are potential of interaction and the relative movement energy of proton and daughter nucleus, $F_l(kr)$ ($G_l(k_cr)$) is the regular(unregular) radial Coulomb function, normalised by δ-function of energy, U_c is the channel function.

Formula (1) is correct for the case of the deep subbarrier proton decay and is not the Born approximation of the disturbed waves method (DWBA), as it is alleged in article [6], because the wave function $(\Psi_{\sigma_i}^{J_iM_i\pi_i})_0$ takes into account of the potential V_{pA-1} for all orders of perturbation theory.

The integral, determining of the proton decay amplitude B_{ipc}, can be represented as the sum of integrals, connected with internal (shell model) ($r \le R_{sh}$), intermediate ($R_{sh} \le r \le R_0$) and external ($R_0 \le r \le R_1$) regions. In contrast to alpha- and cluster-decays for the proton decay amplitude the principal role is played by the shell model region [2], where the parent nucleus wave function $(\Psi_{\sigma_i}^{J_iM_i\pi_i})_0$ coincides with manybody generalized shell model wave function $(\Psi_{\sigma_i}^{J_iM_i\pi_i})_0^{sh}$, taking into account the collective modes of nuclear movement and normal and superfluidity correlations [12,13].

Let us investigate the proton transition from ground states of odd-even and odd-odd parent nuclei to states of ground rotational bands of daughter nuclei. Let us introduce the total orthogonal normalized basis of shell model single nucleon wave functions $f_k(\vec{r}',\vec{\sigma}')$ for the internal coordinate system of deformed nucleus, where multiindex k is presented in the form of $k^{\pi_k}[NN_z\Lambda]$. The function $f_k(\vec{r}',\vec{\sigma}')$ is solution of Shredinger equation, describing of single nucleon movement with energy ε_k in nonspherical shell potential $V^0(\vec{r}',\vec{\sigma}')$.

Representing manybody wave functions of deformed parent and daughter nuclei on the base of the generalised shell model [12] with superfluidity correlations [13] and using formula (1) and methods of article [4], the following expression for the partial proton width Γ_{ipc} can be received:

$$\Gamma_{ipc} = Z_{ipc}\, \Gamma_{ipc}^0,$$

(2)

where the effective single proton width Γ_{ipc}^0 is determined by formula (1) with

$$B_{ipc}^0 = \left\langle \frac{F_{l_p}(k_c r)\Phi_{j_p l_p k_{ip}}(\vec{r}',\vec{\sigma}')}{r} \middle| ((V_p^0(\vec{r}',\vec{\sigma}') - \frac{(Z-1)e^2}{r}) \middle| f_{k_{ip}} \right\rangle.$$

(3)

The proton spectroscopic factor Z_{ipc} is equal to $Z_{ipc} = (\frac{2\hat{J}_f}{\hat{J}_i})\,(C_{J_fJ_p0k_{ip}}^{J_ik_i} u_{k_i}^f)^2$ - for odd-even parent nucleus;

$Z_{ipc} = (\frac{\hat{J}_f}{\hat{J}_i})\,(C_{J_fJ_pk_fk_{ip}}^{J_ik_i} u_{k_{ip}}^f)^2$ - for odd-odd parent nucleus for two sets of possible values k_i, k_f: $(k_i)_1 = k_{ip} + k_{in}, (k_f)_1 = k_{in}$ and $(k_i)_2 = k_{ip} - k_{in}, (k_f)_2 = -k_{in}$. According to Gallagher-Moszkovski rule [14] among the states $(k_i)_1$, $(k_i)_2$ of parent nucleus the minimal energy is connected with the state, where the own spin Z-projections of odd proton and odd neutron have the same signs. This rule is not right only for state $(k_i)_{\frac{\pi_i}{2}} = 0^-$.

In articles [10,11] it was proposed in formula (2) to substitute the true single particle width $\Gamma^0_{j_p l_p}$ of shell model quasistationary state with wave function $f_{k_{ip}}(\vec{r}',\vec{\sigma}')$ and complex energy $\varepsilon_{k_{ip}}$, determinated by Gamov boundary conditions, for the width Γ^0_{ipc}. But the width $\Gamma^0_{j_p l_p}$ in common case differs from the width Γ^0_{ipc} as the proton separation energy $Q_c = E_i^0 - E_f^0$, connected with $F_{l_p}(k_c r)$ function in formula (3), differs from energy $\mathrm{Re}\,\varepsilon_{k_{ip}}$ because of manybody effects. At the same time the method of calculation of proton widths $\Gamma^0_{j_p l_p}$ [10-11] is very complex and can be essentially simplified without loss of accuracy, if instead of Gamov boundary conditions for wave function $f_{k_{ip}}(\vec{r}',\vec{\sigma}')$ the stationary bound conditions, proposed in articles [2,4], were used.

Table 1.

Nucleus	β_2	β_2^0	Nucleus	β_2	β_2^0
105 Sb	0	0	145 Tm	0,1-0,2	0,15-0,2
109 J	0,05-0,1	0,1	146 Tm	0,1-0,2	0,15-0,2
113 Cs	0,1-0,15	0,1-0,2	146m Tm	0,1-0,2	0,15-0,2
131 Eu	0,3-0,35	0,3	147 Tm	0,1-0,2	0,15-0,2
140 Ho	0,3-0,35	0,3	147m Tm	0,1-0,2	0,15-0,2
141 Ho	0,3-0,35	0,3	150 Lu	0,1-0,2	0,15-0,2
141m Ho	0,3-0,35	0,3	151 Lu	0,1-0,2	0,15-0,2

The comparison of calculated on the base of formula (2) proton half life times $T_p^{1/2}$ for deformed nuclei with the corresponding experimental half life times gives the possibility to find the nucleus deformation parameters β_2 and structure characteristics $k^{\pi k}[NN_z \Lambda]$ of odd proton states in the parent nucleus. The calculated [4-5,10-11,15-17] values of β_2 have a good correlation with the values β_2^0, predicted in systematic[18] (see Table 1). Let us remark that in the more late systematic [19] for nuclei ^{147}Tm, ^{151}Lu and ^{150}Lu the negative values of deformation parameters β_2^0 are predicted. It is interesting to repeat the calculations in the case of the oblate shapes of these nuclei.

The Fine Structure Of Proton Spectra.

In article [20] a possibility of observation of the fine structure of proton spectra was discussed. In the proton decay of parent nuclei can be populated not only ground states, but of excited states of daughter nuclei so that spectra of emitted protons can contain several groups of protons with different energies. For strongly deformed daughter nucleus ($\beta_2 \approx 0,3$) the collective structure of the ground rotational band with small excitation energies of the second exited states ($\Delta E_2 \approx 120$ Kev for A ≈ 140) can be the base for the first observation of this fine structure. Recently in Argonne National Laboratory [21] the second line was observed in proton decay spectrum of ^{131}Eu. The experimental value of the ratio α_2 of the partial width Γ_{ip2} of decay on the state $J_f^\pi = 2^+$ to the total width of proton decay is equal to $\alpha_2 = 0,14 \pm 004$. The calculated values of α_2 are equal to $\alpha_2 = 0,15$ и $0,08$ for the configurations of odd proton $3/2^+[411]$ and $5/2^+[413]$ correspondingly. The coincidence of experimental and theoretical values gives the possibility to select the odd proton configuration in ^{131}Eu as $3/2^+[411]$.

Let us investigate the situation with the fine structure of proton decay spectrum for deformed nucleus ^{141}Ho with the values of $\beta_2 \approx 0,3$. As we can see in the Table 2, for proton configuration $7/2^-[523]$ of ^{141}Ho the calculated value α_2 is equal to 0,1. But for configuration $5/2^-[532]$ the partial proton width Γ_{ip2} of decay to the state 2^+ of daughter nucleus is in 2,6 times more than the partial proton width Γ_{ip0} of decay to the ground state 0^+, so that the value of α_2 is equal to 0,73. In this case the total proton width $\Gamma_{ip} = \Gamma_{ip0} + \Gamma_{ip2}$ of decay of is more than the width Γ_{ip0} in 3,6 times. This result is not in discrepancy with experimental data taking into account that the uncertainty in the choice of parameters of shell model potential, giving the uncertainty in the calculated proton widths, estimated by factor 2 [2], and that the calculated proton half life time $T_{p0}^{1/2}$, connected with Γ_{ip0}, is equal to 8 ms and is two times more than experimental total proton half life of ^{141}Ho. It is interesting to continue of investigation of the fine structure proton spectra for ^{141}Ho and for odd-odd nucleus ^{140}Ho. It can be of greate value to investigate the fine structure for isotopes Tm and Lu, for which big negative values of the deformation parameters $\beta_2 \approx -(0,15-0,25)$ are anticipated.

Table 2.

P	$J_i = k_i$	J_f	j_p	l_p	$Z_{J_f j_p l_p}$	$\dfrac{\Gamma_{J_f j_p l_p}}{\Gamma_{0 j_p l_p}}$	α_2
$7/2^-$ [523]	$7/2$	0	$7/2$	3	0,125	1	
$7/2^-$ [523]	$7/2$	2	$7/2$	3	0,29	0,11	0,10
$5/2^-$ [532]	$5/2$	0	$5/2$	3	0,073	1	
$5/2^-$ [532]	$5/2$	2	$5/2$	3	0,13	0,08	0,73
$5/2^-$ [532]	$5/2$	2	$7/2$	3	0,13	2,54	

Proton Decay Of Odd-Odd Nucleus ^{140}Ho.

Using the formula (2) and Gallagher-Moszkovski rule [14], let us investigate the proton decay of odd-odd nuclei on the example of ^{140}Ho, for which the experimental proton half life and the emitted proton energy are equal to

(6 ± 3) ms and $E_p = (1086\pm 10)$ Kev [15,9]. Let us compare found characteristics with characteristics for odd-even nucleus ^{141}Ho, for which the experimental proton half life and the emitted proton energy are equal to $(3,9\pm 0,5)$ ms and $E_p = (1169\pm 8)$ Kev [15,9]. In Table 3 the theoretical and experimental values of the ratio δ of half life times for ^{140}Ho and ^{141}Ho are represented. For odd proton the cofigurations $7/2^-$ [523] and $5/2^-$ [532], fixed for the investigations of the proton decay of ^{141}Ho, are used. For odd neutron the cofigurations $9/2^-$ [514] and $5/2^+$ [402], which lie near the neutron Fermi surface [9] for the values of deformation parameter $\beta_2 \approx 0,3$ and also fixed for the investigations of the proton decay of ^{141}Ho, are used. The interval of theoretical values δ^{th} coincides with interval of experimental values δ^{exp} for proton configuration $7/2^-$ [523] and two used neutron configurations only near borders of these intervals. At the same time the intervals of δth and δ^{exp} completely coincide for the proton configuration $5/2^-$ [532], which was chosen above for the investigation of the fine structure of decay proton spectrum for ^{141}Ho, and for two used neutron configurations.

It is important to continue the experimental and theoretical investigations of proton widths for deformed odd-odd nuclei.

Table 3.

P	N	$J_i = k_i$	$J_f = k_{ni}$	j_p, l_p	δ^{th}	δ^{exp}
$7/2^-$ [523]	$9/2^-$ [514]	8^+	$9/2$	$7/2, 3$	2,26-5,53	0,75-2,25
$7/2^-$ [523]	$5/2^+$ [402]	6^-	$5/2$	$7/2, 3$	2,9-7,1	0,75-2,25
$5/2^-$ [532]	$9/2^-$ [514]	7^+	$9/2$	$7/2, 3$	0,76-1,88	0,75-2,25
$5/2^-$ [532]	$5/2^+$ [402]	5^-	$5/2$	$7/2, 3$	0,92-2,27	0,75-2,25

CONCLUSION

The principal result of this article is demonstration of the high possibility of MTPR for description of different qualities of proton radioactivity phenomenon.

ACKNOWLEDGEMENTS

This work is supported by the Russian Foundation of Fundamental Investigations. The author acknowledges Profs. J.C. Batchelder, C.N. Davids and Doct. A. A. Sonzogni.

REFERENCES

1. Woods P.J., Davids C.N.// Annu. Rev. Nucl. Part. Sci.1997.V.47. P.541.

2. Bugrov V.P., Kadmensky S.G. et.al.// Rus.Nucl. Phys.1985.V.41.P.1123.

3. Bugrov V.P., Kadmensky S.G. et.al.// Rus.Nucl. Phys.1985. V.42.P.57.
4. Bugrov V.P., Kadmensky S.G.// Rus.Nucl. Phys.1989.V.49.P.1562.
5. Kadmensky S.G., Bugrov V.P.// Rus.Nucl. Phys.1996.V.59.P.424.
6. Aberg S.,Semmes P.B., Nazarevicz W.// Phys.Rev. 1997.V.56B. P.1762.
7. Mang H.J., Rasmussen J.O. // Math. -Fys. Medd. Dan. Vid. Selsk. 1962. V.2.№3.
8. Davids C.N. et al. // Phys.Rev.Lett.1998.V.80.1849.
9. Batchelder J.C. et al.// Phys.Rev.1998.V.C57.P.1042.
10. Maglione E., Liotta R.J. ,Vertse T. // Nucl. Phys. 1995.V.A584.P.13.
11. Maglione E., Ferreira L.S., Liotta R.J. // Phys. Rev. Lett. 1998.V.81.P.538; Phys. Rev. 1999.V.C59.P.589.
12. Bohr A., Mottelson B.Nuclear Structure. New-York: W.A.Benjamin, V.1,2 ,1969,1974.
13. Soloviev V.G. The theory of atomic nucleus. Nuclear models.Moscow: Energoatomizdat.1981.
14. Gallagher G.J., Moszkovski S.A. // Phys.Rev.1958.V.111.P.1282.
15. Davids C.N. et al. // Phys.Rev.Lett.1998.V.80.1849.
16. Batchelder J.C., Bingham C.R. et al.// N.York.: Woodbury. ENAM-98.1998.P 264.
17. Rykaczevski K., Batchelder J.C. et al. // Preprint Oak-Ridge Nation.Lab. Phys. and Astron.16 February 1999.
18. Liran S., Zeldes N. // At. Data Nucl.Data Tables. 1976. V.17.P.1.

19. Moller P. et al. // Atom. Data Nucl. Data Tables.1995.V.59.P.185.

20. Kadmensky S.G.// N.York.: Woodbury. ENAM-98.1998.P. 672

21. Sonzogni A.A., Davids C.N. et al. Private communication,1999.

Asymptotic Behavior of the Wave Packet Propagation through a Barrier : the Green's Function Approach Revisited

Bogdan Mihaila[a,b], Shmuel A. Gurvitz[c,d], David Dean[a,e], Witold Nazarewicz[a,e,f]

[a] Physics Division, Oak Ridge National Laboratory, P.O. Box 2008, Oak Ridge, TN 37831
[b] Chemistry and Physics Department, Coastal Carolina University, Conway, SC 29528-6054
[c] Department of Physics, The Weizmann Institute of Science, Rehovot 76100, Israel
[d] Joint Institute for Heavy Ion Research, Oak Ridge, Tennessee 37831
[e] Department of Physics, University of Tennessee, Knoxville, TN 37996
[f] Institute of Theoretical Physics, Warsaw University, Hoża 69, PL-00681, Warsaw, Poland

Abstract. To model the decay of a quasibound state we use the modified two-potential approach introduced by Gurvitz and Kalbermann [1,2]. This method has proved itself useful in the past for calculating the decay width and the energy shift of an isolated quasistationary state [5]. We follow the same approach in order to propagate the wave-packet in time with the ultimate goal of extracting the momentum-distribution of emitted particles. The advantage of the method is that it provides the time-dependent wave function in a simple semi-analytic form. We intend to apply this method to the modeling of metastable states for which no direct integration of the time-dependent Schrödinger equation is available today.

The Two Potential Approximation (TPA) introduced in Refs. [1,2] turned out to be an extremely successful tool for the description of a metastable state. Simple expressions based on the TPA made it possible to obtain a very precise estimate of the life-time of a very narrow resonance without the need of introducing an explicit time dependence. In this work, we use the TPA in order to derive the equations describing the time evolution of the wave function of a particle tunneling through a spherically-symmetric barrier.

Let us consider a particle moving in a central potential $V(r)$ with a barrier. Asymptotically, i.e., at large values of r, we assume that $V(r) \to 0$. In the TPA, $V(r)$ can be decomposed as

$$V(r) = U(r) + W(r), \qquad (1)$$

where

$$U(r) = \begin{cases} V(r) & \text{if } r < R \\ V(R) & \text{if } r > R \end{cases} \quad (2)$$

is an auxiliary potential that produces a bound state at energy E_0 close to the energy of the metastable state, and $W(r)$ is a "closing" potential which is treated perturbatively. The separation radius R should be chosen far from the classical turning points [3].

At $t=0$ the initial state is taken to be the bound eigenstate $\Phi_0(\vec{r})$ of the auxiliary Hamiltonian

$$H_0 = T + U(r) \quad (3)$$

(we take $\hbar=1$). In the following we assume that $\Phi_0(\vec{r})$ is well isolated, i.e., it is well separated from the remaining bound states of $U(r)$ having the same quantum numbers. In such a case, at $t>0$, the wave packet represented by the wave function $\Psi(\vec{r},t)$ can be expanded in the basis $\{\Phi_0(\vec{r}), \Phi_k(\vec{r})\}$:

$$\Psi(\vec{r},t) = b_0(t)\Phi_0(\vec{r})e^{-iE_0 t} + \int \frac{d^3k}{(2\pi)^3} b_k(t)\Phi_k(\vec{r})e^{-iE_k t}, \quad (4)$$

with the initial conditions $b_0(t=0)=1$ and $b_k(t=0)=0$. In Eq. (4) the wave functions $\Phi_k(\vec{r})$ represent the continuum and $E_k = V(R) + k^2/2m$. We shall refer to the first and second terms above, as $\Psi_I(\vec{r},t)$ and $\Psi_{II}(\vec{r},t)$, respectively.

To evaluate the two components, $\Psi_I(\vec{r},t)$ and $\Psi_{II}(\vec{r},t)$, the Laplace transform method can be applied. In terms of the Laplace-transformed expansion coefficients $b(t)$,

$$\tilde{b}(\varepsilon) = \int_0^\infty b(t)e^{i\varepsilon t}\, dt, \quad (5)$$

the Laplace transform of the wave packet $\Psi(\vec{r},t)$ can be written as

$$\tilde{\Psi}_I(\vec{r},\varepsilon + E_0) = \tilde{b}_0(\varepsilon)\Phi_0(\vec{r}), \quad (6)$$

$$\tilde{\Psi}_{II}(\vec{r},\varepsilon + E_0) = \int \frac{d^3k}{(2\pi)^3} \tilde{\bar{b}}_k(\varepsilon_k)\Phi_k(\vec{r}), \quad (7)$$

where $\bar{b}_k(t) = e^{-iV(R)t}b_k(t)$ and $\varepsilon_k = \varepsilon + E_0 + V(R) - E_k$.

Assuming a spherically symmetric potential $V(r)$, the coefficient $\tilde{b}_0(\varepsilon)$ has been calculated as [2]

$$\tilde{b}_0(\varepsilon) = \frac{i}{\varepsilon - \varepsilon_0}, \quad (8)$$

with

$$\varepsilon_0 = \Delta - i\frac{\Gamma}{2} = -\sqrt{\frac{\pi}{2}}\frac{|\phi_0(R)|^2}{2mk_0}\left[\alpha\chi_{lk_0}(R) + \chi'_{lk_0}(R)\right]\left[\alpha\chi^{(+)}_{lk_0}(R) + \chi^{(+)'}_{lk_0}(R)\right], \quad (9)$$

where $\alpha = \sqrt{2m(V_0 - E_0)}$ and $k_0 = \sqrt{2m(E_0 + \varepsilon_0)}$. In this work, $\phi_0(r)$ is the radial wave function of Ψ_0 and $\chi_{lk}(r)$ and $\chi_{lk}^{(+)}(r)$ are, respectively, the regular and outgoing waves of the Hamiltonian with the potential $\tilde{W}(r) = W(r) + V(R)$. (Note, that our radial continuum functions satisfy the orthogonality and completeness relationships

$$\int_0^\infty \chi_{lk}^*(r)\,\chi_{lk'}(r)\,dr = \delta(k - k'), \tag{10}$$

$$\int_0^\infty \chi_{lk}^*(r)\,\chi_{lk}(r')\,dk = \delta(r - r'). \tag{11}$$

Compared with expressions in Refs. [1,2], this results in an additional factor of $\sqrt{\pi/2}$ in the front of every χ [4].)

With the above definitions, the radial part of the first component in Eq. (4) is

$$\psi_I(r,t) = \frac{\phi_0(r)}{r}\,e^{-i(E_0+\varepsilon_0)t}. \tag{12}$$

The coefficients $\tilde{b}_k(\varepsilon_k)$ are determined by solving the system of integral equations

$$\varepsilon \tilde{b}_0(\varepsilon) = i + W_{00}\tilde{b}_0(\varepsilon) + \int \frac{d^3k}{(2\pi)^3}\,\tilde{W}_{0k}\tilde{b}_k(\varepsilon_k), \tag{13}$$

$$\varepsilon_k \tilde{b}_k(\varepsilon_k) = W_{k0}\tilde{b}_0(\varepsilon) + \int \frac{d^3k'}{(2\pi)^3}\,\tilde{W}_{kk'}\tilde{b}_{k'}(\varepsilon_{k'}),$$

with $\tilde{W}_{kk'} \equiv \langle \Phi_k | \tilde{W} | \Phi_{k'} \rangle$. The solution of (13) can be formally written as

$$\tilde{b}_k(\varepsilon_k) = \frac{1}{\varepsilon_k}\,\langle \Phi_k | \left(1 + \tilde{W}\tilde{G}_0 + \tilde{W}\tilde{G}_0\tilde{W}\tilde{G}_0 + \cdots \right) W | \Phi_0 \rangle\,\tilde{b}_0(\varepsilon),$$

where

$$\tilde{G}_0 = \int \frac{d^3k}{(2\pi)^3}\,\frac{|\Phi_k\rangle\langle\Phi_k|}{\varepsilon_k}. \tag{14}$$

The outgoing part of the wave function, $\tilde{\Psi}_{II}(\vec{r},\varepsilon)$, can now be expressed in terms of $\tilde{\Psi}_I(\vec{r},\varepsilon)$ as

$$\tilde{\Psi}_{II}(\vec{r},\varepsilon + E_0) = \int d^3r'\,\tilde{G}(\varepsilon + E_0; \vec{r}, \vec{r}')\,W(r')\,\tilde{\Psi}_I(\vec{r},\varepsilon + E_0), \tag{15}$$

where we now introduce the Green's function

$$\tilde{G}(E) = \tilde{G}_0(E) + \tilde{G}_0(E)\,\tilde{W}\,\tilde{G}(E) = (1 - \Lambda)(E - H + \Lambda\tilde{W})^{-1}, \tag{16}$$

with $\Lambda = |\Phi_0\rangle\langle\Phi_0|$ being the projection operator on Φ_0. The Green's function $\tilde{G}(E)$ is approximated in the spirit of Ref. [2] by neglecting the contribution from

Λ, and then by replacing the potential $V(r)$ by $\tilde{W}(r)$. This gives $\tilde{G}(E) \approx G_{\tilde{W}}(E)$, where

$$G_{\tilde{W}}(E) = (E - H_{\tilde{W}})^{-1} \qquad (17)$$

is the Green's function of $H_{\tilde{W}} = T + \tilde{W}$.

By taking the inverse Laplace transform of (7), one obtains for the the radial wave function

$$\psi_{II}(r,t) = \frac{1}{2\pi} \int_R^\infty r' dr' \, W(r') \phi_0(r') \int_{i\gamma-\infty}^{i\gamma+\infty} d\varepsilon \, e^{-i\varepsilon t} G_{\tilde{W}}(\varepsilon;r,r') \tilde{b}_0(\varepsilon - E_0) \,. \qquad (18)$$

The ε-integral is evaluated using the residue theorem, and results in the sum of the residues corresponding to the two poles of the integrand.

Contribution due to the pole of $\tilde{b}_0(\varepsilon - E_0)$

Using the standard techniques explained in Ref. [2], we obtain

$$\psi_{II,a}(r < R, t) = \sqrt{\frac{\pi}{2}} \, \frac{\phi_0(R)}{k_0 r} \, [\alpha \chi_{lk_0}^{(+)}(R) + \chi_{lk_0}^{(+)\prime}(R)] \, \chi_{lk_0}(r) \, e^{-i(E_0+\varepsilon_0)t} \,, \qquad (19)$$

and

$$\psi_{II,a}(r > R, t) = -\frac{\phi_0(r)}{r} e^{-i(E_0+\varepsilon_0)t}$$
$$+ \sqrt{\frac{\pi}{2}} \frac{\phi_0(R)}{k_0 r} [\alpha \chi_{lk_0}(R) + \chi'_{lk_0}(R)] \chi_{lk_0}^{(+)}(r) e^{-i(E_0+\varepsilon_0)t} \,. \qquad (20)$$

Note that for $r > R$ the contribution from ψ_I is exactly canceled by the first term in (20).

Contribution due to the pole of the Green's function.

The Green's function $G_{\tilde{W}}$ has a continuum of simple poles along the real $E > 0$ axis. After using the spectral representation of $G_{\tilde{W}}(\varepsilon;r,r')$, one can express $\psi_{II,b}(r,t)$ as

$$\psi_{II,b}(r,t) = \frac{2m}{r} \int_R^\infty dr' \, W(r') \, \phi_0(r') \int_0^\infty dk \, \frac{e^{-i\frac{k^2}{2m}t}}{k^2 - k_0^2} \, \chi_{lk}(r) \chi_{lk}^*(r') \,. \qquad (21)$$

The evaluation of the integral (21) represents the corner stone of the present approach.

For now we will restrict ourselves to making some remarks regarding the asymptotic behavior of this integral at large values of r. We shall also assume that the potential $V(r)$ has finite range (i.e., it vanishes at large values of r). While this assumption cannot be used for the case of the Coulomb potential, it is still interesting to investigate the general structure of the solution for the short-range potential. In this limit, the S-matrix is meromorphic in the complete complex k-plane, and [4,6]

$$[S_l(k)]^* = S_l(-k^*), \quad \text{and} \quad S_l(k) = [S_l(-k)]^{-1}. \tag{22}$$

Expressing $\chi_{lk}(r)$ in the asymptotic form:

$$\chi_{lk}(r) = \sqrt{\frac{2}{\pi}} \frac{1}{2i} \left(S_l^{1/2}(k) e^{ikr} - (-)^l S_l^{-1/2}(k) e^{-ikr} \right), \tag{23}$$

one obtains for the k-integral in (21):

$$I(r, r', t) = \int_0^\infty dk \, \frac{e^{-i\frac{k^2}{2m}t}}{k^2 - k_0^2} \chi_{lk}(r) \chi_{lk}^*(r') \tag{24}$$

$$\asymp \frac{1}{2\pi} \int_{-\infty}^\infty dk \, \frac{e^{-i\frac{k^2}{2m}t}}{k^2 - k_0^2} \left[e^{ik(r-r')} - (-)^l S_l(k) e^{ik(r+r')} \right]. \tag{25}$$

The integrand in Eq. (25) is a sum involving two complex functions of complex k,

$$\frac{1}{k^2 - k_0^2}, \quad \text{and} \quad \frac{S_l(k)}{k^2 - k_0^2}, \tag{26}$$

which have common poles at $\pm k_0$. In addition, $S_l(k)$ has an infinite number of simple poles. They are located in the lower half of the complex k-plane, symmetrically with respect to the imaginary axis. Following the notation of van Dijk and Nogami [7], we shall denote the poles in the fourth quadrant with k_ν, $\nu = 1, 2, 3, \ldots$, and the poles in the third quadrant with k_ν, $\nu = -1, -2, -3, \ldots$ It follows from Eq. (22) that

$$\text{Re}(k_\nu) = -\text{Re}(k_{-\nu}), \quad \text{Im}(k_\nu) = \text{Im}(k_{-\nu}). \tag{27}$$

In the following, the residue of the $S_l(k)$ at the pole k_ν is denoted by b_ν.

Since the complex function (26) has no essential singularity at infinity, we can apply the Mittag-Leffler theorem in order to obtain a pole expansion for (26). Consequently, Eq. (26) can be replaced by

$$\frac{1}{2k_0} \left(\frac{1}{k - k_0} - \frac{1}{k + k_0} \right), \quad \text{and} \tag{28}$$

$$\frac{S_l(k_0)}{2k_0} \frac{1}{k - k_0} - \frac{S_l(-k_0)}{2k_0} \frac{1}{k + k_0} + \sum_{\nu = -\infty}^\infty \frac{b_\nu}{k_\nu^2 - k_0^2} \frac{1}{k - k_\nu}. \tag{29}$$

By substituting Eqs. (28) and (29) in (25), the integral (25) becomes

$$I(r,r',t) = \int_{-\infty}^{\infty} dk \, \frac{e^{-i\frac{k^2}{2m}t} e^{ik(r-r')}}{4\pi k_0} \left(\frac{1}{k-k_0} - \frac{1}{k+k_0} \right) \tag{30}$$

$$-\frac{(-)^l}{2\pi} \left[\int_{-\infty}^{\infty} dk \, \frac{e^{-i\frac{k^2}{2m}t} e^{ik(r+r')}}{2k_0} \left(\frac{S_l(k_0)}{k-k_0} - \frac{S_l(-k_0)}{k+k_0} \right) + \sum_{\nu=-\infty}^{\infty} \frac{b_\nu}{k_\nu^2 - k_0^2} \int_{-\infty}^{\infty} dk \frac{e^{-i\frac{k^2}{2m}t} e^{ik(r+r')}}{k-k_\nu} \right].$$

The above can be now expressed in terms of the Moshinsky function

$$M(k, \mathcal{R}, \tau) = \frac{i}{2\pi} \int_{-\infty}^{\infty} dp \, \frac{e^{-ip^2\tau} e^{-ip\mathcal{R}}}{p-k}$$
$$= \frac{1}{2} e^{-ik^2\tau} e^{-ik\mathcal{R}} \, \text{erfc}(y), \tag{31}$$

where

$$y = e^{-i\pi/4} \sqrt{\tau} \left(\frac{\mathcal{R}}{2\tau} - k \right),$$

where $\tau = t/2m$ and $\mathcal{R} = r \pm r'$. The integral $I(r,r',t)$ can now be calculated in the closed form:

$$\frac{1}{2i\,k_0} \left[M\left(k_0, r-r', \frac{t}{2m}\right) + M\left(k_0, r'-r, \frac{t}{2m}\right) - (-)^l S_l(k_0) M\left(k_0, r+r', \frac{t}{2m}\right) \right. \tag{32}$$
$$\left. - (-)^l S_l(k_0) M\left(k_0, -r-r', \frac{t}{2m}\right) \right] + i(-)^l \sum_{\nu=-\infty}^{\infty} \frac{b_\nu}{k_\nu^2 - k_0^2} M\left(k_\nu, r+r', \frac{t}{2m}\right).$$

This concludes our derivation.

ACKNOWLEDGMENTS

This work was supported by the U.S. Department of Energy under Contract Nos. DE-FG02-96ER40963 (University of Tennessee), DE-FG05-87ER40361 (Joint Institute for Heavy Ion Research), and DE-AC05-96OR22464 with Lockheed Martin Energy Research Corp. (Oak Ridge National Laboratory).

REFERENCES

1. Gurvitz, S.A., and Kalbermann, G., *Phys. Rev. Lett.* **59**, 262 (1987).
2. Gurvitz, S.A., *Phys. Rev. A* **38**, 1747 (1988).
3. Gurvitz, S.A., Nazarewicz, W., and Semmes, P.B., in preparation.
4. Baz, A.I., Zeldovich, B., and Perelomov, A.M., *Scattering Reactions and Decay in Nonrelativistic Quantum Mechanics*, Israel Program for Scientific Translations, Jerusalem, 1969.
5. Åberg, S., Semmes, P.B., and Nazarewicz, W., *Phys. Rev. C* **56**, 1762 (1997).
6. Newton, R.G., *Scattering Theory of Waves and Particles*, McGraw-Hill, New York 1966.
7. W. van Dijk and Y. Nogami, Phys. Rev. Lett. **83**, 2867 (1999).

Proton-emitting nuclei in a time dependent formalism

N. Carjan[1], P. Talou[2], M. Rizea[3] and D. Strottman[2]

[1]*Centre d'Etudes Nucléaires de Bordeaux Gradignan, 33175 Gradignan, Fance*
[2]*Theoretical Division, Los Alamos National Laboratory, Los Alamos, NM 87545, USA*
[3]*Institute of Physics and Nuclear Engineering, P.O. Box MG-6, Bucharest, Romania*

Abstract. The time-dependent approach to calculate half-lives for proton-emitting nuclei is reviewed. Doubts about the practicability of this new method are eliminated by showing that it directly leads to the main observable (the average decay rate), that the result does not depend on the prescription used to prepare the initial state or to define the nuclear border and that the method can be applied to low-lying as well as to highly excited states.

The most intuitive and direct way to study proton emission from unbound nuclear states is to start with a proton wave packet localized inside the well of the p-nucleus interacting potential V (**r**) and follow its time evolution by solving numerically the corresponding time-dependent Schrodinger equation (TDSE) :

$$i\hbar\frac{\delta\Psi(\mathbf{r},t)}{\delta t} = \left[-\frac{\hbar^2}{2\mu}\frac{d^2}{d\mathbf{r}^2} + V(\mathbf{r})\right]\Psi(\mathbf{r},t) \quad (1)$$

Then, at any time t, all physical properties of the decaying system can be inferred from the solution $\psi(\mathbf{r},t)$.

Although the suitability of this method was obvious from the very beginning, it became practical only during the last decade [1,2] This was due not only to tremendous progress in computer performances but also to the elimination of three, apparently fundamental, difficulties :

1) There is no simple way the determine the decay rate λ from $\psi(\mathbf{r},t)$.
2) The observables will depend on the choice of the initial quasi-stationary state and there is no unique prescription to prepare it.
3) Due to the huge value of the life-time (~1s) as compared to the nuclear time scale (~10^{-23} s) one needs to calculate a prohibitive number of time steps.

In the following these problems will be tackled and solved :
1) Using $\psi(\mathbf{r},t)$ one can calculate the probability that the proton has tunneled at time t :

$$\rho_\Psi(t,\mathbf{r}_B) = \int |\Psi(\mathbf{r},t)|^2 dV \qquad (2)$$

where r_B defines the border between the inside and the outside of the nucleus. By analogy with the exponential decay law that gives $\rho(t)=1-\exp(-\lambda_0 t)$ one can derive a formula that relates the wave function $\psi(r,t)$, with the decay rate λ that will now depend on time :

$$\lambda_\Psi(t,\mathbf{r}_B) = \frac{1}{1-\rho_\Psi} \frac{d\rho_\Psi}{dt} \qquad (3)$$

2) The TDSE is indeed deterministic : each initial state $\psi(\mathbf{r},0)$, leads to a different time evolution $\psi(\mathbf{r},t)$. We are however interested only in initial quasi-stationary states : a narrow class of states defined by any infinitesimal modification of a given stationary state. Although the dynamical evolution varies in details from one modification to another, all quasi-stationary states lead to the same average asymptotic value λ (∞) of the decay rate [2,3] which is the only characteristic of $\lambda(t)$ experimentally observed so far.

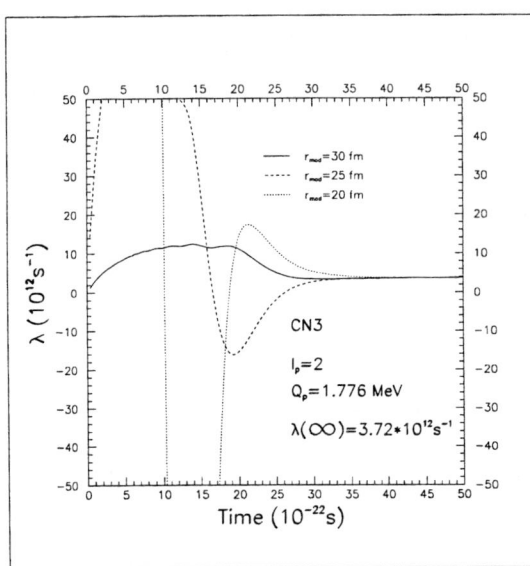

Fig. 1 Time dependent decay rates calculated for an excited state in ^{109}I and 3 initial wave functions defined by different values of the parameter r_{mod}.

3) In fact the time necessary for λ(t) to attain its stationary value (the transient time) increases only very little with the life-time of the state involved [3,4]. Therefore the calculation of proton-decay half-lives using TDSE requires comparable CPU time for long-lived low-lying states as for short-lived excited states. This is also the reason for which it was not possible to measure in nuclear physics the deviations from the exponential decay law during the transient time [5] predicted by quantum mechanics [6].

To illustrate these properties new calculations have been performed for the decay of the metastable p-state $2d_{(5/2)}$ in spherical ^{109}I using the Crank-Nicholson (CN) method to integrate TDSE. The time step and the grid mesh were taken 4.10^{-24}s and 0.125fm respectively. As compared with MSD2 (the previously used method), CN only needs the initial wave function at one starting value of time and it has the advantage of being unconditionally stable. On the other hand it contains a matrix inversion that takes more memory and CPU time.

As usual, the interaction potential V(r) was taken as a sum of Coulomb, nuclear and spin-orbit terms with parameters chosen as in Ref. [3].

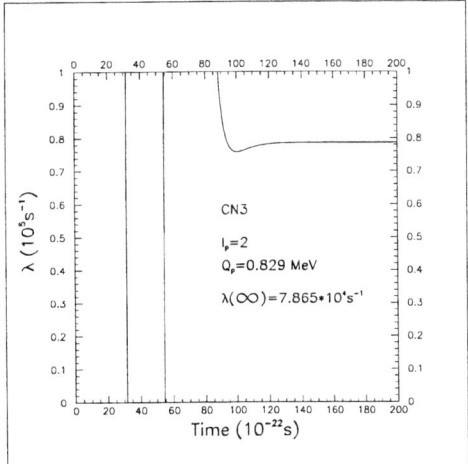 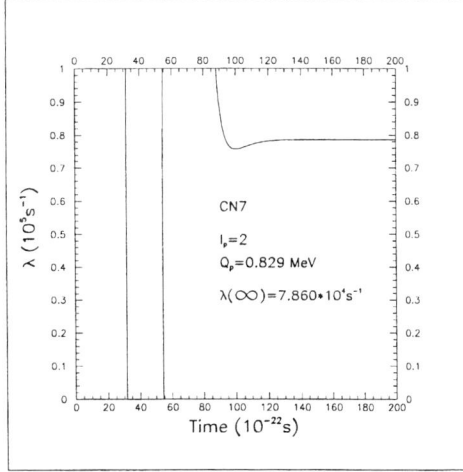

Fig 2. Ground-state decay rate of ^{109}I calculated by numerical integration of TDSE using CN3. A value of r_{mod} = 50 fm was chosen in this example.

Fig. 3. The same as in Fig. 2 but using CN7

The time evolution of the decay rate for 3 different initial quasi-stationary states is represented in Fig. 1. They were chosen as eigenstates of hamiltonians obtained from the original hamiltonian through the replacement of V(r), for r>r_{mod}, by V(r_{mod}).

The 3 states correspond to r_{mod} = 20, 25 and 30 fm. One can notice that, after displaying very different behaviours during the transient time, the 3 curves reach the same asymptotic value λ (∞). Since in this calculation r_B was taken equal with r_{mod}, Fig. 1 also shows the insensitivity of λ (∞) to the choice of the nuclear border in Eq. (2).

To calculate the proton decay from the ground state of ^{109}I, the depth of the nuclear potential, V_0, was slightly varied to obtain a state with an average energy equal to the experimental Q_p-value of 0.892 MeV. The resulting decay rate is plotted in Fig. 2.

The ground-state half-life calculated using TDSE is therefore $t_{1/2} = \ln2/(7.86 \ 10^4 \ s^{-1})$ =8.82µs to be compared with the DWBA estimate of 10µs [7]. To calculate this half-life it was enough to iterate up to $t_{max} = 2.10^{-20}$s, i.e. only 4 limes longer than for the excited state from Fig. 1 that has $t_{1/2} = 1.87.10^{-13}$s. Hence the value of the half-life and the number of time steps necessary to calculate it are not strongly correlated.

So far, in the CN method, the second derivative of the wave function was calculated using 3 grid points (CN3). That this number is enough is shown in Fig. 3 where an identical result was obtain with 7 points (CN7). This feature is encouraging for future use of this method to solve TDSE since extremely fast routines for inversion of tri-diagonal matrices exist.

In conclusion we have demonstrated that our time-dependent approach to calculate half-lives for proton-emitting nuclei is both feasible and reliable. This makes TDSE a powerful tool to study nuclei beyond the proton drip-line.

REFERENCES

1. Serot, O., Ph. D. Thesis, University of Bordeaux (1992)
2. Serot, O., Carjan, N. and Strottman, D. , Nucl. Phys. **A569**, 562-574 (1994)
3. Talou, P., Strottman, D. and Carjan, N., Phys. Rev. **C60**, 054318-1-05318-7 (1999)
4. Talou, P., Carjan, N., Negrevergne, C. and Strottman, D., submitted to Phys. Rev. **C**
5. Norman, E.B., Phys. Rev. Lett. **60**, 2246-2249 (1988) ; Phys. Lett. **B357**, 521-525 (1995)
6. Perez, A., Annals of Physics **129**, 33-46 (1980)
7. Aberg, S., Semmes, P.B. and Nazarewicz, W., Phys. Rev. **C56**, 1762-1772 (1997)

Beta Delayed Proton Emission and High Spin States

Beta-Delayed Proton Emission

J.C. Hardy

Cyclotron Institute, Texas A & M University
College Station, TX 77843

Abstract. Beta-delayed proton emission was first observed thirty-six years ago. At the time, its unique signature opened a window onto a new world of exotic nuclei: on-line isotope separation was not yet available to isolate newly produced nuclei, but a readily separable decay mode was a fine alternative. In the first two decades after its discovery, beta-delayed proton decay was a rich source of spectroscopic information on nuclei far from stability, yielding everything from analog states and atomic masses to level densities and excited-state lifetimes in the femtosecond region. However, with the advent of more sophisticated methods for the production and isolation of exotic nuclei, the importance of beta-delayed protons as an identifier of new nuclei effectively disappeared. Nevertheless, the potential spectroscopic value of the process remains undiminished. In fact, as the number of known beta-decaying nuclei steadily increases, beta-delayed proton emission is seen to be a widely occurring phenomenon. Its usefulness needs to be rediscovered. This paper presents an overview of beta-delayed proton emission, its occurrence, its usefulness and its potential for the future.

INTRODUCTION

As I look around the audience at this conference, I realize that I go back farther in this field than anyone else here. In fact, I think I go back farther than anyone else anywhere who is still active in nuclear physics. I find this to be a very sobering thought to contemplate. The only advantage that I can think of to my status as graybeard is that it should give me the right to indulge in a bit of history. I will begin my talk, therefore, with a brief history of the first observation of proton-emitting nuclei — the β-delayed proton precursors. I will go on to demonstrate the rapid growth in the number of known precursors and their applications, and describe the current status of the field with illustrations taken from very recent experiments. I will conclude by looking at possible future directions for study.

HISTORICAL DEVELOPMENT

The more neutron-deficient a nuclide is, the higher the energy of its β^+-decay. If a nuclide's energy is high enough that its decay populates states unbound to proton emission, then its radioactivity is characterized by the presence of energetic protons with the same half-life as that of its β^+-decay. This phenomenon, referred to as β-delayed proton emission, is illustrated in Fig. 1.

In 1959 and 1960, Baz, Zeldovich and Goldansky (see, for example ref. [1], and

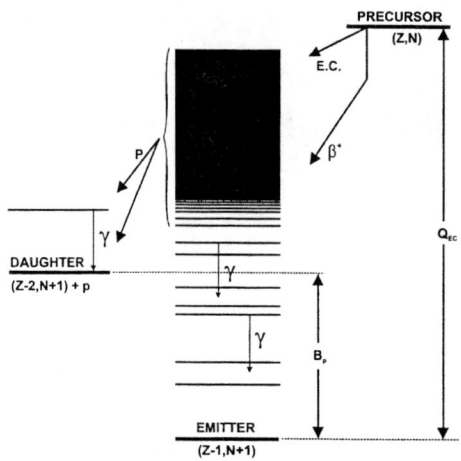

FIGURE 1. Decay scheme for a typical β-delayed proton precursor, illustrating the energetics and defining the terms used.

references therein) began using isobaric symmetry relations to calculate the masses of as-yet-unknown neutron-deficient light nuclei. Their results predicted the likely existence of nuclides that decay by β-delayed proton emission, as well as some that might be candidates for one- and two-proton radioactivity.

Following a visit to the Soviet Union, in which he discussed these calculations with Goldansky, R.E. Bell of McGill University, Canada, initiated a program to search for candidate β-delayed proton precursors using the 97-MeV proton beam from the McGill synchrocyclotron. The graduate student assigned the task, R. Barton, had remarkable and rapid success, reporting in 1963 the first positive identification of a β-delayed proton precursor, ^{25}Si, and the tentative identification of several others nearby [2]. Meanwhile, Karnaukhov et al [3] in Dubna had also seen evidence for delayed protons following ^{20}Ne bombardment of Ni but had, as yet, been unable to identify their source. Both experimental groups used what were then called "silicon junction" detectors, one of the first applications of this type of detector in nuclear physics.

I was a very junior graduate student at McGill while all this was going on but, with Barton's experiment completed, it became my assignment to work with R. McPherson and later R.I. Verrall, other graduate students, to improve the experimental efficiency so that we could positively identify more precursors and begin to use delayed protons as a real spectroscopic tool. This we did by mounting our target and detector on a probe that was inserted into the cyclotron, where we could make use of several orbits of the more intense internal proton beam. With a factor of three better resolution and ten times the count rate, by 1965 we had positively identified six more precursors in the same $T_z = -3/2$ series using (p,3n) and (p,p3n) reactions, and had tied down their decay schemes, including in most cases the first observation of the lowest analog state in the emitter, which is populated via superallowed β-decay from the precursor.

Soon, other groups became involved in addition to those at McGill and Dubna — first Poskanzer's at Brookhaven, later Cerny's and also Nitzsche's at LBL, Hansen's at ISOLDE in CERN, and my own at Chalk River. The number of known precursors, which was 11 in 1967, became 42 by 1977, and 97 by 1987. Today, with β-delayed proton emission a routine feature of the observed radioactivity from exotic nuclei, there are 134 identified precursors (Fig. 2). The useful applications of the process have grown apace.

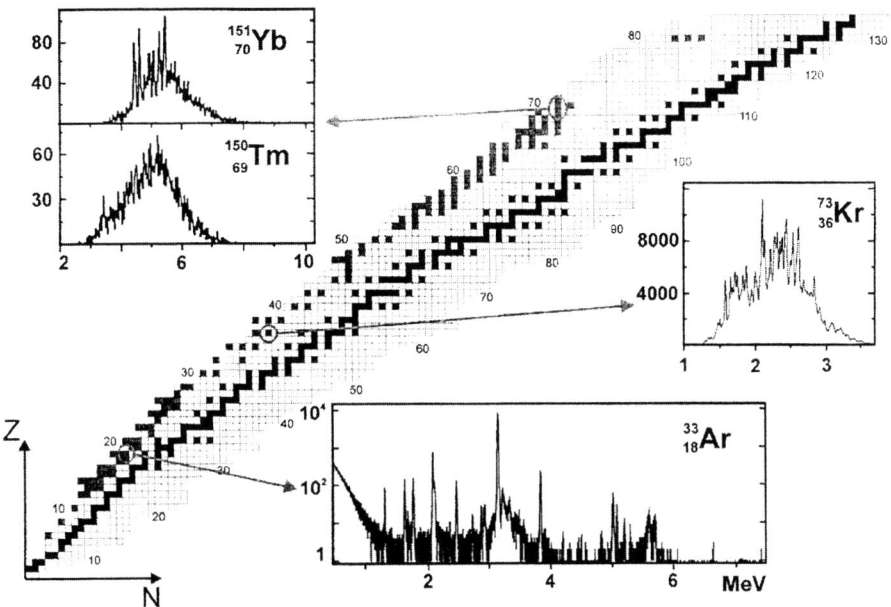

FIGURE 2. This chart of the nuclides shows the stable nuclei (in black) and the known β-delayed proton precursors (in gray, on the neutron-deficient side of stability). The four experimental delayed-proton spectra are identified by precursor and by an arrow from the precursor's location on the chart; these spectra are all of comparable experimental resolution and illustrate how the proton spectra differ qualitatively from one mass region to another. The spectra are taken from references [4] (for ^{33}Ar), [5] (^{73}Kr) and [6] (^{150}Tm and ^{151}Yb).

OCCURRENCE AND PROPERTIES

Beta-delayed proton precursors are observed in regions of the nuclear chart where Q_{EC}-values exceed the corresponding proton separation energies, B_p, by an amount that exceeds the energy of the Coulomb barrier for protons (see Fig. 1). Thus, for each element below about $Z = 70$, there is (or is predicted to be) a band of neutron-deficient isotopes that are β-delayed proton precursors. Each band is bounded on the right by pure β-emitters and on the left by the proton drip line, the latter sometimes being characterized by proton-radioactivity. In addition to the known precursors shown in Fig. 2, reliable mass predictions leave little doubt that at least 75 more neutron-deficient nuclei will be found to exhibit delayed proton emission as a component of their decay.

An examination of Fig. 1 shows that the energies of proton groups emitted in delayed-proton decay are determined by the energies of levels in the emitter and, in some cases, by those in the daughter. The intensity of each group, however, depends on the product of two factors: the intensity with which a state in the emitter is populated by β-decay from the precursor, and the probability that the emitter state decays by proton, rather than gamma, emission. If we neglect the perturbations introduced by the details of nuclear structure and consider only the general trends, the first factor decreases with increasing proton energy (*i.e.* with decreasing β energy), while the second factor increases (from 0 to a maximum of 1) with increasing proton energy. The resulting spectral shape is shown Fig. 3; it was specifically calculated for the precursor ^{117}Xe but is generally illustrative of all delayed proton emitters.

The spectra actually observed from several precursors are shown in Fig. 2: their features are characteristic of spectra generally observed in their different regions of the nuclear chart. In the case of the relatively light nucleus ^{33}Ar, the level density in its β-decay daughter, ^{33}Cl, is low enough that its spectrum is marked by individual proton peaks, each one corresponding to the decay of an individual state. Although the spectrum

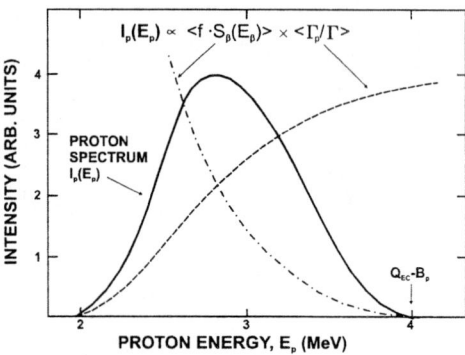

FIGURE 3: The delayed-proton spectrum calculated for ^{117}Xe assuming $l_p=0$ proton emission. The first factor in the expression for $I_p(E_p)$ corresponds to the β-decay feeding of states in the emitter: f is the statistical rate function for allowed decay and $S_β(E_β)$ is the beta strength function, taken to be constant in this case. The second factor corresponds to the relative probability for proton emission from states in the emitter.

does follow the general behavior illustrated in Fig. 3, cutting off on the low-energy end at the Coulomb barrier around 1 MeV (the exponential rise at low energy is due entirely to beta particles) and tapering off at high energy, its main features reflect the local details of nuclear structure, and its resemblance to the bell-shaped spectrum in Fig. 3 is not immediately evident.

In contrast, the other three precursors' spectra are predominantly bell-shaped, with a superimposed peak structure that is more or less pronounced depending on the case. In all three cases, however, the level density in the emitter is high enough that the peak structure can be attributed [7] to Porter-Thomas fluctuations: in general the peaks do not correspond to individual strong transitions but rather to clusters of several [8]. The pair of spectra measured following the decays of ^{150}Tm and ^{151}Yb is particularly illustrative of the effects of level densities. The first precursor is an odd-odd nucleus, which β-decays to an even-even emitter, ^{150}Er. Thus, both Q_{EC} and B_p are large, and the proton-emitting states in erbium are necessarily at rather high excitation energy (in fact, 6-10 MeV). The high level density at that excitation leads to a rather smooth spectrum without prominent fluctuations. The second precursor, ^{151}Yb, is an even-odd nucleus, which β-decays to an odd-even emitter, ^{151}Tm. In this case, both Q_{EC} and, particularly, B_p are smaller and the protons are emitted from states in thulium as low as 3 MeV in excitation. There, the level density is relatively low and the fluctuations are considerably more pronounced.

There is a further important feature evident only in one spectrum in Fig 2. By far, the strongest peak in the ^{33}Ar spectrum corresponds to proton emission from the excited state in ^{33}Cl that is analog to the ground state of ^{33}Ar; this state is fed by a superallowed β-transition, which is more than an order of magnitude faster than any other decay branch. Only precursors with $T_z \leq -3/2$ exhibit this important feature. In all other cases, the analog state in the emitter either is not energetically available to beta decay or else is bound against proton emission. Several important applications of β-delayed proton emission make use of this feature.

APPLICATIONS

Among light nuclei, β-delayed proton emission has been used extensively to obtain the energies of excited states in various emitters, measure their relative β-decay feeding, determine limits on their spins and parities (via β-decay selection rules) and, where possible, to identify analog states. In heavier nuclei, analysis of the characteristic bell-shaped proton spectra has provided more limited spectroscopic information but it has offered a ready means to determine experimental values for $Q_\beta - B_p$ (from the "end-point" energy — see Fig. 3) in regions of the chart where little experimental mass information is available. All these applications have been valuable, but they are straightforward enough that I need not elaborate on them here. I will focus instead on some of the applications that take unique advantage of β-delayed proton emission.

Beta-Neutrino Angular Correlations

The β-ν angular correlation observed in a nuclear β-transition depends on the make-up of the transition, the correlation coefficient being different for vector, axial vector and — if such were to occur — for scalar and tensor components. However, no realistic measurement of the β-ν angular correlation can determine the neutrino's motion directly, but instead requires it to be inferred from the motion of the daughter nucleus. In most cases, even that is a prohibitively difficult task. Not so if the daughter nucleus is a particle emitter. As illustrated in Fig. 4, if a particle is emitted in flight from the recoiling daughter nucleus following β decay, then the energy of that particle, as well as that of the final nucleus, will undergo a kinematic shift that reflects the motion of the daughter nucleus. A measurement of the kinematic shift can thus be

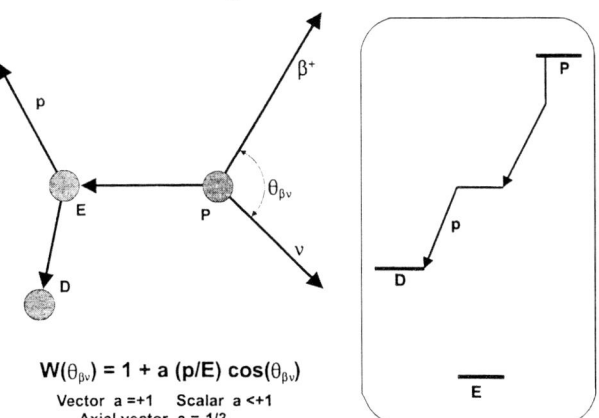

FIGURE 4: Schematic representation of a β-delayed particle decay. The left half of the figure shows the kinematics involved, while the right gives the simplified decay scheme. The letters P, E and D designate the precursor, emitter and daughter, respectively; the emitted particle is denoted by a lower-case p. Note the different possible values for the correlation coefficient, a, depending on the β transition.

233

used to determine the β-ν angular correlation, $W(\theta_{\beta\nu})$. One advantage of β-delayed particle experiments is that the particle-emitting states populated by β decay are so short-lived that the daughter nucleus does not collide with neighboring atoms before the particle emission takes place. Another is that reasonably energetic heavy particles are observed, in contrast to any case in which the recoiling β-decay daughter is detected directly. Not only does this facilitate the detection of the relevant particles but it also permits the investigation of non-gaseous activities because of the reduced sensitivity to scattering and energy losses in the source material.

The first use of this technique was forty years ago in a classic experiment [9] that established Gamow-Teller transitions as axial vector in nature. The β decay employed was that of ^{8}Li, which populates states in ^{8}Be that immediately break up into two α particles. Subsequently [10,11], the technique was applied to a more typical β-delayed alpha precursor, ^{20}Na, where it was used to determine the ratio of vector to axial-vector components in a number of β transitions; this produced [11] the first measurement of the vector coupling constant from a $2^+ \rightarrow 2^+$ transition and also set limits on possible isospin mixing in states in ^{20}Ne. Very recently, Adelberger et al [12] have used it for the first time with β-delayed proton decay to study the superallowed $0^+ \rightarrow 0^+$ β-decay branch of ^{32}Ar. If a scalar component were present in this otherwise pure vector transition, the correlation coefficient would be less than 1.0 (see angular correlation relation in Fig. 4). The experimental result obtained, a = 0.9989 ± 0.0065, sets rather tight constraints on any contribution from a scalar current.

PXCT Measurements of 10^{-16} s nuclear lifetimes

The lifetimes of proton-emitting states populated by the β decay of a precursor are usually of the order of 10^{-16}s or less, a time scale not normally accessible to experiment. However, the unique properties of this decay process have made it possible in some cases to measure such lifetimes directly via the particle-X-ray coincidence technique (PXCT). The technique is described by the cartoon in Fig. 5. Any precursor with atomic number Z that decays by electron capture to excited states in a Z-1 emitter produces simultaneously a vacancy in an atomic shell. If those excited states are unstable to proton emission, then the energy of the X-ray emitted with the filling of the atomic vacancy will depend upon whether the proton has already been emitted (in which case the X-ray would be characteristic of a Z-2 element) or has not yet been emitted (characteristic of a Z-1 element). If the nuclear and atomic lifetimes are comparable, then the K_α X-rays observed in coincidence with protons will lie in two peaks, the relative intensities of which uniquely relate one lifetime with the other. Because the K-shell vacancy lifetimes are well known both experimentally and theoretically, ranging from $\tau \approx 2 \times 10^{-15}$s for carbon down to $\tau \approx 6 \times 10^{-18}$s for uranium, excited-state lifetimes can be established with reasonable accuracy in this range of very short times.

The PXCT was initially proposed and developed in Chalk River, where it was first applied [13] to the measurement of lifetimes of proton-unstable excited states in ^{69}As, populated by β decay of the precursor ^{69}Se. In this case, excited states could not be isolated individually since their density was too high, but their *average* lifetimes were determined and information was also obtained on the level density parameter, a. Similar data were obtained [14] for states in ^{73}Br (from decay of the delayed-proton precursor ^{73}Kr), which yielded mean lives ranging from 4×10^{-15}s at 4.0 MeV of excitation to 4×10^{-16}s at 6.5 MeV. Less detailed information has been obtained on several other cases as well [15]. Very recently [5], new high-resolution PXCT data on ^{69}Se and ^{73}Kr decays have been obtained and are refining our understanding of the statistical properties of exotic nuclei in this region.

Statistical Beta-Decay Properties

Delayed protons have an important advantage over β-delayed γ rays in the measurement of β-decay branching ratios in a complex decay scheme. If excited states populated in the β-decay daughter de-excite by γ emission, a number of γ-decay paths are usually possible from each state, and the multiplicity of γ rays in each path can be quite large. If many β branches are involved, it is an impossible task to extract reliable β branching ratios from a plethora of γ rays, many of them too weak to identify and measure [16]. However, in the case of β-delayed proton decay, every β decay is followed by the emission of only a single proton. Furthermore, since the proton-decay daughter nucleus (see Fig. 1) is usually populated near or in its ground state, only one or two decay paths occur from each state in the emitter. These paths can even be distinguished if data on proton-γ-ray coincidences are obtained. Thus, reliable β-decay schemes can be extracted both when individual peaks are present in the proton spectrum (*e.g.* ^{33}Ar in Fig. 2) or when the spectrum is characterized by Porter-Thomas fluctuations (*e.g.* ^{73}Kr).

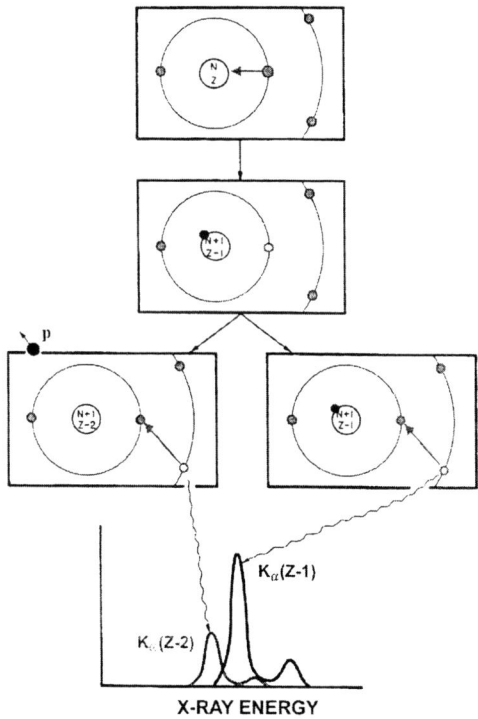

FIGURE 5: Cartoon illustrating the proton-X-ray coincidence technique. The first panel shows a K-shell electron being captured; the second indicates that a K-shell atomic vacancy and a proton-unstable nuclear state (designated by a proton at the surface of the nucleus) have been created simultaneously. The remaining panels show how the emitted X-rays depend on the subsequent order of events.

In the past, this property of β-delayed proton decay has been exploited to study the β strength function in the tail of the Gamow-Teller (GT) giant resonance as observed in medium-mass nuclei [15]. Recently, though, it has also become interesting in the study of light precursors with $T_z \leq -3/2$. These are the nuclei with large Q_{EC} values, whose β-decays populate the analog state in their daughters. They also populate a much larger fraction of the GT giant resonance than is accessible to β decay in any other region of nuclei. In a recent experiment [17], in which delayed protons (and a few γ transitions) were measured following the β decays of ^{36}Ca and ^{37}Ca, the GT strength function was determined over 7 MeV of excitation in ^{36}K and 9 MeV in ^{37}K. The results were used to test the efficacy of large-scale shell-model calculations.

Isospin Mixing

Obviously, such a detailed study of the decay of a light nucleus with $T_z \leq -3/2$ also yields information on the strength of the superallowed β-decay branch to the analog state. This can be used to probe the extent of isospin mixing in that state. To take ^{33}Ar ($T_z = -3/2$) as a specific example: the T=3/2 analog state in its β-decay daughter, ^{33}Cl ($T_z = -1/2$), must experience a certain amount of mixing with nearby T=1/2 states since the state decays via that channel into a proton (T=1/2) and a T=0 state in ^{32}S. However, for the decay of the analog state to be dominated by proton (rather than γ-ray) emission requires very little isospin mixing to be present. Delayed-proton-based measurements of the intensity of the superallowed β branch have indicated that the amount of mixing may actually be considerably larger — up to ten percent.

The first measurement of this type was done nearly 30 years ago [18]. It indicated a possible T=1/2 impurity in the T=3/2 analog state in ^{33}Cl of ≈10%. Within the last two years, two new measurements on similar $T_z = -3/2$ precursors have reached similar conclusions. The study of ^{37}Ca decay [17], already mentioned in the last section, claims ≈4% impurity in the analog state in ^{37}K, and another independent measurement concludes that the T=3/2 analog state in ^{41}Sc has ≈10% T=1/2 component admixed. In all three cases, the experiments could not distinguish between the vector and axial-vector components of the superallowed decay branch. Thus, any conclusions about isospin mixing — revealed by reduced intensity in the vector component — still dependeded upon a model calculation to determine the magnitude of the axial-vector matrix element. Definitive model-independent conclusions will have to await experiments that are sensitive to the β-ν angular correlation. As already described, such experiments are now feasible.

FUTURE DIRECTIONS

I began this talk on β-delayed proton emission with some history dating from the early 1960's. I have ended it by describing four unique applications of the decay mode, each leading to significant publications that have appeared within the last two years. Can there be any better demonstration that this is a field with a past, a present and very definitely a future?

It is always risky to *predict* the future but I do not hesitate to suggest some directions that I believe *should* be pursued in the future:

1) *Unique applications.* I have already described the unique applications that are now being explored. As "rare-ion accelerators" increase the production of exotic nuclei, and experimental techniques improve, these applications should be exploited more fully. Undoubtedly, new applications will appear as well. We are perhaps witnessing the birth of one of them at this meeting with the paper by Rikovska and Stone [19] promising a test run at ISOLDE to measure the angular distribution of β-delayed protons from a precursor oriented at low temperatures.

2) *Beta spectroscopy.* Wherever β-delayed proton precursors occur, their beta spectroscopy will become accessible to study. In fact, as I have stated already, β-delayed proton precursors offer the *only* means available among neutron-deficient nuclei to do reliable β spectroscopy on a complex decay scheme. In this context, I would also argue that such decay schemes will be truly understood only within the framework of a statistical model. Here again, the sensitivity of the delayed proton spectrum to each emitter's level density, lifetime profile and (β and γ) strength functions, will make it a valuable tool as we strive to characterize and understand nuclei far from stability.

3) $N \leq Z$ *nuclei.* Among the β-delayed proton precursors with $N \leq Z$, the density of levels accessed in each nuclear β-decay is usually low enough that individual-state spectroscopy is possible. In those cases, the precision afforded by the process is high enough that tests of symmetry and isospin mixing can be improved and extended to nuclei heavier that $A \approx 40$, which is effectively where they end today.

4) *Necessary evil to be accounted for.* In some cases where extreme precision is required in measurements of the decay of a neutron-deficient nucleus, β-delayed proton emission simply cannot be ignored. One example is in the measurement of superallowed $0^+ \rightarrow 0^+$ β-decay from odd-odd, $T = 1$ ($T_z = 0$) nuclei above $A \approx 70$. In principle, such nuclei could have non-negligible delayed proton branches — perhaps a total of 0.1% of all decays for ^{74}Rb, 1.0% for ^{86}Tc, and greater for even heavier cases. Though not particularly significant in their own right, these branches must be accounted for before an *ft*-value can be reliably determined for the superallowed branch.

Thus, whether as a sensitive probe of the physics of exotic nuclei or as a source of unavoidable "background", there seems little doubt that β-delayed proton emission will be an important factor in a future dominated by rare ion beams and nuclei ever more remote from stability.

ACKNOWLEDGEMENTS

This work was supported by the U.S. Department of Energy under Grant number DE-FG03-93ER40773 and by the Robert A. Welch Foundation.

REFERENCES

1. Goldansky, V.I., *Nucl. Phys.* **19**, 482 (1960).
2. Barton, R., McPherson, R., Bell, R.E., Friskin, W.R., Link, W.T., and Moore, R.B., *Can. J. Phys.*

41, 2007 (1963).
3. Karnaukhov, V.A., Ter-Akopian, G.M., and Subbotin, V.G., in *Proc. Asilomar Conf. Reactions between Complex Nuclei,* edited by A. Ghiorso *et al,* U. Cal. Press, 1963, p 434.
4. Schardt, D., and Riisager, K., *Zeit für Physik A,* **345**, 265-271 (1993).
5. Giovinazzo, J., Dessagne, Ph., and Miehé, Ch., to be published.
6. Nitschke, J.M., Wilmarth, P.A., Gilat, J., Möller, P., and Toth, K.S., in *Nuclei Far From Stability-1987* edited by I.S. Towner, AIP Conference Proceedings 164, New York: American Institute of Physics, 1988, pp 697-707.
7. MacDonald, J.A., Hardy, J.C., Schmeing, H., Faestermann, T., Andrews, H.R., Geiger, J.S., Graham, R.L., and Jackson, K.P., *Nucl. Phys.* **A288**, 1-22 (1977).
8. Hardy, J.C., Jonson, B., and Hansen, P.G., *Nucl. Phys.*, **A305**, 15-28 (1978).
9. Barnes, C.A., Fowler, W.A., Greenstein, H.B., Lauritsen, C.C., and Nordberg, M.E., *Phys. Rev. Lett.* **1**, 328-330 (1958).
10. MacFarlane, R.D., Oakey, N.S., and Nickles, R.J., *Phys. Lett.* **B34**, 133-134 (1971).
11. Clifford, E.T.H., Hagberg, E., Hardy, J.C., Schmeing, H., Azuma, R.E., Evans, H.C., Koslowsky, V., Schrewe, U.J., Sharma, K.S., and Towner, I.S., *Nucl. Phys.* **A493**, 293-322 (1989).
12. Adelberger, E.G., Ortiz, C., García, A., Swanson, H.E., Beck, M., Tengblad, O., Borge, M.J.G., Martel, I., and Bichsel, H., *Phys. Rev. Lett.* **83**, 1299-1302 (1999).
13. Hardy, J.C., Macdonald, J.A., Schmeing, H., Andrews, H.R., Geiger, J.S., Graham, R.L., Faestermann, T., Clifford, E.T.H., and Jackson, K.P., *Phys. Rev. Lett.* **37**, 133-136 (1976).
14. Asboe-Hansen, P., Hagberg, E., Hansen, P.G., Hardy, J.C., Jonson, B., and Mattsson, S., *Nucl. Phys.* **A361**, 23-34 (1981).
15. Hardy, J.C., in *4th Int. Conf. On Nuclei far from Stability,* edited by P.G. Hansen and O.B. Nielsen, CERN report 81-09, 1981, pp 217-228.
16. Hardy, J.C., Carraz, L.C., Jonson, B., and Hansen, P.G., *Phys. Lett.*, **71B**, 307-310 (1977).
17. Trindner, W., Adelberger, E.G., Brown, B.A., Janas, Z., Keller, H., Krumbholz, K., Kunze, V., Magnus, P., Meissner, F., Piechaczek, A., Pfützner, M., Roeckl, E., Rykaczewski, K., Schmidt-Ott, W.-D., and Weber, M., *Nucl. Phys.* **A620**, 191-213 (1997).
18. Hardy, J.C., Esterl, J.E., Sextro, R.G., and Cerny, J., *Phys. Rev.,* **C3**, 700-718 (1971).
19. Rikovska, J., and Stone, N.J., these proceedings.

The Effects of β-delayed Proton Emission on the Path of the rp-Process

R.N. Boyd

Department of Physics, Department of Astronomy
The Ohio State University, Columbus, OH 43210, USA

Abstract. The rp-process occurs in a hot hydrogen-rich stellar environment. Its trajectory passes through the most proton-rich nuclides in the periodic table. It has long been thought to be responsible for synthesizing at least the light p-process nuclides. Thus these nuclides can provide signatures for rp-process nucleosynthesis. Difficulties with various rp-process scenarios often focus on ^{92,94}Mo and ^{96,98}Ru p-nuclides, as their anomalously large abundances are difficult to produce in any model of nucleosynthesis. However, it now appears that they might be produced in the rp-process resulting from accretion onto a neutron star. If the rp-process does synthesize these nuclides, β-delayed proton emission might well resolve some of the difficulties made evident by the model calculations.

INTRODUCTION

The astrophysical rp-process is expected to occur in high-temperature proton-rich environments. It does not occur in the quasi-static burning stages through which a massive star evolves, since the required combination of temperature and hydrogen richness does not occur. Rather, the rp-process is thought to occur in situations in which a compact star—a white dwarf or a neutron star—accretes matter from a less evolved companion. The accreted matter comes from the periphery of the star, so will be hydrogen rich. It will also be hot, as it will have fallen into a deep gravitational potential well as it approaches the surface of the compact star. Thus the requisite rp-process conditions are fulfilled. Another possible site can occur in the shock wave of a supernova, as the energy deposited in the existing nuclei as the shock wave passes can provide the requisite temperature, while the required proton density is supplied by (γ,p) reactions on pre-existing nuclei or by mixing with material from hydrogen rich zones. In any event, the rp-process produces explosive nucleosynthesis, as it occurs rapidly, in tens of seconds or less, and produces a large amount of energy in that time which will emerge in some form. If the accretion is onto a white dwarf, the temperature achieved will be several hundred million K, and the result may be a nova. If accretion occurs onto a neutron star, the temperature achieved will exceed 1 billion K, and the result may be an x-ray burst. General descriptions of the nuclear astrophysics of this process have been given by Wallace and Woosley, [1], Champagne and Wiescher [2], and Schatz et al. [3].

SPECIFICS OF THE rp-PROCESS

Because the rp-process involves a rapid succession of proton captures, it will drive its seed nuclei to near the proton drip line. Occasional (rapid) β^+-decays will allow the progression from light seed nuclei to the heavier nuclei produced by the rp-process. Occasional (p,α) reactions may close cycles, but the general progression from light to heavy nuclides is expected to proceed none the less. The reactions relevant to the rp-process are indicated in Fig. 1.

The rp-process drives its seed nuclei to nuclides at which further (p,γ) reactions are prohibited by negative Q-values, or inhibited by small positive Q-values. If this were the case for nucleus (N,Z) in Fig. 1, then the reaction (N,Z+1)(γ,p) would dominate over (N,Z)(p,γ), as the high-temperature environment would destroy the

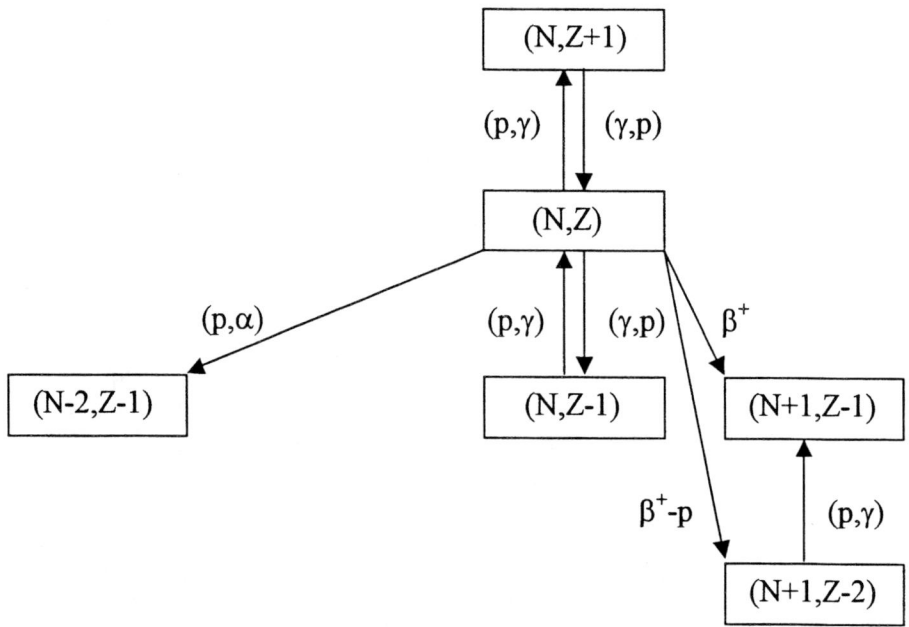

FIGURE 1. Various reactions that can occur on nucleus (N,Z) in rp-processing. If (γ,p) dominates over (p,γ), and both β^+ and β^+-p decays are slow, nucleus (N,Z) will become a waiting point.

nuclei made by (p,γ) almost as rapidly as they were produced. If the nucleus (N,Z) β^+-decayed rapidly, the resulting nucleus, (N+1,Z-1) would quickly undergo a (p,γ) reaction, and the rp-process would continue almost without pause. If, however, the half-life were long, the rp-process would have a "waiting point," at nucleus (N,A), and its abundance would build up during rp-processing. After the high-temperature that drove the rp-process subsided, the subsequent β^+-decays would yield the ultimate products of rp-process nucleosynthesis. The stable nucleus to which nucleus (N,Z) ultimately decayed would be expected to have a large abundance.

Of course, β^+-delayed proton emission could affect the trajectory of the rp-process. If β^+-p emission occurred rapidly, the resulting nucleus would be considerably closer to stability than its parent nucleus, so that a subsequent (p,γ) reaction would almost immediately return the nucleus to that which would have been produced by simple β^+ decay. In this case the β^+-p emission would have had essentially no effect on the rp-process path. If, however, the nuclide from which the β^+-p emission occurred was long-lived, then abundance will build up at nucleus (N,Z), in which case the β^+-p decay mode would impact the nucleosynthesis of the rp-process, as some of those decays would occur after the rp-process conditions ended. I return to this below.

SIGNATURES OF rp-PROCESSING

Are there key nuclides that might provide signatures of rp-processing? Clearly one would like to select nuclides that cannot be synthesized in any other way, but such "smoking gun" nuclei probably do not exist. However, the p-process nuclei (see section by Boyd in [4]) come as close as one can get. These nuclides lie to the proton-rich side of stability, and are blocked by stable isotones from being synthesized by the r- or s-processes, the neutron capture processes that synthesize virtually all of the nuclei more massive than iron. Thus, the p-nuclides must be synthesized by processes having trajectories through the nuclides along the proton-rich side of stability; one such process is the rp-process. Of particular interest in this context are those p-nuclides around mass 90 to 100 u, the Mo and Ru p-nuclides. These stand out because of their high abundances. While p-nuclear abundances in this mass region are typically around 1% of their respective elemental abundances, those for ^{92}Mo, ^{94}Mo, ^{96}Ru, and ^{98}Ru are 14.8%, 9.3%, 5.5% and 1.9% respectively [5]. The explanation of these anomalously high abundances has represented a longstanding problem for nuclear astrophysicists.

OTHER PROCESSES THAT SYNTHESIZE THE p-NUCLIDES

However, there are other processes as well, most notably the γ-process, that produce p-nuclides. Its current description was formulated in 1978 by Woosley and Howard

[6], and updated in 1990 [7]. It processes abundant high-mass nuclei, e.g., lead, at high temperatures via (γ,n), (γ,p), and (γ,α) reactions to lighter nuclei. Because of the Coulomb barrier, the trajectory of the γ-process will tend to lie among the proton-rich nuclei. As with the rp-process, the progenitors that occur during the period of high temperature will decay back to stability after the high temperature subsides. This process is inevitable in high temperature environments, so it always accompanies the rp-process. Some of the stable nuclei it ultimately produces will be p-nuclei.

An adaptation of the rp-process is the repetitive-rp-process, or (rp)^2-process [8]. This occurs under the same conditions as the rp-process, but alternates rp-processing bursts and quiescent periods. Thus the very proton-rich nuclides synthesized in each of the processing bursts have time to decay back to stability before the next processing burst begins. In this way the waiting point nuclei of the rp-process are circumvented.

Finally, another process, the ν-process [9], is thought to synthesize some of the rarest nuclides in the periodic table, ^{180}Ta and ^{138}La. This process operates by (neutral-current induced) neutrino spallation on ^{181}Ta in the former case, and by both neutrino spallation on ^{139}La and charged current reactions on ^{138}Ba in the latter. This process is also thought to synthesize a few other nuclides, e.g., ^{7}Li and ^{19}F, but they are not a part of the rp-process story.

SUCCESSES AND FAILURES OF rp-PROCESS MODELS

How well do the different models of rp-processing work? It has been shown [10] that the γ-process does explain the nucleosynthesis of most of the heavy p-nuclides, those heavier than 100 u, quite well, but that the rp-process appears to be essential for fitting the lower mass p-nuclides, Se to Sr. However, no theory has successfully reproduced the abundances of the Mo and Ru p-nuclides without some negative features. One fairly successful description arises from the rp- plus γ-process with high temperatures (T = 1.5x10^9 K) and densities (10^6 g cm^{-3}), and very long processing times (100 s) [3]. The sorts of conditions used in that study might be characteristic of accretion of matter onto a neutron star. However, that host's deep gravitational well would allow very little matter to escape into the interstellar medium, thus necessitating a very high enhancement of the abundance of the Mo and Ru isotopes from rp-processing. This is achieved in the model of Schatz et al. [3]; their enhancement factors are in excess of 10^7 for the nuclides of interest. The downside of that model, though, was the production of large amounts of ^{80}Kr, a nuclide that is made in the s-process, so is potentially overproduced when all processes of synthesis are included.

Not surprisingly, the (rp)^2-process also is able to achieve high enhancements of the Mo and Ru p-nuclides. However, it does not require such long processing times; ~1 s works well, or such high density; ~10^4 g cm^{-3} is sufficient to achieve the same abundance enhancements. The downside of the (rp)^2-process, however, is similar to that of the rp-process. It produces some non-p-nuclides, most notably, ^{93}Nb, with abundance enhancements comparable to those of the Mo and Ru p-nuclides. Since ^{93}Nb is made in the s-process, this might present the same sort of problem that ^{80}Kr does for the rp-process.

The ν-process does successfully produce the two rare nuclides mentioned in the previous section, but has difficulty producing large quantities of the four p-nuclides on which we are focusing. It should be noted, though, that neutrino induced reactions might be able to synthesize some very specific p-nuclides from very abundant seed nuclei formed in the s-process. The most notable case is ^{92}Mo, which could be produced from (abundant) ^{92}Zr by ^{92}Zr(ν,e$^-$)^{92}Nb(ν,e$^-$)^{92}Mo reactions. However, even this process does not appear to be capable of producing appreciable amounts of the other three very abundant p-nuclides in this mass region.

THE EFFECTS OF β-DELAYED PROTON EMISSION

The path of the rp-process through the mass 90 to 100 u nuclides is shown in Fig. 2. This figure is similar to one in Schatz et al. [3]. As can be seen in that figure, the abundances produced in that model are determined by the abundances and decay modes of the progenitors that exist at the end of the processing period. A very similar picture occurs in the (rp)^2-process, at least in this mass region. The progenitor nuclei for ^{92}Mo, ^{94}Mo, ^{96}Ru, and ^{98}Ru are expected to be ^{92}Pd, ^{94}Pd, ^{96}Cd, and ^{98}Cd respectively, all waiting points along the high-temperature rp-process path. The progenitor of ^{93}Nb in the rp- and (rp)^2-processes would be ^{93}Pd. However, if this nucleus decayed a significant fraction of the time by β$^+$-delayed proton emission, it would ultimately populate ^{92}Mo, not ^{93}Nb, which would improve the ability of these processes to produce ^{92}Mo, and would also solve the possible problem in the (rp)^2-process with overproducing ^{93}Nb. The energetics would suggest that β$^+$-delayed proton emission is very likely for ^{93}Pd, but the actual fraction has not been measured at this time. Because the other progenitors are all even-even, their β$^+$-delayed proton emission would be expected to be inhibited by energetics.

A similar situation exists for ^{97}Cd. If it decayed by simple β$^+$ emission, it would populate ^{97}Mo. It is known [7] to decay to some extent by β$^+$-p, so some of the ^{97}Cd abundance will end up as ^{96}Ru, a benefit to the rp- and (rp)^2-processes. Although some of the proton-rich Ag nuclides are also known to decay by β$^+$-p emission, none of them is a waiting point in the rp-process, so their decay modes will have little effect on its nucleosynthesis.

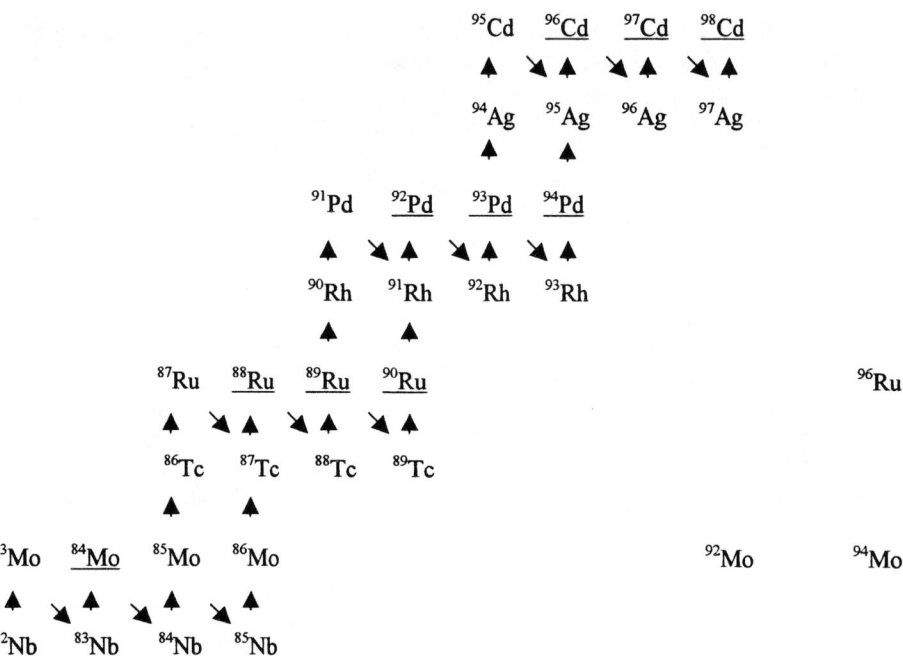

FIGURE 2. Nuclides along the primary trajectory of the rp-process that populate the Mo and Ru p-nuclides, three of which are shown (to the lower right). The underlined nuclides are waiting points, in that they have half-lives of about 1 s or longer (although in some cases the half-lives have not been measured) and (p,γ) Q values less than 1 MeV. Upward arrows represent (p,γ) reactions, whereas arrows pointing downward and to the right represent β^+ decays.

CONCLUSIONS

If β-delayed proton emission does affect the rp- and (rp)^2-processes, it must satisfy two conditions. First, the nuclei that undergo the β^+-p decay must be long-lived with respect to the time scales of the rp-process, of order 10 seconds. Secondly they must lie along the main trajectory of the rp-process, which means that they are close to the proton drip line. These two conditions are rarely simultaneously fulfilled until the nuclei that lie along the rp-process trajectory attain masses in the 90 to 100 u region.

However, the possibility that progenitor nuclides such as ^{93}Pd and ^{97}Cd do decay by β^+-p emission may well impact the nucleosynthesis that is performed by the rp- and (rp)^2-processes, and may even help to solve some of the difficulties with current models of the synthesis of the Mo and Ru p-nuclides. Further data that determine the β-delayed proton emission probabilities of the key nuclei will be important in resolving these issues.

ACKNOWLEDGEMENTS

The support of the National Science Foundation through grant PHY9901241 is gratefully acknowledged.

REFERENCES

1. Wallace, R.K. and Woosley, S.E., Astrophys. J. Suppl. Series 45, 389 (1981)
2. Champagne, A.E. and Wiescher, M., Ann. Rev. Nucl. Part. Sci. 42, 39 (1992)
3. Schatz, H., Aprahamian, A., Görres, J., Wiescher, M., Rauscher, T., Rembges, J.F., Thielemann, F.-K., Pfeiffer, B., Möller, P., Kratz, K.-L., Herndl, H., Brown, B.A., and Rebel, H., Phys. Reports 294, 167 (1998)
4. Wallerstein, G., Iben, I., Jr., Parker, P., Boesgaard, A.M., Hale, G.M., Champagne, A.E., Barnes, C.A., Kappeler, F., Smith, V.V., Hoffman, R.D., Timmes, F.X., Sneden, C., Boyd, R.N., Meyer, B.S., and Lambert, D.L., Rev. Mod. Physics 69, 995 (1997)
5. Anders, E. and Grevesse, N., Geochim. Cosmochim. Acta 53, 197 (1989)
6. Woosley, S.E. and Howard, W.M., Astrophys. J. Suppl. 36, 285 (1978)
7. Woosley, S.E. and Howard, W.M., Astrophys. J. 354, L21 (1990)
8. Boyd, R.N., Hencheck, M., and Meyer, B.S., *Origin of Matter and Evolution of Galaxies 97*, ed. by S. Kubono, T. Kajino, K.I. Nomoto, and I. Tanihata, (World Scientific, Singapore) 350 (1998)
9. Woosley, S.E., Hartmann, D.H., Hoffman, R.D., and Haxton, W.C., Astrophys. J. 356, 272 (1990)
10. Howard, W.M., Meyer, B.S., and Woosley, S.E., Astrophys. J. 373, L5 (1991)
11. Schmidt, K., Divari, P.C., Elze, Th.W., Grzywacz, R., Janas, Z., Johnstone, I.P., Karny, M., Keller, H., Kirchner, R., Klepper, O., Plochocki, A., Roeckl, E., Rykaczewski., K., Skouras, L.D., Szerypo, J., and Zylicz, J., Phys. A624, 185 (1997)

A Distribution of GT-Strength and βp-Emission near ^{100}Sn.

M.Karny

Institute of Experimental Physics, University of Warsaw, PL-00681 Warsaw, Poland
Joint Institute for Heavy Ion Research, Oak Ridge, Tennessee 37831, USA
Department of Physics, University of Tennessee, Knoxville, TN 37996, USA

Abstract. Results on $^{103-107}$In strength function measurements are presented. The domination of the core $(\pi g_{9/2})^8$ over the valence proton decay is confirmed. Contribution of a βp decay channel to the total β strength function of 0.5% and 43% for ^{102}In and ^{100}In, respectively, is expected based on the B_{GT} systematics, available βp spectra and statistical model calculations.

β - DECAY NEAR ^{100}Sn

The region of ^{100}Sn has been the subject of intense experimental and theoretical study for many years. Although ^{100}Sn was observed [1–3], its detailed spectroscopy appears to be still out of experimental range. However, for nuclei near ^{100}Sn both in-beam and, in particular, decay spectroscopy is already feasible today, and some results are available for nuclei as close as ^{100}In or ^{101}Sn.

One particularly interesting feature of decay studies in the ^{100}Sn region is the occurrence of fast β transitions related to the Gamow-Teller (GT) transformation of a $\pi g_{9/2}$ proton into a $\nu g_{7/2}$ neutron. A measurable quantity suited for comparison with theoretical predictions is the β strength B_{GT} of this decay mode, defined as:

$$| < J_i|\sigma\tau|J_f(E) > |^2 = B_{GT}(E) = \frac{D \cdot I(E)}{f(Q_{EC} - E) \cdot T_{1/2} \cdot 100}, \quad (1)$$

where $D = 3860(18)$ s is a constant, corresponding to the value of the axial vector weak interaction coupling constant g_A for the decay of the free neutron [4,5], I the β intensity normalized to 100% per decay, E the excitation energy in the daughter nucleus, f the statistical rate functions, Q_{EC} the total energy released in electron–capture (EC) decay, $T_{1/2}$ the β–decay half–life, $< J_i|$ and $|J_f >$ are wave functions of the initial and final states respectively and $\sigma\tau$ is a free GT operator. The $B_{GT}(E)$ distributions, deduced from measurements of $I(E)$, Q_{EC} and $T_{1/2}$, can be compared to the calculated square of the GT transition matrix element. The quenching of the experimental GT transition rates with reference to model predictions has been a puzzle for many years. A renormalization of g_A (or of the GT operator) has been applied (see e.g. [6,7]) in order to account for the missing GT strength in the ^{100}Sn region. This led to a picture with the GT strength from shell model calculations being quenched by about a factor of 4-5 compared to experiment [8–10].

For nuclei close to the proton drip line a high level density and small β feeding to particular state make γ transitions very weak and thus difficult to detect. In addition, deexcitation of highly excited state can proceed also via β-delayed proton emission (if energeticly possible). Therefore, while measuring the $I(E)$ distribution great care (by taking into account two possible β delayed processes) should be taken in deriving the β feeding to the highly excited states in the daughter nucleus. This is because the statistical rate function (f) becomes very small for excitation energies close to Q_{EC}. Thus a small β feeding can result in a large $B_{GT}(E)$ value. Although experimental information for β delayed gammas from highly excited states in proton emitters are not available so far, some information can be derived based on the systematics of strength distribution in heavier systems (see Table 1). This work will present some of the existing experimental information on $\beta\gamma$ and βp proton studies in the vicinity of ^{100}Sn and then focus on the predictions of strength distribution for ^{100}In and ^{102}In based on measured β delayed protons and systematic studies of B_{GT} distribution in heavier indium isotopes.

EXPERIMENTAL TOOLS.

β delayed γ measurements: standard set-up for measurement of β delayed γ's in ^{100}Sn region would consist of a few (2-4) high resolution germanium detectors, placed after a mass separator. Very low efficiency of such devices allows only for measurement of relatively strong gammas rays. In practice, this results in reliable measurement of the GT strength of even-even nuclei (e.g.^{98}Cd [8]) and significant underestimation of the total GT strength and its distribution in case of non even-even isotopes. (e.g. ^{103}In compare [11] and [9]).

Another device used for $\beta\gamma$ measurement is Total Absorption Spectrometer (TAS) [12,13] used at the GSI on-line mass separator to study $^{103-107}$In [9] and ^{97,98}Ag [10]. This large NaI(Tl) detector is characterized by sufficient efficiency to detect very weak gamma transitions, and thus reconstructs β decay feeding pattern and B_{GT}, correctly. Poor energy resolution of NaI(Tl) (in comparison to HPGe detectors), makes it impossible to resolve excited states on level-by-level basis, nevertheless TAS measurements are sufficient to determine the value of the integrated GT strength function.

An array of 42 germanium detectors (6 EUROBALL cluster detectors), was used to study ^{102}In and ^{97,98}Ag. This device (Cluster Cube) was characterized by the energy resolution of germanium detectors and significantly improved photo-peak

TABLE 1. The latest references on the β delayed γ and p data in ^{100}Sn region.

Element	Mass	Ref.[a] βp	$\beta \gamma$ [b]
Tin (Sn)	101	[15]	no
	102	no	no
	103	[16]	no
	104	no	[17]
Indium (In)	100	[18]	no
	101	no	[19]
	102	[18]	[18]
	103	no	[9]
Cadmium (Cd)	97	[21]	no
	98	[22] [c]	[8]
	99	[23]	[24]
	100	no	[25]
Silver (Ag)	94	[20]	no
	95	[20]	[21]
	96	[21]	[21]
	97	no	[10]

[a] References taken from the NNDC-online service
[b] Some limited information on $\beta \gamma$ can be derived from in-beam studies (references not included.)
[c] Upper limit for βp branching given

efficiency (up to 17% for 1.3MeV line), which allowed detection of very weak $\beta \gamma$ branchings [10]. Although, available data for ^{97}Ag show good over-all agreement between TAS and Cluster Cube data, up to 30% of total B_{GT} is still missed by the germanium array. Note that this 30% of B_{GT} corresponds to only 9% of missed β feeding [10].

β delayed p measurements: β delayed protons as charged particles can be efficiently detected with the use of ΔE-E silicon detector telescopes. For example, the setup used by Szerypo et al., to measure ^{102}In and ^{100}In consisted of two ΔE-E telescopes mounted close to the carbon foils in which mass separated beam was implanted [18]. Detector thicknesses were about 16-32μm and 530-750μm for ΔE and E wavers respectively, while corresponding sensitive areas ranged to $\approx 150 mm^2$ and $450 mm^2$. In order to study possible βp branches to the excited states a Ge detector was added to the setup.

Figure 1 presents a schematic view of ^{100}Sn region with marked isotopes where the decays via $\beta \gamma$ (measured with TAS or CLUSTER CUBE) or βp were studied.

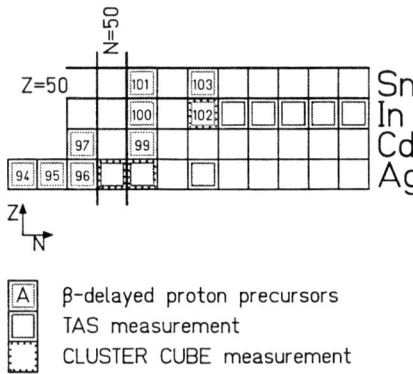

FIGURE 1. A schematic view of the ^{100}Sn region with marked nuclei of interest to this work.

SINGLE PARTICLE MODEL PREDICTIONS AND EXPERIMENTAL RESULTS

The EC/β^+ decay in the ^{100}Sn region proceeds via the $\pi g_{9/2} \to \nu g_{7/2}$ transformation. For even-even nuclei one expects transformations of $J = 0^+$ initial state to a couple of $J = 1^+$ states in the daughter nuclei. Final states are expected to be at relatively low excitation energies (between 1.7 and 2.5 MeV for ^{98}Cd [8]). In case of odd-odd nuclei the simple single particle shell model predicts two possible scenarios: (i) transformation of the valence proton from the $g_{9/2}$ orbital to the neutron $g_{7/2}$ state, or (ii) decay of one of the protons from the $(g_{9/2})^8$ core to the neutron on the $g_{7/2}$ orbital. The unpaired $g_{9/2}$ valence proton acts then as a spectator for this process. For both scenarios there is a significant difference in the excitation energy of the populated state. In (i) the final state can be described as a broken neutron pair with one of the nucleons excited to the $g_{7/2}$. This would yield the excitation energy of approximately 2Δ, where Δ is a pairing energy (for ^{100}Sn region $\Delta \approx 1.5$). The transformation of type (ii) populates the states with the 4 particles being unpaired, two protons (one valence and one unpaired after core decay) and two neutrons (like in case of (i)). Those states would result in the excitation energy of about $4\Delta \approx 6~MeV$. (More accurate theoretical description can be found in [7,6,26].) For nuclei close to the valley of β stability, where Q_{EC} energies are small, only transitions of type (i) will be possible, while close to the proton drip line both types of states can be fed. In addition, for nuclei far away

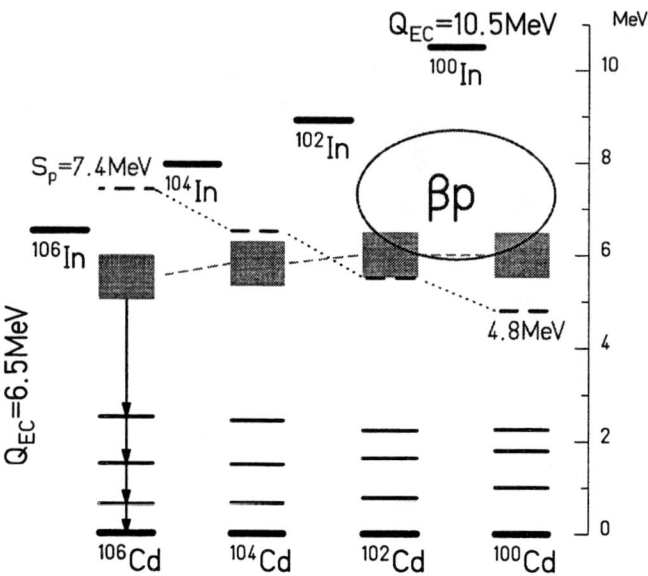

FIGURE 2. Systematics of Q_{EC} values of even mass indium isotopes (A=100-106) together with proton separation energies (S_p) in daughter systems. Rectangles show measured (^{104,106}In) or expected (^{100,102}In) position of the 4-quasiparticle levels. First ($2^+, 4^+$ and 6^+) exited states in even-even cadmium isotopes are also shown.

from β-stability the proton separation energy becomes small and lower then Q_{EC} allowing the proton emission from excited states. As can be seen in Fig.2 for ^{102}In and ^{100}In the S_p drops below 6 MeV making proton emission possible from states which are β fed via scenario (ii). This very simplified description gives an important conclusion that the relative amount of β delayed protons associated with the precursor decay will strongly depend on the ratio between transformations of type (i) and (ii). Systematic investigation of the β decay strength near ^{100}Sn has been undertaken at the GSI on-line mass separator with the use of the Total Absorption Spectrometer. Figure 3 shows strength function of neutron deficient indium isotopes measured with TAS. Maxima on the plots for both odd-even as well as odd-odd isotopes correspond to the transformation of a proton from the $(\pi g_{9/2})^8$ core, showing domination of the (ii) process over the (i). A similar result was found for ^{97}Ag decay [10].

The β delay proton distribution for ^{100}In and ^{102}In is known from work of Szerypo et al. [18], a computer code - DELPA, was design to calculate within the statistical model [27,28] βp distribution based on the knowledge of the strength

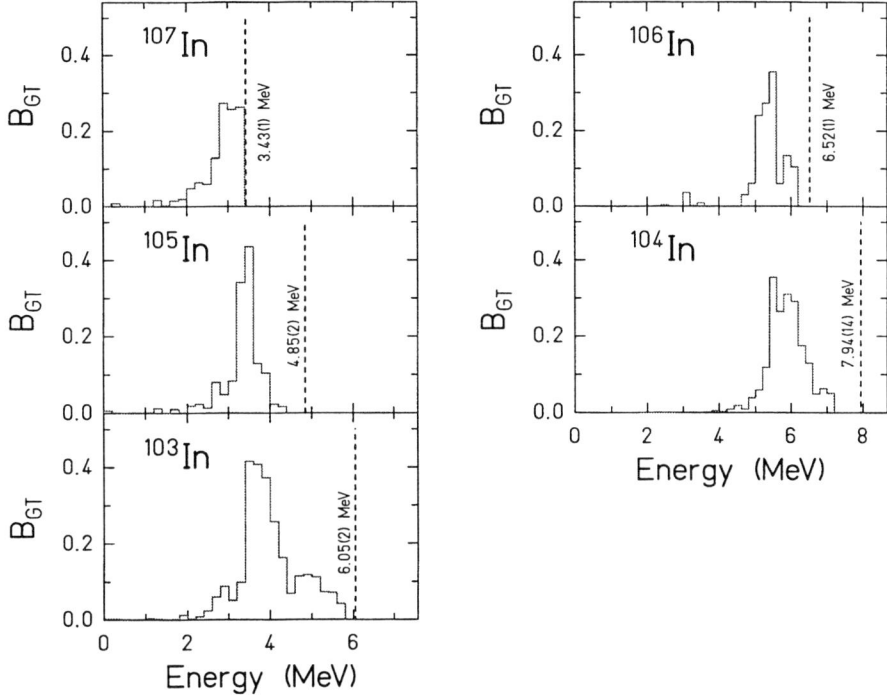

FIGURE 3. $^{103-107}$In B_{GT} distribution derived by means of Total Absorption Spectrometry. Vertical dashed lines show Q_{EC} limits.

TABLE 2. Branching ratios for βp as of experiment [18] and calculation.

Isotope	βp^{exp}	βp^{DELPA}
^{102}In	$9.3 \cdot 10^{-3}$%	$8.6 \cdot 10^{-3}$%
^{100}In	> 3.9%	3.9%

function. These can be combined, in order to propose a strength function which will reproduce the βp experimental results. The quality of the fit is shown in figure 4, while the branchings are compare in table 2. Figure 5 presents a B_{GT} distribution of ^{100,102}In adjusted to get good agreement between calculated and measured β delayed protons and βp branching ratios. In case of ^{102}In only ≈ 0.5% of β strength is carried away with β delayed protons, while remaining the 99.5% can be detected by measurement of β delayed γ rays. The βp and $\beta \gamma$ branches (in terms of strength) are almost equal for ^{100}In (43% and 57% for βp and $\beta \gamma$, respectively). The result for ^{100}In shows that for nuclei close to the proton drip line both types of measurements: βp and highly efficient $\beta \gamma$, are necessary to reconstruct B_{GT} correctly.

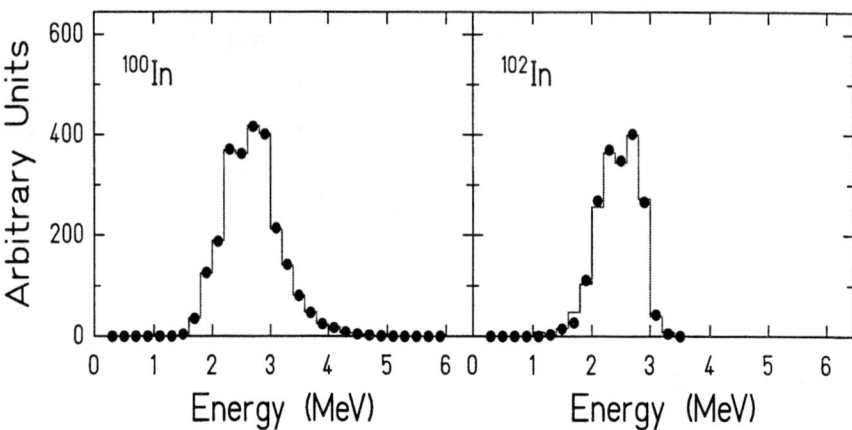

FIGURE 4. βp distribution for ^{102}In and ^{100}In. Experimental results are shown based on [18] (solid line) together with the calculated values (full circles). Note, that the input parameters of the statistical model calculation, namely strength function, were adjust in order to get the best agreement.

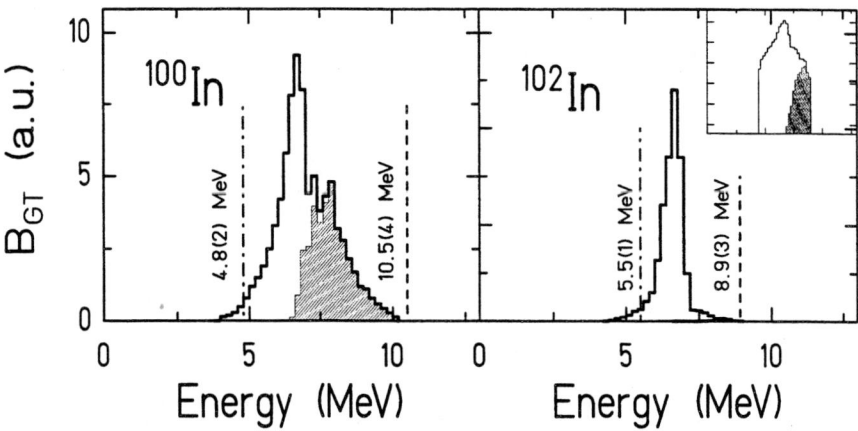

FIGURE 5. B_{GT} distribution for ^{102}In and ^{100}In used to reproduce with the statistical model calculation an experimental spectrum of β delayed protons. Filled areas show amount of strength carried out by βp. Inset shows ^{102}In B_{GT} with a logarithmic scale.

SUMMARY

It was shown that β decay of non even-even nuclei in the vicinity of ^{100}Sn is dominated (if energeticly possible) by a transformation of a proton from the $(\pi g_{9/2})^8$ core to a neutron in the $\nu g_{7/2}$ orbital. Results of the calculation of β strength, needed to reproduce experimental β delayed proton spectra for ^{102}In and ^{100}In, are consistent with the systematics of the measured β strength distributions of heavier indium isotopes. Since the calculations may be model dependent the direct measurement of $\beta\gamma$ will not only give the result for β strength distribution but should also yield (together with known βp branches) information on $\frac{\Gamma_p}{\Gamma_\gamma}$ ratio which can be used to verify theoretical predictions.

ACKNOWLEDGMENTS

The author would like to acknowledge partial support from the Polish Committee of Scientific Research, grant no. KBN 2 P03B 086 17. Nuclear physics research at The University of Tennessee is supported by the U.S. Department of Energy through Contract No. DE-FG02-96ER40983.

REFERENCES

1. R. Schneider et al., Z.Phys. **A348** (1994) 241.
2. M. Lewitowicz et al., Phys.Lett. **332B** (1994) 20.
3. K. Rykaczewski et al., Phys.Rev. **C52** (1995) R2310.
4. E. Klempt et al., Z. Phys. **C37** (1988) 179.
5. S.J. Freedman, Comments Nucl. Part. Phys. **19** (1990) 209.
6. B.A. Brown and K. Rykaczewski Phys. Rev. **C 50** (1994) R2270.
7. I.S. Towner, Nucl. Phys. **A444** (1985) 402.
8. A. Płochocki et al., Z. Phys. **A342** (1992) 43.
9. M. Karny et al., Nucl.Phys. **A640** (1998) 3.
10. Z. Hu et al., Phys. Rev. **C60** 024315 (1999).
11. J. Szerypo et al., Z. Phys. **A259** (1997) 117.
12. M. Karny et al., Nucl. Instr. and Meth. Phys. Res. **B 126** (1997) 320.
13. D. Cano et al., Nucl. Instr. and Meth. Phys. Res. **A430** (1999) 333.
14. Z. Hu et al., Nucl. Instr. and Meth. Phys. Res. **A 419** (1998) 121-131.
15. Z. Janas et al., Phys.Scr. **T56** (1995) 262.
16. P. Tidemand-Petersson et al., Z.Phys. **A302**, (1981) 343.
17. J. Szerypo et al., Nucl.Phys **A507**, (1990) 357.
18. J. Szerypo et al., Nucl.Phys **A584**, (1995) 221.
19. M. Huyse et al., Z,Phys **A330**, (1988) 121.
20. K. Schmidt et al., Z.Phys. **A350**, (1994) 99.
21. K. Schmidt et al., Nucl.Phys **A624**, (1997) 185.
22. M. Hellström et al., Z.Phys **A356**, (1996) 229.

23. T. Elmorth et al., Nucl. Phys. **A304,** (1978) 493.
24. A.W.B. Kalshoven et al., Nucl.Phys. **A337,** (1980) 120.
25. K. Rykaczewski et al., Z.Phys **A332,** (1989) 275.
26. K. Rykaczewski, GSI Report, GSI-95-09.
27. P. Hornshøj et al., Nucl.Phys. **A187** (1972) 609.
28. B. Jonson et al., CERN 76-13, (1976) 277.

Spectroscopy Of β-delayed Charged Particles At Projectile Fragment Separators

Zenon Janas

Gesellschaft fur Schwerionenforschung mbH, D-64291 Darmstadt, Germany
Warsaw University, 00-681 Warszawa, Poland

Abstract. The combination of projectile fragmentation reactions and in-flight separation has proved to be a powerful tool to produce nuclei at the limits of stability. Decay studies of very neutron-deficient projectile fragments led to the discovery of several new β-delayed particle emitters. Basic principles of the method are described and various aspects of extracting interesting spectroscopic information from β-delayed particle studies at projectile fragment separators are discussed.

INTRODUCTION

Studies of β-delayed particle emission gain their importance with the increased distance from the valley of stability. More exotic neutron-deficient nuclei exhibit both higher β-decay Q-values and lower charged particle separation energies. Far enough from stability β-delayed proton or α-particle emission turn into the dominating decay mode. Moreover, exotic β-delayed multiparticle decay modes become energetically possible.

Different types of spectroscopic information can be extracted from the β-delayed particle studies. The observed decay rate of β-delayed particles has been frequently used to determine the lifetime of the precursor. The energy spectrum of the delayed particles contains Q-value information. In light nuclei, where the decay of the isobaric analog state (IAS) can be clearly identified in the spectrum of β-delayed particles, the precursor ground state mass can be precisely determined with the isobaric multiplet mass equation. Many studies of β-delayed particle emitters have focused on the detailed determination of the Gamow-Teller (GT) strength distribution. Results of these studies allowed one to measure the GT-strength distribution over a large range of excitation energies in the daughter nuclei, to investigate the quenching of the GT-strength in the β-decay, and to probe isospin symmetry in β-transitions of mirror decays.

In recent years the in-flight projectile fragmentation has been successfully used to produce nuclei at the limits of stability. The observation of the very neutron-deficient doubly-magic nucleus ^{100}Sn at GSI [1] and GANIL [2] and the very recent identification of ^{48}Ni at GANIL [3] are the most spectacular results, demonstrating the sensitivity of the method. Decay studies of very neutron-deficient projectile fragments lead to the discovery of several new β-delayed proton emitters [4-6], and permitted detailed studies of the β-decays of very neutron-deficient nuclei [7-11].

In this contribution we restrict our discussion to the experimental aspects related to the spectroscopy of β-delayed particles at the projectile fragment separators. In the following the basic principle of projectile fragment isotope separation will be described. The further discussion will concentrate on the specific problems related to β-delayed particle detection in conditions encountered at fragment separators.

PRODUCTION AND IDENTIFICATION OF ISOTOPES AT PROJECTILE FRAGMENT SEPARATORS

In the fragmentation reaction projectiles, accelerated to the energies between 50 and 1000 MeV/u, are stripped of several nucleons in collision with the target and form pre-fragments. Due to the small momentum transfer between the projectile and the target, pre-fragments move with the velocity of the projectile and lose their excitation energy by the particle evaporation and γ-ray emission.

The fragmentation cross-sections are well described by the EPAX formula [12] based on the parametrization of measured values. Experiments at the projectile-fragmentation facilities at GANIL [13], GSI [14], MSU [15] and RIKEN [16] have demonstrated that the nuclear fragmentation process is a very efficient and universal way of producing radioactive isotopes over the entire chart of nuclei, including the most exotic species.

The FRS spectrometer [14] employing beams with relativistic energies delivered by the SIS synchrotron at GSI-Darmstadt, Germany, is an example of an in-flight separator based on projectile-fragmentation kinematics. The layout of the apparatus is schematically shown in Figure 1.

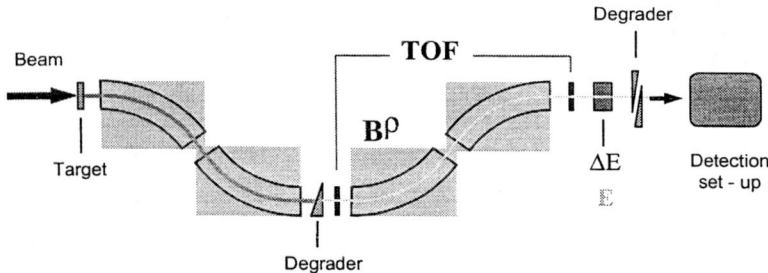

FIGURE 1. Schematic view of the Fragment Separator (FRS) at GSI-Darmstadt [14].

The first part of the spectrometer plays the role of a A/Z separator of the produced fragments. In addition to the A/Z separation, the isotopes undergo a selection according to $A^{2.5}/Z^{1.5}$ due to their different slowing-down in the profiled degrader placed in the central dispersive plane of the spectrometer. The selectivity of the separation can be further enhanced by using the velocity filter (LISE3) or additional degrader after the first dipol (FRS). Beams of projectile fragments with the purity higher than 95% [7,8] have been separated for spectroscopic studies.

The set-up of the in-beam detectors is used to measure the time of flight (TOF) of ions through the spectrometer, their magnetic rigidity (Bρ) and the energy loss (ΔE) spectrum. For completely stripped atoms standard TOF-Bρ-ΔE analysis allows an unambiguous identification of the atomic and mass number of every ion passing through the spectrometer.

Ions arriving at the end of the spectrometer have energies of the order of 200 MeV/u and have to be slowed down and stopped in the radiation detection set-up for decay studies. The sensitivity of these studies can be significantly enhanced by the possibility of time and position correlation between the identified ion and observed radiation.

As the in-flight separation method does not involve any chemical processes, it may be applied to study isotopes of any element. The half-life limit for the isotopes to be studied by this technique is determined only by the time of flight through the ion optical system, which is of the order of few hundred nanoseconds. This time is much shorter than the time scale characteristic for the fastest β–decays but it may become critical for studies of nuclei beyond the particle driplines, e.g. in searches for direct proton emitters.

The fast, selective and element independent separation favours the in-flight separation technique over the on-line isotope separators (ISOL) method. The drawback of the projectile-fragmentation method from the point of view of β-delayed particle spectroscopy, is the broad stopping range distribution of the separated isotopes. Due to the fragmentation reaction mechanism and the energy straggling in the materials it may extend up to few hundred micrometers, depending on the beam energy on spectrometer setting. This is in contrast to the ISOL technique where very thin, point-like, surface deposited sources are available for spectroscopic studies.

SPECTROSCOPY OF β-DELAYED CHARGED PARTICLES

A combination of magnetic separation and energy-loss selection methods allows for separation of very pure beams of selected projectile fragments. An influence of possible contaminants can be further suppressed by applying the time and position correlation between the ion of interest and detected radiation [8-10]. High purity of the separated beams is particularly favorable for studies far from stability, where β-delayed particle emission becomes a dominating decay mode, but the intensities of separated projectile fragments beams are limited by the very low reaction cross-ections and available intensity of the primary beam.

To adequately address physics questions related to β-delayed particle emission studies, the detection set-up for spectroscopy of β-delayed particles should enable efficient, precise and complete measurements of the decay process. These requirements imply high efficiency, good energy resolution, capability of particle identification, and the possibility of detection of β-delayed multiparticle decay events in the detection set-up.

So far, the implantation of projectile fragments in silicon detectors is commonly used as a simple and efficient way of observing β-delayed particle emission. Figure 2 shows, as an example, the detection set-up used in the precise decay studies of ^{36,37}Ca at the GSI Fragment Separator [7]. To obtain as narrow as possible an implantation profile along the beam direction (measured FWHM ≈ 100 μm), the FRS was operated in the "monoenergetic" mode. At the exit of the spectrometer, the Al degrader was used to adjust the energy of the fragments and to implant ^{37}Ca ions in the middle of the central, 500 μm thick, Si detector.

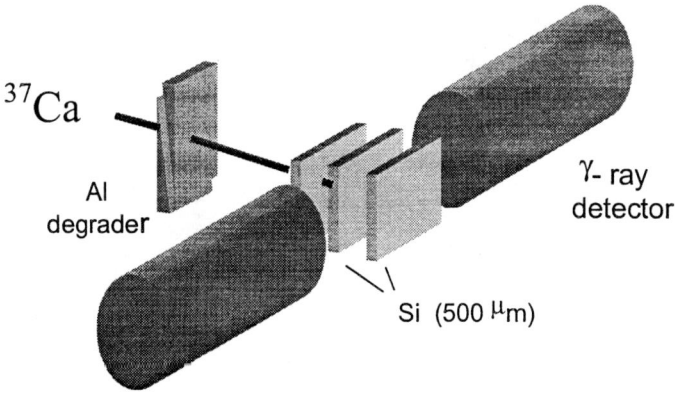

FIGURE 2. Detection set-up used at GSI Fragment Separator for decay spectroscopy of ^{36,37}Ca [7] The isotopes were implanted into a central 30 mm × 30 mm × 500 μm silicon detector. Two neighboring silicon counters were used for detecting β-rays, while large volume germanium detectors registered γ-rays emitted by implanted nuclei.

This assures that protons with energies up to about 4 MeV were fully stopped in the implantation detector while the full-energy detection efficiency for 6 MeV protons amounts to about 85 %.

The upper histogram in Figure 3 shows the raw energy spectrum of β-delayed protons registered in the implantation detector during the ^{37}Ca measurement. Signals registered by the implantation detector represent the sum of three components:(i) the energy of the recoil atom, (ii) the energy of the emitted β-delayed particle(s), and (iii) the energy deposited by the β-particle.

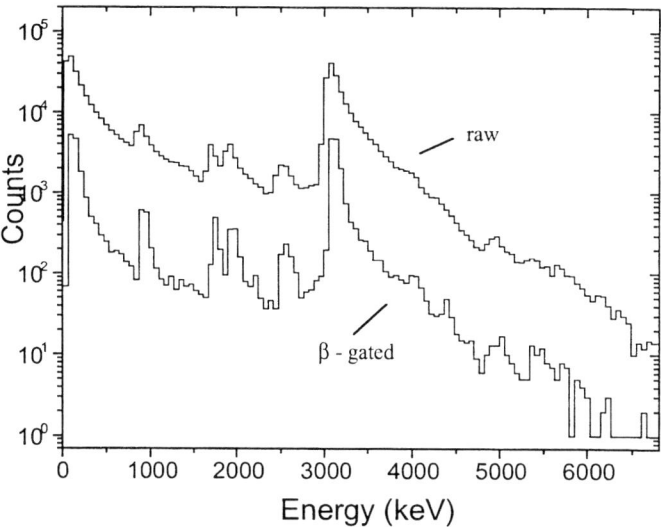

FIGURE 3. Energy spectra of β-delayed protons from ^{37}Ca decay: raw spectrum, spectrum in coincidence with a β-ray detected with the energy smaller than 450 keV in the β-counter [7].

The kinetic energy of the recoil is determined by the masses of the emitted particle and recoil, and by the transition energy. Due to the pulse-height defect [17], only part of the recoil energy is registered by the detector. For instance, the kinetic energy of ^{36}Ar recoil after proton decay (Q_p = 3.1MeV) of IAS in ^{37}K amounts to 84 keV, but only 28 keV will be recorded by the silicon detector.

The energy deposited by β-particles has a continuum spectrum, its shape depends on the detector thickness, implantation profile and the end-point energy of the β spectrum. Summing with the β-rays energy-loss spectrum influences both - the position and the shape of β-delayed particle line. This is illustrated in Figure 4 which shows the simulation of the effect of β-summing for the 3.1 MeV proton transition deexciting IAS in ^{37}K, after implantation in the middle of the 500 μm silicon detector.

The effect of β-summing can be reduced by restricting the path length of β-particles in the implantation detector. In the detection set-up shown in Figure 2 this was accomplished by accumulating proton spectra in coincidence with β-ray energy losses in the β-detector. As it is shown in Figure 3, this requirement reduces the summing effect in the proton line shape (the FWHM of 3.1 MeV line in the β-gated spectrum amounts to 70 keV) but at the same time the spectrum statistics is decreased by a factor corresponding to the efficiency of the β-counter.

The effect of β-summing can be diminished by decreasing the thickness of the implantation detector. The lower limit is given by the increase of capacity of the detector which increases the noise level and deteriorates the energy resolution. In a typical situation, the thickness of the detector is comparable or smaller than the width of the implantation profile. In experiments at the FRS, usually a stack of implantation detectors has to be used to cover the full implantation range of selected isotope.

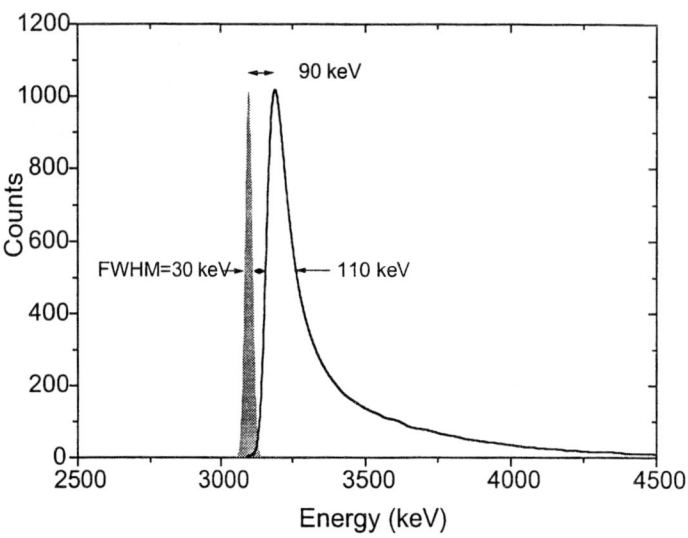

FIGURE 4. The simulation of β-summing effect for the 3.1 MeV proton line (shaded histogram) deexciting the IAS state in ^{36}K, after implantation (FWHM = 100 μm) in the middle of the 500 μm Si detector. The observed line shape (blank histogram) reflects the energy-loss spectrum of β-rays with the end-point Q_β=5 MeV: 90 keV energy shift corresponds to the most probable energy-loss of β-particles, the total width of the line (FWHM= 110 keV) results from the intrinsic detector resolution (30 keV) and broadening due to the β-summing. Both lines were normalized to the same height.

A broad, compared to the detector thickness, implantation profile results in the proton's escapes from the detector. This effect is illustrated in Figure 5 which shows the calculated full-energy detection efficiency as a function of energy of protons emitted from the source homogeneously distributed in the 150 and 500 μm thick implantation detectors, respectively. In the case of the thin detector only 50% of protons with the energy of about 4 MeV is registered with the full energy.

Decay studies at projectile fragment separators offer the possibility of measurements of absolute intensities of transitions, without invoking the intensity balance in the decay scheme. In this case one benefits from the identification procedure which allows to determine the number of ions of selected isotope implanted in the detector. This number has to be corrected for losses due to the secondary reactions in the final degrader and in the detector itself. The actual value of the correction can be calculated by knowing the thickness of the stopping material and the total reaction cross-sections [18,19]. Typical correction factors are of the order of 10 % [10] at FRS and about 2 % [9] at LISE energies.

Precise determination of the number of counts in the lines identified in the spectrum of β-delayed particles is usually difficult due to poor energy resolution and complex line shapes. Moreover, the correction for the full-energy detection efficiency, which depends e.g. on the implantation profile, has to be applied to obtain true intensities of the observed transitions. Finally, the absolute intensities are determined

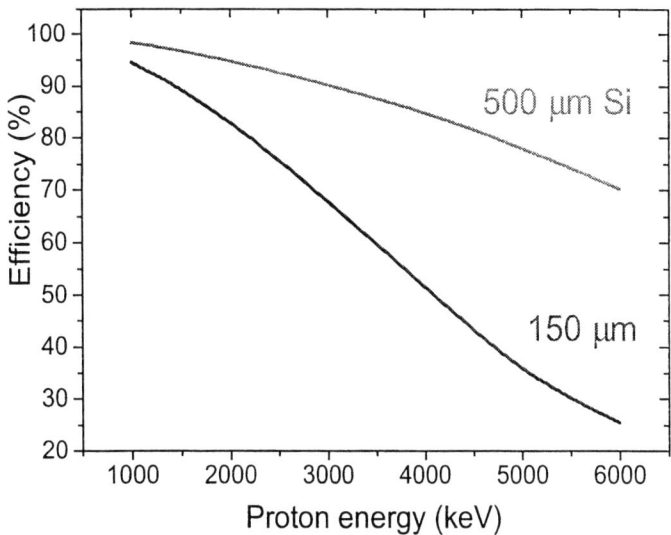

FIGURE 5. The full-energy detection efficiency for protons emitted from the source homogeneously implanted in the 150 and 500 μm thick silicon detector.

as the ratios of the efficiency-corrected peak intensities and the actual number of implanted ions.

For β-transitions to excited states lying high enough above the particle separation threshold one can assume that the partial width for particle emission dominates the total decay width. Under this condition, β-decay feedings can be determined from the absolute intensities of particle transitions deexciting these states.

For the levels close to the particle emission threshold, where γ-ray emission may compete with particle decay, supplementary γ-rays measurements are essential [7]. Measurements of β-delayed particle-γ-rays coincidences are required to identify transitions to excited states in the final nucleus.

The implantation of nuclei in solid state detectors causes severe limitations in studies of β-delayed multiparticle decay modes. Unambiguous identification of processes like β-delayed two-proton ($\beta 2p$), proton-alpha ($\beta p\alpha$) or three-proton ($\beta 3p$) decay requires the decay multiplicity to be measured. The energies and emission angles of individual particles contain information on the decay scheme and the process mechanism.

To be detected separately, the emitted particles have to escape the volume of silicon surrounding the implanted nucleus. However, since the total transition energy is shared by the emitted particles, their energies will constitute only a fraction of the decay Q-value. This enforces the use of very thin implantation detectors and results in losses in the implantation efficiency, since, even at the lowest energies used for projectile fragmentation, only a small fraction of the range distribution will be covered by the detector.

In the search for the β-delayed multiparticle decay of ^{31}Ar at GANIL [20], the ions were implanted in the 20 μm silicon detector placed between two 4-fold segmented, 500 μm silicon counters. The latter detectors were used to register β-delayed particles, which escaped from the implantation counter. The geometry of the telescope determined the complexity of the detector response. In particular, due to several detector and particle dependent factors, there was no simple correspondence between the detected event fold and the particle multiplicity. Therefore, the identification of the multiparticle decays had to be based on the simulated response of the detector and decay energetics considerations. Among the energetically allowed β-delayed multiparticle decay channels of ^{31}Ar, $β2p$ decay was identified and the first observation of the $β3p$ decay was reported [20]. The claim of the existence of a 2 % $β3p$ branch in the decay of ^{31}Ar was not confirmed by recent decay studies of this isotope at ISOLDE [21].

This example indicates difficult to overcome limitations of the solid state implantation technique for studies of multiparticle decays. Even with a dedicated detection set-up and under optimized implantation conditions complete information on the decay process remains unaccessible. Consequently, the interpretation of the data is complicated and unfirm.

An alternative to the solid state implantation technique has been proposed by Blank et al. [20]. In the method investigated, nuclei produced in the projectile fragmentation are implanted in a gas detector and deposited by electrostatic collection at the surface of the silicon counter. The detector yielded a good energy resolution, the possibility of particle identification and multiplicity measurement but it suffered from a low collection efficiency.

SUMMARY

The combination of projectile fragmentation reactions and the in-flight separation technique has proved to be a powerful method to study very exotic nuclei. Owing to the good production and superior separation performances of fragment separators, very neutron-deficient isotopes, up to the border lines of particle stability, have been identified for elements below Z=50. For many isotopes, first characteristics of their radioactive decays and nuclear structure have been found.

In studies of β-delayed particle decays at projectile-fragment separators, the technique of deep implantation of nuclei in the solid-state detectors has been commonly used. This method guarantees high efficiency for the detection of β-delayed particles but it introduces severe limitations in precision of the energy determination and the achievable energy resolution. The complicated response of the detector requires extended on-line calibrations involving implantation of well-known activities in the detection set-up. Moreover, the implantation of nuclei in the detector volume excludes the particle identification and practically disables measurement of multiplicity as well as individual particle energies in the case of multiparticle decays. As these disintegration modes become widely open for nuclei at the limits of stability, to study these exotic processes new techniques for particle detection have to be developed. Another challenge for the spectroscopy of charged particles at the projectile-fragment separators represent studies of nuclei beyond the particle driplines. Short (submicrosecod) lifetimes of these nuclei require identification of a decay signature in the presence of the high-energy fragment signal.

ACKNOWLEDGMENTS

This work is partially supported by the Polish Committee of Scientific Research under grant KBN 2 P03B 086 17.

REFERENCES

1. Schneider, R., et al., Z. Phys. A348, 241 (1994).
2. Lewitowicz, M., et al., Phys. Lett. B332, 20 (1994).
3. Giovinazzo, G., contribution to this Symposium.
4. Faux, L.,et al., Phys. Rev. C49, 2440 (1994).
5. Faux, L., et al., Nucl. Phys. A602, 167 (1996).
6. Blank, B., et al., Phys. Lett. B364, 8 (1995).
7. Trinder, W., et al., Nucl. Phys. A620, 191 (1997).
8. Trinder, W.,et al., Pys. Lett. B459, 67 (1999).
9. Bhattacharya, M., et al., Phys. Rev. C58, 3677 (1998).
10. Liu., W., et al., Phys. Rev. C58, 2677 (1998).
11. Piechaczek, A., et al., Nucl. Phys. A584, 509 (1995).
12. Summerer, K., et al., Phys. Rev. C42, 2546 (1990).
13. Muller, A. C., Nucl. Instr. and Meth. B56/57, 559 (1991).
14. Geissel, H., et al., Nucl. Instr. and Meth. B70, 286 (1992).
15. Sherrill, B. M., et al., Nucl. Instr. and Meth. B56/57 1106 (1991).
16. Kubono, S., et al., Nucl. Instr. and Meth. B70 583 (1992).
17. Wilkins, B. D., et al, Nucl. Instr. and Meth. 92, 381 (1971).
18. Kox, S., et al., Phys. Lett. B159, 15 (1985).
19. Shen, W., et al., Nucl. Phys. A491, 130 (1989).
20. Bazin, D., et al., Phys. Rev. C45, 69 (1992).
21. Fynbo, H. O. U., et al., Phys. Rev. C59, 2275 (1999).
22. Blank, B., et al. Nucl. Instr. and Meth A330, 83 (1993).

Beta-Delayed Two-Particle Emission

M. J. G. Borge

Instituto de Estructura de la Materia, CSIC, Serrano 113bis, E-28006 Madrid, Spain

INTRODUCTION

The main characteristic of beta-decay far from stability is the number of decay channels open. This is due to the quadratic increase of the isobaric mass differences and the reduction in the separation energies for emitting nucleons or clusters of nucleons when going away from the valley of stability.

The process of β-delayed particle emission has been the object of much study during the last few decades, as it allows one to uniquely determine the decay-pattern from the energy of the emitted particle, if the final state is known. A summary of this subject with a beautiful description of the first steps in the field is presented elsewhere in these Proceedings (see J. Hardy).

Ground state proton emission was predicted in 1914 [1] when there was not even a definition of proton and it was first observed in ^{151}Lu [2] and ^{147}Tm [3] in 1982. From then on many cases have been observed both in the spherical and deformed region of masses around A = 150 (see contributions to this proceedings). The observation of these decays is enabled by the long proton half-lives when traversing the large Coulomb barriers.

Two proton radioactivity as well as beta-delayed two-neutron emission were predicted in 1960 [4] as a consequence of the pairing force. The latter process was observed for first time in 1979 in the decay of ^{11}Li [5] and inspired the prediction of the mirror process, beta-delayed two-proton emission ($\beta 2p$) by Gol'danskii [6]. Near the proton drip line it should be easier for an even-Z nucleus to eject a pair of protons than to break them apart, and therefore decay by two-proton radioactivity. The proton pair will escape via tunneling through the Coulomb barrier and one can, by detecting their energies and angular correlations, obtain information about the nucleon-nucleon interaction in the region where the nuclear field is gradually replaced by the free nucleon-nucleon force. The ocurrence of this decay mode is thus a result of the odd-even staggering in the single proton separation energies which results in situations were $S_p > 0$ and $S_{2p} < 0$. In spite of the intense experimental efforts the 2p decay mode has not been identified, the lightest candidates, (estimated S_{2p} < 1 MeV) ^{22}Si [7], ^{31}Ar [8] and ^{39}Ti [9] have been observed to dominantly beta decay. Remaining candidates would be nuclei with predicted [10] $S_{2p} > 1$ MeV and therefore having half-lives in the observable range (1 μs - 10 ms). They are ^{45}Fe and ^{48}Ni. The observation of the latter isotope (3 events) and, with higher intensity, of the former one recently in GANIL (see contribution of J. Giovinazzo in this proceedings) gives new hope for the observation of this decay mode in the near future. Nevertheless 2p-emission has been observed from the unbound ground states of ^{6}Be [11] and ^{12}O [12] and in several cases from excited states populated by superallowed beta decay.

In the case of β-delayed multi-particle emission the nucleus breaks up into more than two particles. The kinematics of two-particle breakup is fully determined by momentum and energy observation. The main interest in β-delayed multi-particle emission is the fact that the mechanism of the break-up is not fully determined by energy and momentum conservation. In three body break-up there are three binary subsystems and each subsystem have resonances controlling the break-up. Either the break-up proceeds via each of these resonances sequentially or the beta-daughter breaks up directly into the three body continuum; as occured in β-delayed deuteron emission from ^{6}He.

If we use the traditional "knocking" picture as in α-decay, the width of the resonance in the sequential decay is small, and the Δt is large so the third particle has time to penetrate the barrier if $\Delta t \leq 10^{-21}$s. These resonances are large in binary subsystem which thus leads to simultaneous emission. This picture indicate that the process is determined by the width of the renonance in the binary subsystem and the height of the barrier but this line of argumentation neglects the importance of the structure within the decaying state that could single out a specific channel. *Learning about this structure is one of the main goals in studying β-delayed multi-particle emission.*

Beta-Delayed Two-Neutron Emission

The study of extremely neutron-rich isotopes remain a major experimental challenge in the current research on far-unstable nuclei. Twenty years has elapsed since the first observation of β-2n emission but the progress has been slow due to the technical problems associated with the detections of neutrons. The β-2n precursors found in the literature are summarized in Table 1. The multiplicities of the processes have been established either by time correlation using a 4π neutron long counter, by identifying the growing activity of the βn daughter, or by looking indirectly to the multiplicities. Due to the technical problems of measuring delayed neutrons nothing is known about the mechanism of these decays. The most extreme case is the decay of ^{17}B ($Q_\beta = 24$ MeV), where the ^{17}C excited states are formed by a core of ^{12}C and five loose neutrons, all accesible to the Q_β window, four of them were detected in the experiment at GANIL [13].

As delayed neutron and multineutron emission dominates the decay of neutron rich nuclei far from stability, one needs new detectors with high efficiency and good energy and angular resolution. This is fulfilled by the new generation of scintillator arrays such as the ones existing at MSU, RIKEN or the newly built array TONNERRE at GANIL, comparison of their characteristics are summarized in reference [14]. These neutron detector arrays will give more insight into the problem in the coming decade.

TABLE 1. Beta delayed multi-neutron precursors

Nuclide	$T_{1/2}$ (ms)	Decay Mode	P_{0n} (%)	P_{1n} (%)	P_{2n} (%)	P_{3n} (%)	P_{4n} (%)	Ref.
$^{11}_{3}$Li	8.5(2)	β, βn, β2n, β2α3n, $\beta\alpha$n, βt, βd	6.3(6)	87.6(8)	4.2(4)	1.9(2)		[15]
$^{14}_{4}$Be	4.35(17)	βn, β2n, β3n		~100	< 2.4^a			[16]
$^{17}_{5}$B	5.08(5)	β, βn, β2n, β3n, β4n	21(2)	63(1)	11(7)	3.5(7)	0.4(3)	[13]
$^{19}_{6}$C	49(4)	β, βn, β2n	46(3)	47(3)	7(3)			[13]
$^{30}_{11}$Na	50(3)	β, βn, β2n	69(4)	30(4)	1.17(16)			[17]
$^{31}_{11}$Na	17.0(4)	β, βn, β2n	62(5)	37(5)	0.9(2)			[17]
$^{32}_{11}$Na	13.2(4)	β, βn, β2n	68(7)	24(7)	8(2)			[17]
$^{33}_{11}$Na	8.2(4)	β, βn, β2n	36(21)	52(20)	12(5)			[17]
$^{34}_{11}$Na	5.5(10)	β, βn, β2n		115(20)^b				[17]
$^{53}_{20}$Ca	90(15)	β, βn, β2n		< 30^b				[18]
$^{98}_{37}$Rb	114(5)	β, βn, β2n	86.4(5)	13.6(5)	0.051(7)			[19]
$^{100}_{37}$Rb	51(8)	β, βn, β2n		6(3)	0.16(8)			[20]
$^{136}_{51}$Sb	820(20)	β, βn, β2n		24				[21]

^{a}limit given for β(2n+3n) ^{b} value stated for β(n+2n)

Beta-delayed two-proton emission

Due to the elusive character of the 2p-radioactivity the possibility of observing correlated emission of two protons from excited states fed in β-decay has caused quite some interest in the β2p decay mode. Experimentally this mode is more accesible than that of its partner (β2n) as the detection of delayed protons with good energy resolution is relatively easy over a broad energy range using ordinary Si detectors. Secondly for nuclei with Z > N the superallowed Fermi transition collects a large feeding to the IAS. Close to the drip line this state is placed at large excitation energy and therefore open to 2p-emission. As the escape of the delayed proton pair require tunnelling through the potential barrier, the observation of energy and angular correlation between the protons could give information on the interaction between these protons in the subbarrier region, where the attraction due to the nuclear forces is gradually replaced by Coulomb repulsion. However, this decay mode will compete with sequential emission where the first proton feeds a proton emitting state. The question of the mechanism of β2p concerns the relative importance of sequential and direct emission of the two protons. Direct emission depends upon the degree of correlations between the protons. When the protons are fully correlated in a L = 0, S = 1 state it is referred to as *"diproton"* emission. Experimentally the different mechanisms can be distinguished by looking at the individual proton energies and their angular distributions. In sequential emission the individual proton energies depend on the intermediate state. Thus, the spectrum will peak at a certain energy while in direct emission the proton spectrum is continuous. The angular distribution in direct emission is far from being isotropic while the angular dependence introduced by the momentum coupling in the sequential case is negligible.

The fact that the IAS collects a large part of the β^+-strength explain that the first observation of β2p was done from the IAS of ^{22}Al [22,23], ^{26}P [23,24] and ^{35}Ca [25] at the Lawrence Berkeley Laboratory. In their studies the radioactivity was deposited in a catcher foil that was viewed by a set of Si telescopes at two different angles. This pioneering work, described in detail elsewhere in this Proceedings (J. Cerny), established that the β2p emission mechanism is in these cases sequential. Since then, other β2p-emitters have been identified at GANIL: ^{23}Si [26], ^{27}S [27], ^{31}Ar [8], ^{39}Ti [9] and ^{43}Cr [28]. The detector system used here consisted of a set of 4-6 Si detectors forming a telescope. The radioactive nuclei produced, identified by their time of flight and energy loss, were implanted in a thick detector and the total energy of the emitted particles was measured in the telescope.

These studies were not sensitive to the mechanism of the two proton emission since the energies of the protons were not measured individually due to the high energy of the radiactive beam. The beam was slowed down and finally stopped in the Si-detectors, the straggling produced in this process gives an energy resolution not better than 100 keV. Due to chemical reasons, out of these nuclei only ^{31}Ar has been produced for study at ISOLDE.

Beta-Delayed Multi-Particle Emission

Due to the stability of the α-particle $\beta\alpha$ has been observed in very light neutron rich nuclei as ^{11}Li and ^{8}He. This fact combined with the unbound character of ^{8}Be dominates the multiparticle break-up of the very light proton rich nuclei. As an example ^{9}C β-decays to states in ^{9}B that are all well above the p$\alpha\alpha$ threshold. Moreover, this decay mode is more favoured than that of α^5Li. The mirror partner, ^{9}Li has also large decay branches (50 %) to n$\alpha\alpha$ final states. The latter has been fully explained in terms of sequential decays through both the ^{5}He and ^{8}Be channels [29], although components directly to n$\alpha\alpha$ states have also been suggested [30]. In the A = 12 isobars, the β-decays of ^{12}N and ^{12}B have both small branches (3.5(5) % and 1.6(3) % respectively) to states in ^{12}C lying above the 3α-threshold. The decay mechanism is not known although it is assumed to proceed via the ^{8}Be ground state. The difficulty in this study rests on the fact that the main decay branches populate a 0^+ state in ^{12}C only 287 keV above the ^{8}Be + α threshold. In the beta decay of ^{17}Ne, pα-decay has been observed from the IAS in ^{17}F to ^{12}C ground state, decaying sequentially through states in ^{16}O and ^{13}N [31].

Although several light neutron deficient nuclei have been observed to decay both by βp and $\beta\alpha$ (^{36}K, ^{40}Sc, ^{48}Mn), the βpα decay mode had never been observed before. Nuclei with A $\sim$ 100 and heavier, decay simultaneously via βp, $\beta\alpha$ as well as via alpha and proton radioactivity.

Theoretical approach

Little theoretical effort has been made towards a description of multiparticle breakups. Shell model has been used to calculate spectroscopic factors for di-proton emission from the decay of ^{22}Al [32] and for the possible sequential decays of ^{9}C and ^{9}Li [33]. The calculation of Brown [32] indicated that the ^{2}He branch is hindered by two orders of magnitud with respect to the 1p-branch due to the reduction in penetration factor caused by the doubling in charge of the emitted particle, This calculation did not take into account the final-lifetime of the di-proton. For the case of the A=9 decays, the spectroscopic factors were calculated using four interactions giving quite different predictions.

The case of no narrow resonances compared with the energy available in any of the two-body subsystems has been discussed in [34] and references there in. The authors argue that in this case the breakup of the binary system proceeds so fast that the presence of the third particle is still felt. They have developed a formalism for expanding the decay amplitude in a set of functions (hyperspherical harmonics) that fully describes the final state of three particles. This formalism has been applied to the break-up resonances in ^{6}Be, ^{6}He, ^{6}Li, ^{9}Be and ^{9}B [34]. Its potentiallity and modus operandi has been described by I. Mukha in this conference where the application to new cases is presented.

THE CASE OF ^{31}AR

In this section I report on two recent experiments done at ISOLDE (CERN) on the decay of ^{31}Ar aiming to study the mechanisms of β-delayed two proton emission.

The progress in the understanding of the decay of ^{31}Ar has come from the development and improvement in the detection system of the produced radioactivity. Two main features of ISOLDE have been used in this work. Firstly the fact that the radioactivity is extracted and transported to the detection setup with low energy and high beam quality. Secondly the pulsed structure of the beam with a frequency at maximum of 1.2 s/pulse. The latter allows, in case of short lived species, a determination of signal and background in the same pulse. The former makes it possible to stop the beam in a thin carbon foil resulting in a point like source and to use thin Si detectors to measure the particles emitted in the decay. The use of thin detectors allows a beta-proton discrimination well below 1 MeV.

Detecting very exotic species a high efficiency of the setup is mandatory in order to compensate for the low yield. Furthermore to increase the efficiency of multiparticle detection it is neccessary to use several detectors. To distinguish between the different possible mechanisms in the distribution of the individual energies, granularity as well as large solid angle are required. The efficiency of the two proton detection in a set-up with N detectors with solid angles $\Omega_1, \Omega_2, ..., \Omega_N$ is,

$$\epsilon_{2p} = (\sum_{i=1}^{N} \Omega_i)^2 = \sum_{i=1}^{N} \Omega_i^2 + 2 \sum_{i,j(i \neq j)} \Omega_i \Omega_j \quad (1)$$

Where the first term represents the 2p summing and the second corresponds to the two protons hitting different detectors. It is this last term that give us the individual proton energies and the angular distribution. As this term goes quadratically with the number of detectors it is attractive to have a highly segmented set-up.

In the following, I will briefly describe the setup used at ISOLDE to study the decay of ^{31}Ar For the production of ^{31}Ar a calcium oxide target (CaO) connected by a water cooled transfer line to a plasma ion source was used. The target was bombarded by a pulsed 1 GeV proton beam of 3.0×10^{13} protons coming from the PS-Booster (PSB). The resulting ^{31}Ar yield was of 2-3 atoms/s depending upon the number of pulses (6-9 out of 12 in a 14.2 s repetition cycle) delivered to ISOLDE by the PSB. The production of Ar ions was determined, in both experiments, by normalising the branching ratio of the most intense proton peak, 2.1 MeV, to 29(3) % determined as the average mean of the values obtained in the previous experiments done at GANIL [27,35]

1995: General Features of the ^{31}Ar Decay Scheme

The first experiment done in 1995 aimed to gain a better understanding of the β-decay of ^{31}Ar by the study of the decay modes βp, β2p, β3p, $\beta\alpha$, βpα and $\beta\gamma$ which are all energetically allowed. The first three had been identified in previous works [8,27,35,36]. So the setup should allow to separate and identify low energy proton and alphas, high energy protons and gammas. The beam passed through the hole of a thick annular Si detector (Ω=4.3 %) and was stopped in a 40 μg/cm^2 C-foil tilted with respect to the beam direction. A 70 % coaxial HPGe γ-detector (Ω=7.4 %) was located behind the foil. In a perpendicular plane two telescopes were placed one with Si in the front and back (Ω=3.3 %) and the other consisted in a CF$_4$ filled front detector and a Si in the back (Ω=2.3 %). For details see reference [37].

This experiment , covering a wide energy range (from 200 keV to more than 12 MeV) allowed to give a strict limit in the 2p-radiactivity , 6.0×10^{-4}, from ^{31}Ar [38]. Due to the pulsed structure of the beam, the study of the temporal behaviour of the γ-rays allowed to identify four γ-ray following the decay connecting states in ^{30}S, ^{30}P (^{30}S daughter) and from ^{31}S (from the β-decay of ^{31}Cl). No γ-ray was seen connecting states in ^{31}Cl. Therefore we used the β-delayed proton events selected in the telescopes to determine the ^{31}Cl ground state feeding to 23(8) % in good agreement with the expected 25% obtained by SM calculations [35].

The method to analyse the β2p data is the same in both experiments and will be described later. In this first experiment we suffered from low solid angle coverage (Ω_{tot}=10 % of 4π, ϵ_{2p}= 0.9 %) and low granularity only 65 % of the events goes

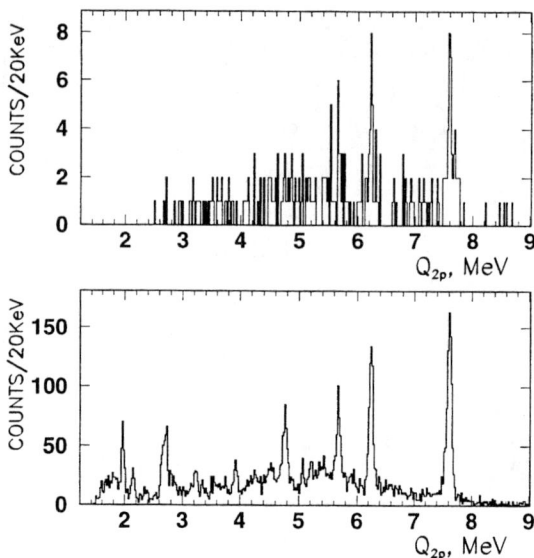

FIGURE 1. The recoil corrected two-proton sum spectrum. Upper half shows the data from 1995 and the lower half the improved data from 1997. Notice that both spectra correspond to the same number of deposited ^{31}Ar ions.

to different detectors, plus an energy cut-off at 1.3 MeV due to the use of a thick annular detector. The resulting 2p-spectrum show in the upper half of Fig. 1 correspond to a total of 9×10^5 ^{31}Ar ions deposited in the foil. Three peaks could be identified connecting the IAS with the ground state and the two lowest excited states in the two proton daughter ^{29}P.

But in order to study the mechanism of the two-proton one has to go beyond the Q_{2p} spectrum. The individual proton projection for each of these peaks was not conclusive. The decay could be interpreted as sequential through many intermediate states but the observed spectrum was also compatible with the continuous energy distribution expected for simultaneous two proton emission, see details in [38].

1997: Mechanisms of the $\beta 2p$ resolved for the decay of ^{31}Ar

Aiming for high granularity and large solid angle a new setup was assembled. It consisted of a hemispherical mount holding 15 Si p-i-n diodes (FUTIS) ($\Omega_{tot}=14$ %) developed in Jyväskylä having a central hole to allow the ^{31}Ar beam to reach and be stopped in the C-foil placed at the center of the hemisphere. On the opposite side of it a double sided Si strip detector with 16x16 strips and total dimension 5x5 cm^2 was placed. Behind the strip detector a thick large area Si-detector was placed to fully stop high energy protons and to allow for a good discrimination between β-particles and protons. FUTIS was operated using standard electronics whereas the strip channels were analyzed using specially designed preamplifiers and amplifiers from RAL of the type shown by P. Woods in his contribution to this Symposium.

With a total geometric solid angle of 25 % of 4π divided in 271 segments and 95 % of the total 2p-efficiency for events with two protons hitting different detectors, this set up combines excellent efficiency with good angular resolution.

To analyse the β-delayed two-proton data the multiplicity-two events were selected and the energy and positions of the two detected protons determined. Using energy and momentum conservation we then derive the recoil energy of the daughter nucleus and thereby reconstruct the full decay energy (Q_{2p}) of the event

$$Q_{2p} = E_1 + E_2 + \frac{m_p}{M_{29P}} \left(E_1 + E_2 + 2\sqrt{E_1 E_2} \cos \Delta\Theta \right), \quad (2)$$

where E_i is the detected energy of the ith proton, m_p is the proton mass, M_{29P} is the mass of the recoiling two-proton daughter and $\Delta\Theta$ is the angle between the two protons. Note that this calculation is independent of the mechanism of the two-proton emission, but the quality of the recoil correction depends on the accuracy in determining $\Delta\Theta$. Due to the large size of the individual detectors, the effect of finite solid angles was included, when determining the Q_{2p} values of the IAS. In the last setup each detector segment was much smaller making this step unnecessary. Fig. 1 shows the Q_{2p} spectrum obtained in this way for the two experiments. In the last experiment the threshold could be lowered to 600 keV which allowed us to detect several new unexpected low energy transition from states fed in Gamow-Teller (GT) decay. Notice that this is the first time two-proton emission has been observed and resolved from states fed in GT-decay (the assignment in [36] was wrong, see [38]).

Due to the low efficiency of β-particles plus the energy cut-off at 600 keV, the contribution of βp events to multiplicity-two events is negligible. The analysis of the multiplicity three events give an upper limit for the β3p branch of 1.1×10^{-3} (99 % C.L.) for an energy cut-off of 500 keV, see details in [39]. Thus there is no "background" from higher multiplicities.

Fig. 2 shows a scatter plot displaying Q_{2p} versus the energy E_i of the individual protons. Each event is represented by two points lying in the same horizontal line, one for each proton. The right part is the projection onto the Q_{2p}-axis. We can identify diagonal lines between the lower left and upper right corners of the scatter plot that correspond to several states in ^{31}Cl all feeding by proton emission the same state in ^{30}S as schematically shown on the right in Fig. 2. To each diagonal line corresponds a vertical line in the left hand side from the emission of the second proton. The recoil of ^{30}S, from the emission of the first proton, broadens these vertical lines increasingly with Q_{2p}. It is therefore possible, already from this figure, to conclude that the mechanism of 2p-emission is dominantly sequential. In addition, horizontal lines are clearly identificable corresponding to the peaks observed in Fig. 1 and defined by their Q_{2p}-values. The Fermi part of the β-decay corresponds to the high energy peaks from 5.22 to 7.67 MeV connecting the IAS with the lowest states in ^{29}P. One can identify other peaks at lower Q_{2p} corresponding to 2p-transitions from other states populated in GT-tansitions. The events outside the peaks in Fig. 1 follow mainly the pattern of the diagonal lines corresponding to states in ^{30}S. This fact gives us confidence that most of them

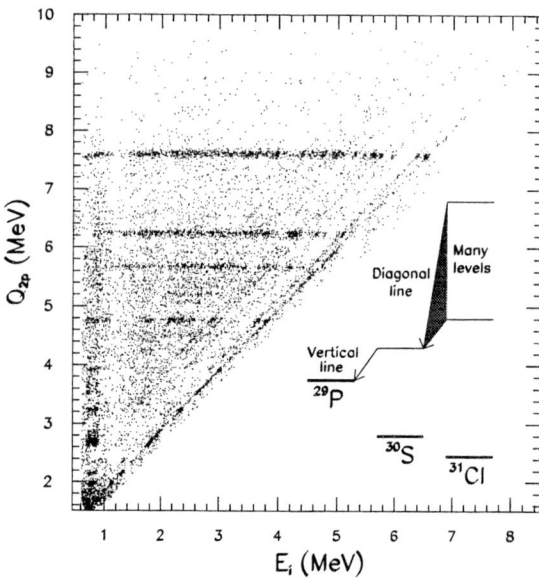

FIGURE 2. Scatter plot showing Q_{2p} vs individual proton energy. Each event corresponds to two points in the same horizontal line, one for each proton. The diagram gives a schematic view of the main features of the data, see text.

corresponds to real 2p-events and that the lack of statistics plus the increase in state density with increasing energy in the 1p-daughter ^{30}S prevent from distinguishing individual peaks in the Q_{2p} spectrum. Certainly the existence of a broad level in ^{31}Cl responsible of the bump around Q_{2p}=5-6 MeV (9.7-10.7 MeV excitation energy in ^{31}Cl) cannot be excluded. As already mentioned, the mechanism of two proton emission should be studied via the individual proton energy distribution and by looking to the angular dependence of the two-proton events.

Fig. 3 shows the distribution of individual proton energies for two of the peaks seen in Fig. 1. The upper half display the distribution for the transition between the IAS and the ground state in ^{29}P. The bottom half corresponds to the 2p-transition from a GT-fed state in ^{31}Cl at 7.36 MeV also ending in the g.s. of ^{29}P. In both cases the first proton is the one of higher energy, and is matched in the left hand side of the plot by a recoil broadened second proton peak, see cosine dependence in eqn. (2). This effect is clearly seen in the bottom part of the figure where due to the small energy window (Q_{2p}) only one intermediate state in ^{30}S is involved.

The individual proton distributions confirm the mechanism as sequential and allow us to derive information about excited states in ^{30}S. This subject goes beyond this presentation, but I should mention that this method has allowed to identify more than 20 levels in ^{30}S up to 8 MeV excitation energy, several of them seen for first time, see [40] for more details. It is remarkable that at the drip line one can obtain information on nuclear spectroscopy comparable or better than the one obtained in reaction studies on stable nuclei.

It is also interesting to note that both states, the IAS and the 7.36 MeV state in

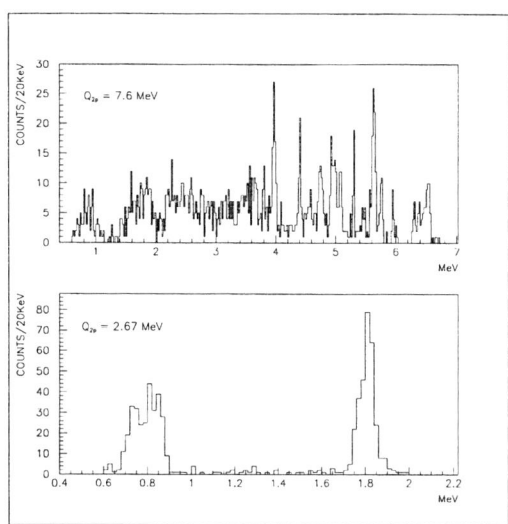

FIGURE 3. Two individual proton projection spectra. The top one corresponds to the 2p transition between the IAS and the g.s. in ^{29}P. At the bottom, the 1p-projection for the transition between a GT fed state in ^{31}Cl at 7.36 MeV excitation energy also ending in the g.s. of ^{29}P.

^{31}Cl decay by 1p and 2p. The ratio of B_{2p}/B_{1p} is almost a factor of 100 between these two states. This factor depends of the ratio of unbound to bound states in ^{30}S as well as on the increase in level density in this nucleus.

The high granularity of the setup has also been used to study the βp coincidences extracting the angular distribution of the β-particle and the proton or the energy of the proton as a function of the βp angle. Both the recoil shift of the proton and the βp angular distribution depend on the spin sequence of the involved states and the angular momentum of the proton and therefore it can be used to extract spectroscopic information. The recoil shift is easier to measure and less dependent upon the experimental conditions, therefore it was used in this experiment to measure the recoil shift of the strongest proton peak at 2.08 MeV. This proton comes from a state of spin parity $5/2^+$ (established by comparison with the mirror nucleus) and it has allowed to determine the spin-parity of ^{31}Ar to be $5/2^+$ [41].

SUMMARY AND OUTLOOK

In this contribution a panorama of beta-delayed multiparticle emission has been sketched. An example of application of new technologies to disentangle the mechanism of β-delayed two-proton emission was presented by the case of ^{31}Ar.

This example shows the potentiallity of the new technologies that allow to design setups with high efficiency for multiparticle detection. This fact in combination with high purity sources (ISOL-technique) and the use of low energy beams to produce point-like sources allows to extract information at the drip line of comparable quality to the one obtained near stability. A new era opens up with the low-energy radioactive beam facilities, where the same technique can be applied to other nuclei with difficult ionization properties.

REFERENCES

1. Marsden, E., *Philos. Mag.* **27**,824 (1914)
2. Hofmann, S. et al., *Z. Phys.* **A305**, 111-123 (1982)
3. Klepper, O. et al., *Z. Phys.* **A305**, 125-130 (1982)
4. Gol'danskii, V.I., *Nucl. Phys.* **19**, 482-495 (1960)
5. Azuma, R.E. et al., *Phys. Rev. Lett.* **43**, 1652-1654 (1979)
6. Gol'danskii, V.I., *JETP Lett.* **32**, 554-556 (1980)
7. Saint-Laurent, M.G. et al., *Phys. Rev. Lett.* **59**, 33-35 (1987)
8. Borrel, V. et al., *Nucl. Phys.* **A473**, 331-341 (1987)
9. Détraz, C. et al., *Nucl. Phys.* **A519**, 529-547 (1990)
10. Brown, B.A., *Phys. Rev.* **C43**, R1513-R1517 (1991)
11. Bochkarev, O.V. et al., *Nucl. Phys.* **A505**, 215-240 (1989); *Sov. J. Nucl. Phys.* **55** 955-969 (1992)
12. Kryger, R.A. et al., *Phys. Rev. Lett.* **74**, 860-863 (1995)
13. Dufour, J.P. et al., *Phys. Lett.* **206B**, 195-198 (1988)
14. Orr, N. et al., *Nouvel. Ganil* **63**, 4-7 (1998)
15. Borge, M.J.G. et al., *Phys. Rev.* **C55**, R8-R11 (1997)
16. Bergmann, U.C. et al., *Nucl. Phys.* **A658**, 129-145 (1999)
17. Langevin, M. et al., *Nucl. Phys.* **A414**, 151-161 (1984)
18. Junde, H. and Dailing, H. *Nucl. Dat. Sheets* **61**, 47-91 (1990)
19. Singh, B. et al., *Nucl. Dat. Sheets* **84**, 565-716 (1998)
20. Singh, B. et al., *Nucl. Dat. Sheets* **81**, 1-181 (1997)
21. Kitao, K., *Nucl. Dat. Sheets* **75**, 99-198 (1995)
22. Cable, M.D. et al., *Phys. Rev. Lett.* **50**, 404-406 (1983)
23. Cable, M.D. et al., *Phys. Rev.* **C30**, 1276-1285 (1984)
24. Honkanen, J. et al., *Phys. Lett.* **133B**, 146-148 (1983)
25. Äystö, J. et al., *Phys. Rev. Lett.* **55**, 1384-1387 (1985)
26. Blank, B. et al., *Z. Phys.* **A357**, 247-254 (1997)
27. Borrel, V. et al., *Nucl. Phys.* **A531**, 353-369 (1991)
28. Borrel, V. et al., *Z. Phys.* **A344**, 135-144 (1992)
29. Nyman, G. et al., *Nucl. Phys.* **A510**, 189-208 (1990)
30. Langevin, M. et al., *Nucl. Phys.* **A366**, 449-460 (1981)
31. Chow, J.C. et al., *Phys. Rev.* **C57**, R475-R483 (1998)
32. Brown, B.A., *Phys. Rev. Lett.* **65**, 2753-2756 (1990)
33. Mikolas, D. et al., *Phys. Rev.* **C37**, 766-780 (1988)
34. Korsheninnikov, A.A., *Sov. J. Nucl. Phys.* **52**, 827-835 (1990)
35. Bazin, D. et al., *Phys. Rev.* **C45**, 69-79 (1992)
36. Borge, M.J.G. et al., *Nucl. Phys.* **A515**, 21-30 (1990)
37. Axelsson, L. et al., *Nucl. Phys.* **A634**, 475-496 (1998), Errata **A641** 529 (1998)
38. Axelsson, L. et al., *Nucl. Phys.* **A628**, 345-362 (1998)
39. Fynbo, H.O.U. et al., *Phys. Rev.* **C59**, 2275-2277 (1999)
40. Fynbo, H.O.U. et al., *Nucl Phys. A* In preparation
41. Thaysen, J. et al., *Phys. Lett.* **B467**, 194-198 (1999)

Isospin-Forbidden β-delayed Proton Emission

W. E. Ormand

Physics Directorate, Lawrence Livermore National Laboratory[1]
L-414, P.O. BOX 808, Livermore, CA 94551

Abstract. The effects of isospin-symmetry breaking on proton emission following β-decay to the isobaric analog state are discussed in detail. Of particular importance is the mixing with a dense background of lower isospin states, whose properties are not well known. The possibility of observing T=4 states in even-even, N=Z nuclei, which is viable if the decay proceeds via isospin-forbidden particle emission, is also discussed.

INTRODUCTION

The study of nuclei at the extreme limits of stability has recently become a central focus in nuclear structure research. A powerful tool used in these studies is β-delayed proton emission. Near the proton drip-line, the β-endpoint is quite large, and the β-decay proceeds through a variety of excited states in the daughter nucleus. In turn, these states are generally proton-unbound. Because of the relative ease in detecting and measuring the energy of the emitted protons, it is possible to construct a detailed picture of the structure of the daughter nucleus. In addition, it is also possible to determine the branching ratios for the β-decay, and, hence, measure the B(GT) values. Some particularly important applications are ^{37}Ca [1] and ^{40}Ti [2,3] (the analogs of ^{37}Cl and ^{40}Ar, respectively), which provide a mechanism for calibrating solar-neutrino detectors.

Because of its approximate conservation, isospin is a powerful spectroscopic tool that can often be exploited to map out the structure of nuclei. In nuclei near the proton drip-line, the β-decay to the isobaric analog state is permitted because of the large β-endpoint energy (~ 8-15 MeV). In many cases, however, the analog state is bound relative to the emission of a proton into the T'=T-1/2 state. If isospin were a conserved quantity, the isobaric analog state could not decay by particle emission at all. However, because of the presence of the Coulomb interaction and other weaker isospin-nonconserving (INC) components of the nucleon-nucleon interaction, isospin symmetry is broken and each state can have admixtures of up to ΔT= ± 2. Consequently, the analog state can decay via the emission of protons to T=T-3/2 states. A schematic of this type of decay is illustrated for ^{40}Ti in Figure1, where the J=0, T=2 ground state of ^{40}Ti β-decays to J=1, T=1 (via Gamow-Teller transitions) states in ^{40}Sc that are all unbound relative to the T=1/2 ground state of ^{39}Ca. In

---

[1] Lawrence Livermore National Laboratory is managed for the Department of Energy by the University of California under contract No. W-7405-ENG-48.

addition, the β-decay proceeds to the analog J=0, T=2 (Fermi transition) state in ^{40}Sc, which is bound relative to the J=7/2, T=3/2 excited state in ^{39}Ca. Since isospin is not a conserved quantity, the isobaric analog state in ^{40}Sc has admixtures of T=1 states,

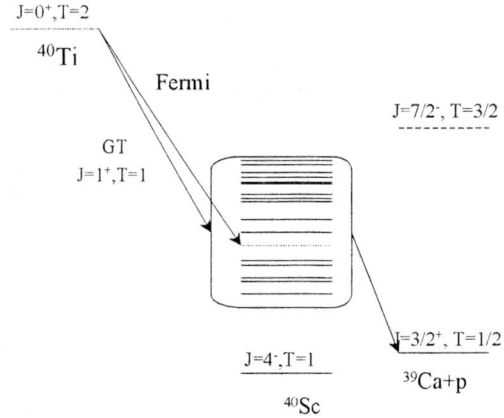

FIGURE 1. Schematic of the decay of ^{40}Ti.

while the low-lying T=1/2 states in ^{39}Ca have admixtures of T=3/2. Hence, proton emission in the analog proceeds via two processes: (1) the decay of the small T=1 admixtures in the analog to the T=1/2 component of the final state; and (2) the decay of the T=2 analog to small admixtures of T=3/2 in the final state. This is schematically illustrated in Figure 2, where the dotted lines represent the analog T=2 state and the T=3/2 state in ^{39}Ca. The solid lines represent the T=1 states in ^{40}Sc and the T=1/2 ^{39}Ca ground state.

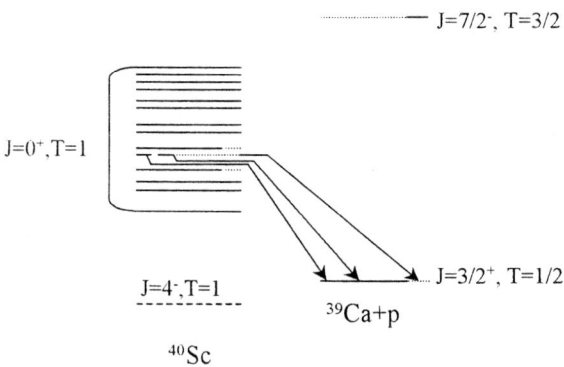

FIGURE 2. Schematic illustration of the isospin-forbidden decay of the analog state.

In the next section, a detailed description of the process behind isospin-forbidden proton emission is given. It will be seen that of particular importance is the role played by the background T=1 states, especially their excitation energy relative to the analog state. Afterwards, is a short discussion on the effects of isospin mixing on the β-decay. In addition to applications regarding β-delayed proton emission, the possibility of observing high-lying T=4 states in even-even, N=Z nuclei, such as ^{40}Ca, which is possible if proton and neutron emission is isospin forbidden, is discussed. Concluding remarks are gathered in the final section.

DETAILED PICTURE OF ISOSPIN-FORBIDDEN PROTON EMISSION

In this section, I will discuss the physics required for a quantitative description of isospin-forbidden particle emission. Instead of a specific example relating to β-delayed proton emission, I will focus on a systematic study of isospin-forbidden resonances carried out on even-even, N=Z *sd*-shell nuclei [4]. In these experiments, the compound nucleus shares the same feature exhibited in Figure 2; namely that the state under investigation lies at a relatively high excitation energy and is embedded in a background of T-1 states. From first-order perturbation theory, the mixing amplitude between the analog and the T-1 states is given by

$$\alpha = \frac{\langle \psi_{T-1} | V_{INC} | \psi_T \rangle}{E_T - E_{T-1}}, \quad (1)$$

where V_{INC} represents the isospin-nonconserving (INC) interaction, and E_T is the excitation energy of the state with isospin T. In general, the matrix elements of V_{INC} are found to be of the order 10-50 keV. From Eq. (1), it is immediately clear that the background T-1 states play a crucial role; especially those with excitation energies within 100 keV of the analog state. Unfortunately, even the best theoretical estimates of excitation energies (e.g., from large-basis shell-model calculations) have an uncertainty of the order 200-500 keV. Consequently, in the absence of experimental information about the excitation energies of the T-1 states, a quantitative description of the effects of isospin mixing is difficult to assess. The uncertainty imposed on a quantitative picture by the lack information about the background T-1 states was investigated in Ref. [5], and here I will recount the main features.

The experimental observable was the nuclear-structure spectroscopic factor Θ extracted from the resonance width Γ via *R*-matrix theory

$$\Gamma = 2 P \gamma_{sp}^2 \Theta^2, \quad (2)$$

where $\gamma_{sp}^2 = \hbar^2 / R_c^2$, with $R_c \approx 1.4 (A_1^{1/3} + A_2^{1/3})$ representing the channel radius. The penetrability P_l for a final state with relative orbital angular momentum *l* is given by

$$P_l = \frac{kR_c}{F_l^2(kR_c) + G_l^2(kR_c)}, \quad (3)$$

where *k* is the wave number of the emitted particle and F_l and G_l are regular and irregular free-particle solutions to the Schrödinger equation.

In the theoretical study of Ref. [5], the spectroscopic factor Θ was evaluated via perturbation theory. Hence, Θ represents a sum of the contributions due to the mixing of T=1/2 states in the compound nucleus as well as to contributions due to mixing with T=1 and 2 states in the target nucleus. The situation is similar to that in Figure 2, and is illustrated in Figure 3 for the specific case of ^{20}Ne. As mentioned above,

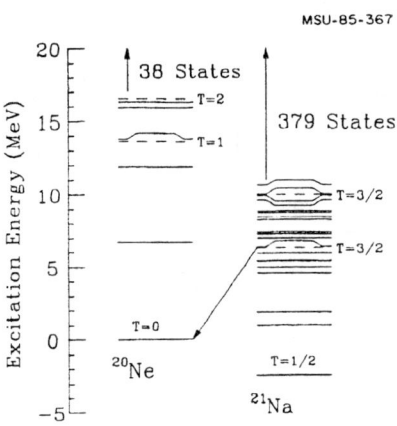

FIGURE 3. Level scheme in ^{20}Ne and ^{21}Na illustrating the relative location of T=1 and 2 states in the target nucleus and the T=1/2 states in the compound nucleus.

the excitation energy of the background T=1/2 states is unknown. The effect of the location of the T=1/2 states is illustrated in Figure 4, where Θ is evaluated while shifting the entire T=1/2 spectrum by an amount δε. In Figure 4, it is seen that over a reasonable range of the shift δε, say ± 300 keV, the magnitude of Θ changes by approximately a factor of ten, and reaches a maximum when one of the background T=1/2 states is "accidentally" degenerate with the T=3/2 state. Given this sensitivity, in Ref. [5] the best estimate of Θ was obtained by taking the average over the range δε = ± 500 keV, with a "theoretical" uncertainty given by the variance over this range. The calculated spectroscopic factors are compared with experimental values on the right-side of Figure 4. In the figure, the solid and open squares represent the experimental data [4] (the experimental errors are approximately the size of the symbols), while the crosses and error bars represent the "best" theoretical estimates. In the figure, it is seen that although in each case, the spectroscopic factor cannot be estimated to better than a factor of two, the range generally encompasses the experimental data. The noted exception being for A=37, where the shell-model space used in the calculation is most likely inadequate.

A reasonable estimate of the lifetime for the decay of the analog state by isospin-forbidden proton emission can be obtained using Eq. (2) and estimating the spectroscopic factor to be of the order 10^{-2}. With a separation energy of approximately

5 Mev, we have $2P\gamma_{sp}^2 \approx 1 - 2$ MeV and $\Gamma \sim 10$ eV (or $T_{1/2} \sim 10^{-16}$ s). On the other hand, typical widths associated with γ-emission are of the order $10^{-7} - 10^{-3}$ eV. Hence, even though it is isospin forbidden, the proton emission is often the dominant decay mode.

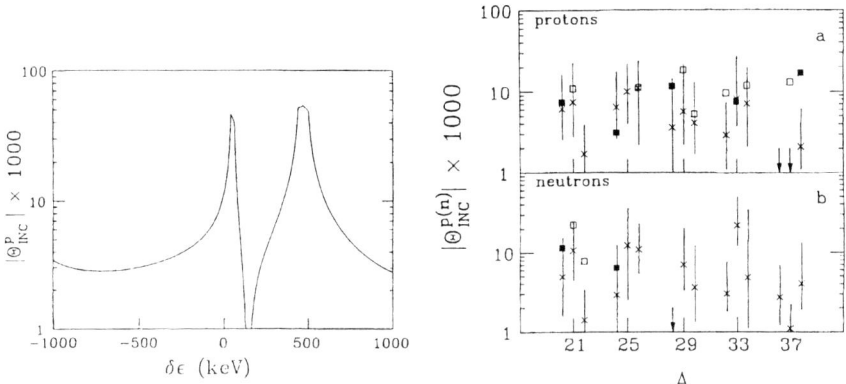

FIGURE 4. On the left, Θ evaluated as a function of the shift δε in the background T=1/2 spectrum. On the right, the theoretical estimates of Θ (crosses with error bars) are compared with experimental data (solid and open squares) [4].

EFFECTS OF ISOSPIN MIXING ON BETA-DECAY

Another question that can be raised is whether isospin-symmetry violation will significantly alter the decay rate for the Fermi transition. Because of isospin mixing, the Fermi strength will generally be distributed over a range of background T-1 states. Given that the decay rate, Γ_β, for allowed β-decay is proportional to W^5, where W is the β-endpoint, a distribution of the strength over an energy range could lead to a change in the total Fermi decay rate, and affect estimates for the half-life. Typically, the mixing matrix elements are generally less than 50 keV. Hence, substantial mixing will occur over a narrow range of background states. In Ref. [2] the affect of isospin-mixing was estimated using a two-level mixing model. It was found that the *total* Fermi decay rate differed from the "pure" transition by no more than 0.2% even in the case of maximal mixing with a large mixing-matrix element of 100 keV. It should be noted, however, that in this case the decay rate, and, hence, the branching ratio, to each of the individual states are from the "pure" transition. In general, one expects that the Fermi strength will be distributed over a range of only ± a few hundred keV around the analog state.

I conclude the discussion of isospin mixing on β-decay studies involving proton emission with some remarks regarding the Fermi transition in odd nuclei. In the example in shown in Figure 1, the Fermi and Gamow-Teller transitions are separated by angular momentum selection rules, i.e., the transitions to J=1, T=1 states are pure Gamow-Teller. The same is not true for odd-A nuclei, where because of isospin mixing, the Fermi strength will be distributed to all states with the same angular momentum as the analog state. As a consequence, some β-transitions to T-1 states will contain both Gamow-Teller and Fermi components. Since the Fermi matrix element is generally larger than the Gamow-Teller matrix element even small admixtures of the Fermi transitions can enhance the decay rate to these states. This can be very important for experiments comparing measured and theoretical B(GT) values to assess the magnitude of the quenching of the Gamow-Teller strength.

T=4 STATES IN EVEN-EVEN, N=Z NUCLEI

A goal in physics is to probe the response of nuclear systems at the extreme limits of existence; in particular, at high spin, high excitation energy, or a large proton or neutron excess (i.e., high isospin). Of interest is also the location and formation of the analog of these high isospin states in stable nuclei, for example in even-even, N=Z nuclei. The observation of these states poses several challenges in that they are expected to lie at a high excitation energy, e.g., 20-35 MeV. The viability of observing these high-isospin states lies in the lifetime of the state, and, in particular, whether they are bound relative to isospin-allowed particle emission. If they are, the fact that the decay proceeds through isospin-forbidden channels offers the promise that the states will be long-lived, and, hence, have a width narrow enough to be resolved experimentally. This is especially important given that one is searching for a single state buried in a dense background, whose total density is expected to exceed several thousand per MeV.

Here, I address the question as to whether it is feasible to observe T=4 states in even-even, N=Z nuclei. These T=4 states can be formed with the (α,^{8}He) transfer reaction on T_z=-2 nuclei [here T_z=(Z-N)/2], and can be performed with the S800 spectrometer at the National Superconducting Cyclotron Laboratory at Michigan State. Viable targets include, ^{36}Ar, ^{44}Ca, ^{46}Ti, ^{52}Cr, ^{56}Fe, ^{60}Ni, and ^{64}Zn. In this work, I will focus on ^{40}Ca (^{44}Ca target) to illustrate the principal features.

Predictions for the Binding Energy of the T=4 State

The primary questions to be answered are: (1) what is the excitation energy of the T=4 state? and (2) is it bound relative to isospin-allowed particle emission? As mentioned above, shell-model predictions for excitation energies generally have uncertainties of several hundred keV. The primary source of this uncertainty lies in the treatment of the isospin-conserving part of the strong interaction. On the other hand, the isospin-nonconserving (INC) components (the Coulomb and isotensor part of the strong interaction) are considerably weaker, and can be reasonably addressed with perturbation theory. In order to minimize the uncertainties due to the isospin-conserving part of the strong interaction, I make use of the isobaric-mass-multiplet equation (IIME), namely that for the elements of an isospin-multiplet, the binding energy (BE) in each member is given by [6]

$$BE(T_z) = a + bT_z + cT_z^2, \qquad (4)$$

where the a-coefficient is due to the isospin-conserving part of the strong interaction, and the b- and c-coefficients are due to the INC interaction.

The utility of Eq. (4) is that if the binding energy of any one member of the isospin multiplet is known experimentally, the binding energy of all the other members can be predicted using theoretical estimates for the b- and c-coefficients. In Ref. [7], shell-model V_{INC} Hamiltonians were developed for a variety of model spaces ranging from the p-shell to the fp-shell. These V_{INC} interactions were found to reproduce experimental b- and c-coefficients at the level of 25 keV and 15 keV, respectively.

In order to estimate the binding energy of the T=4 state in ^{40}Ca, a shell-model calculation was carried out for the b- and c-coefficients using the INC interaction of Ref. [7] within the $0d_{3/2}$-$0f_{7/2}$ model space. These were then added to the experimental binding energy [8] of the T_z=-4 analog state, i.e., ^{40}S. The same procedure was then followed to determine the binding energies of the T=5/2 and 7/2 states in both ^{39}Ca and ^{39}K. The binding energies and relative excitation energies are shown in Table 1. The quoted uncertainties arise from experimental errors ($\sim$ 230 keV for ^{40}S) and the theoretical uncertainties of 25 keV and 15 keV for the b- and c-coefficients, respectively. In order to provide a further overall check, the same procedure was followed to predict the excitation energies of the experimentally known T=1 and 2 states in ^{40}Ca and the T=3/2 state in ^{39}K. Agreement was achieved at the level of 10-20 keV for each of these levels.

TABLE 1. Predicted binding and excitation energies for states in ^{40}Ca, ^{39}Ca, and ^{39}K and the experimental binding energy of the T=3 state in ^{38}Ar.

Nucleus	T	Binding Energy (MeV)	Excitation Energy (MeV)
^{40}Ca	4	306.936(412)	35.116(412)
^{39}Ca	5/2	318.086(150)	15.638(150)
^{39}Ca	7/2	306.134(271)	27.589(271)
^{39}K	5/2	310.926(150)	15.485(150)
^{39}K	7/2	298.982(271)	27.429(271)

With the predicted binding energies, we are now in the position to determine if the T=4 state will be sufficiently narrow so as to be observable. From Table 1, it is apparent that the T=4 state is bound relative to isospin-allowed proton and neutron emission. By comparing the predicted binding energy with the experimental binding and excitation energies tabulated in Refs. [8,9], the J=0, T=4 state in ^{40}Ca is found to be unbound to the emission of two protons by 1.6(4) MeV to the T=3 state in ^{38}Ar. Note that this decay mode is isospin allowed.

Estimates of the Width

The width of the T=4 state will have two components. The first is associated with the lifetime for particles to escape, which, following the notation used in descriptions of the giant-dipole resonance, is denoted by $\Gamma^{\uparrow}$. The second component, $\Gamma^{\downarrow}$, is due to the fact that the T=4 state will mix with nearby T=2 and 3 states, thereby spreading over an energy window comparable with the mixing matrix elements.

Spreading width $\Gamma^{\downarrow}$

The spreading width can be estimated following the arguments in Ref. [10] giving

$$\Gamma^{\downarrow} \approx 2\pi v^2 \rho_{T=2,T=3}(E_{T=4}), \quad (5)$$

where v^2 is the square of the average matrix element of V_{INC}, $\rho_T(E_{T=4})$ is the density of T=2 and 3 states at an excitation energy $E_{T=4}$ that mix with the T=4 state. In general, the mixing matrix element is of the order 10 keV. In practical application of Eq. (5), ρ_T is the most difficult quantity to estimate reliably. One possible recourse is to perform a shell-model calculation, and simply count the number of levels in the vicinity of the T=4 state. Unfortunately, a calculation utilizing the full 4p-4h excitations in the *sd-pf* shell is not feasible. Consequently, a corrected Fermi-gas estimate will be employed with the proviso that it most likely represents a significant over counting the relevant levels. The Fermi-gas formula for the density of levels as a function of excitation energy E is [10]

$$\rho(E) = (2J+1)\frac{1}{48}\sqrt{\frac{6}{\pi}}a\left(\frac{a\hbar^2}{I_{rig}}\right)^{3/2}\frac{\exp(2\sqrt{aE'})}{[aE']^{5/4}}, \quad (6)$$

where $a \sim A/10$ is the level density parameter, I_{rig} is the rigid body moment of inertia, and $E' = E - J(J+1)\hbar^2/2I_{rig}$. As mentioned above, at the excitation energy of the T=4 state, approximately 35 MeV, Eq. (6) vastly overestimates the density of 4p-4h states that can mix. To account for this overcounting, the excitation energy E' in Eq. (6) is further shifted by 18.5 MeV, which corresponds to the difference in the excitation energies between the T=4 state and 4p-4h, T=2 states obtained from a shell-model calculation. The shell-model calculation was carried out using the interaction of Ref. [11] while limiting the 4p-4h excitations to the $0d_{3/2}, 1s_{1/2}, 0f_{7/2}, 1p_{3/2}$ model space. This leads to $\rho \sim 880$ MeV^{-1}, and, hence, $\Gamma^{\downarrow} \sim 500$ keV.

Escape width $\Gamma^{\uparrow}$

A simple estimate of the escape width can be obtained from Eq. (2). The maximum energy for an emitted proton is of the order 9.5 MeV, which for an $l=1$ proton gives $2P\gamma_{sp}^2 \approx 6$ MeV. Note that since the T=4 and 5/2 states are of different parity, the l of the emitted proton must be odd. As was seen in the previous subsection, the T=4 state spreads over an energy range of approximately 100-200 keV, therefore, a reasonable estimate of the spectroscopic factor associated with the emission is the average Θ for the background T=2 and 3 states that can mix into the T=4 state. For this estimate, a shell-model calculation was again carried out by allowing 4p-4h excitations within the $0d_{3/2}, 1s_{1/2}, 0f_{7/2}, 1p_{3/2}$ model space using the interaction of Ref. [11]. In the energy region near the T=4 state, the average allowed spectroscopic factor was found to be $\langle\Theta\rangle \sim 0.03$, which gives an escape width to the lowest T=5/2 state of the order 5 keV. Because of the large separation energy, however, the decay can also occur to higher-lying T=5/2 states. Although the penetrability factor decreases rapidly with the separation energy, this can easily compensated for by the density of final states. The escape width is then summed over the possible final states, giving

$$\Gamma^\uparrow = \langle\Theta\rangle^2 \int_0^{E_{max}} dE \rho(E_{max} - E)\Gamma^\uparrow(E) \approx \langle\Theta\rangle^2 74\,\text{MeV} \approx 70\,\text{keV}, \quad (7)$$

where E is the energy of the emitted proton and a Fermi-gas model was used to estimate the density of final states $\rho(E)$.

In addition to proton emission, the T=4 state can also decay via the isospin-allowed emission of two-protons to the T=3 level in ^{38}Ar. The escape width for this channel can also be estimated using Eq. (2). Using the separation energy 1.6(4) MeV to this level and assuming a spectroscopic factor of unity (worst-case scenario) leads to an escape width of approximately 1 eV. Consequently, the J=0^+, T=4 state in ^{40}Ca, with an excitation energy of 35.116(412) MeV, is expected have a decay width of the order 500 keV or less.

CONCLUSIONS

The effects of isospin-symmetry breaking on the emission of protons following β-decay and the possibility of observing high-lying T=4 states in even-even, N=Z nuclei was discussed in the present work. For β-delayed proton emission, it was shown that the decay widths for the isospin-forbidden proton decay of the analog state are strongly dependent on the exact locations of the background T-1 states that mix into the analog. Consequently, in the absence of experimental information about these states, it is possible to present only a qualitative picture of the decay process. In particular, comparisons were made between experimental values and theoretical estimates for isospin-forbidden spectroscopic factors, which were found to be in agreement within the limits imposed on theory due to uncertainties in the excitation energies of the mixed T-1 states. In addition, due to the phase space available to isospin-forbidden proton emission, this decay mechanism is found to be generally favored over the emission of photons.

The excitation energy of the T=4 analog state in ^{40}Ca was estimated making use of the IMME, and was found to be bound relative to isospin-allowed proton and neutron emission. Estimates were presented for the spreading and escape widths for this state, leading to a total width of approximately 500 keV.

ACKNOWLEDGEMENTS

I wish to thank J. A. Nolen (Argonne National Laboratory) for bringing the topic of the T=4 state in ^{40}Ca to my attention.

REFERENCES

1. Garcia, A., et al., Phys. Rev. Lett. **67**, 3654 (1991); Garcia, A., et al., Phys. Rev. **C51**, R439 (1995).
2. Ormand, W. E., Pizzochero, P. M., Bortignon, P. F., and Broglia, R. A., Phys. Lett. **B345**, 343 (1995).

3. Liu, W. *et al.*, Phys. Rev. **C58**, 2677 (1998); Bhattacharya, M., *et al.*, Phys. Rev. **C58**, 3677 (1998).
4. Wilkerson, J. F., Anderson, R. E., Clegg, T. B., Ludwig, E. J., and Thompson, W. J., Phys. Rev. Lett. **51**, 2269 (1983).
5. Ormand, W. E. and Brown, B. A., Phys. Lett. **B174**, 128 (1986).
6. Wigner, E. P., Proceedings of the Robert A. Welch Conference on Chemical Research, Houston: R. A. Welch Foundation, 1957, Vol. 1, p. 67.
7. Ormand, W. E. and Brown, B. A., Nucl. Phys. **A491**, 1 (1989).
8. Audi, and Wapstra, Nucl. Phys. **A491**, 1 (1989).
9. "Table of Isotopes, " edited by Firestone, R. B., and Shirley, V. S., eigth edition , New York: Wiley & Sons, 1996.
10. Bohr, A. and Mottelson, B., "Nuclear Structure" Vol. 1, New York: Benjamin, 1969
11. Warburton, E. K., Becker, J. A. and Brown, B. A., Phys. Rev. **C41**, 1147 (1990).

Prompt Particle Decays From Deformed High-Spin States

D. Rudolph*

Department of Physics, Lund University, S-22100 Lund, Sweden

Abstract. Well deformed and superdeformed rotational structures in the second well of nuclei near doubly-magic ^{56}Ni were recently identified by means of heavy-ion fusion-evaporation reactions. Some of these bands were found to decay by prompt particle emission in competition to the expected decay via γ radiation. Subsequently, dedicated experiments were performed to investigate this exotic decay mode in more detail. First results from these studies are reported.

I INTRODUCTION

The combination of contemporary 4π Germanium detector arrays and powerful ancillary detector systems has led to many exciting and unexpected results in nuclear structure physics [1]. One example is the so-called prompt particle emission of discrete energy, which connects high-spin states in the deformed second well with spherical states in the daughter nucleus. Hence, different from the majority of the experimental contributions to the workshop, this paper deals with combined *in-beam* particle- and γ-ray spectroscopy close to the proton drip line at high excitation energies and spins, rather than direct proton decay studies at or near the ground state of nuclei at or beyond the drip line. Up to now, the exotic and unprecedented prompt particle decays have only been observed in nuclei in the vicinity of the $N = Z = 28$ doubly-magic isotope ^{56}Ni.

The first case is the prompt proton decay established in ^{58}Cu [2] [Fig. 1(a)]. The irregularly spaced states in the first, spherical minimum and the rotational band in the second minimum are schematically shown in the upper left part. The spin values of the states in the first minimum were measured while those in the second minimum were inferred from the assigned configuration of the band, i.e., $\pi(g_{9/2}) \times \nu(g_{9/2})$, as well as the best estimates originating from in part measured spins and parities of initial and final states of both γ and proton decay out. Angular distribution and correlation measurements prove quadrupole character for the γ-ray transitions within the band [3]. Despite the low mass, the band in ^{58}Cu is relatively regular, and its configuration assignment is often used as benchmark in the $A \sim 60$ region [4]. The γ-ray spectrum at the bottom of Fig. 1 implies that in essence only the $I^\pi = 9/2^+$ state at 3701 keV excitation energy in the daughter nucleus ^{57}Ni is populated by the main branch observed at 2.3(1) MeV proton center-of-mass energy (cf. spectrum in the upper right part of Fig. 1). Many other levels with spin values close to 9/2 were observed in ^{57}Ni at similar excitation energies, and shell-model calculations indicate a high degree of mixing within these negative-parity states [3]. This fact plus the apparent selectivity of the proton decay support the positive parity assignment of the 3701 keV level and its interpretation as the neutron $1g_{9/2}$ single-particle level with respect to the doubly-magic core ^{56}Ni. The parity of the daughter state was established experimentally be means of a dedicated study (see below). The intriguing decay scenario is such that along with the drastic change in shape the $1g_{9/2}$ proton is emitted from the 8915 keV 9^+ band-head in ^{58}Cu, which leaves the spherical ^{57}Ni daughter with a single neutron in

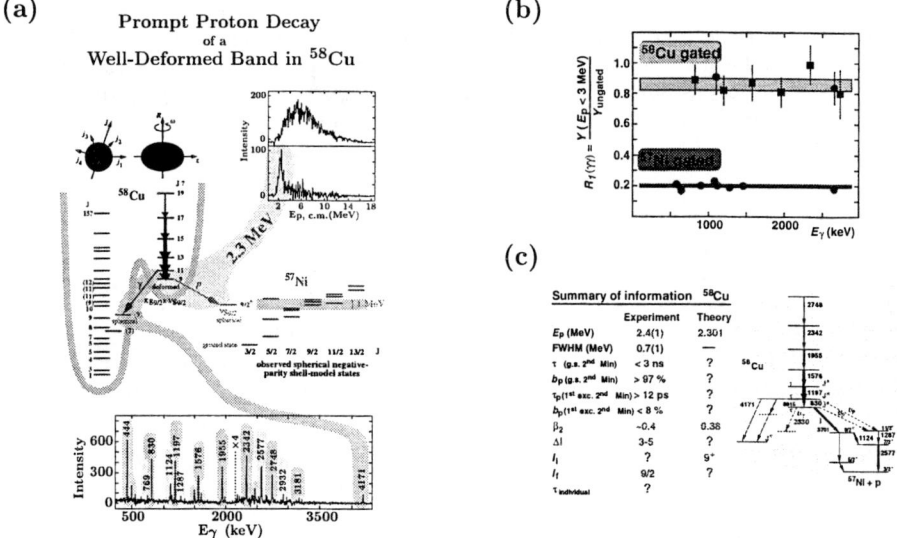

FIGURE 1. (a): Overview of the prompt proton decay from the deformed band in the second potential well of ^{58}Cu into a spherical daughter state in ^{57}Ni. The γ-ray spectrum from the reaction ^{36}Ar+^{28}Si was gated by one detected α-particle, proton, and neutron, and measured in coincidence with the second and third transition in the rotational band of ^{58}Cu (1197 and 1576 keV). It shows the other transitions in this band (830, 1955, 2342, 2748, and 3181 keV), the discrete linking transition at 4.2 MeV, and γ rays between states in the spherical minima of both ^{58}Cu (e.g., 444 keV) and ^{57}Ni (e.g., 1124 and 2577 keV). In the upper right part of the figure proton center-of-mass energy spectra are shown gated by one α-particle, one or two protons, and one neutron, in prompt coincidence with the 444 keV ground-state transition in ^{58}Cu (top) and the 830 keV transition which feeds the proton decaying ^{58}Cu band head. (b) Ratio of yields of γ-ray transitions in ^{58}Cu and ^{57}Ni. An overall particle gating of one α particle, two protons, and one neutron was applied. See text for details. (c) Summary of the early results on the prompt proton decay in ^{58}Cu.

the $1g_{9/2}$ orbit. The high angular momentum character is supported by an angular distribution measurement of the proton, which indicates $\Delta l = 3\text{-}5\ \hbar$. A more precise measurement of the angular distribution may in fact shed more light on this two-dimensional tunneling process, as discussed in more detail in the contributions of N. Carjan [5] and P. Talou [6] and references therein. It should also be noted that the measured FWHM of 0.7(1) MeV of the proton peak can be attributed to kinematic broadening rather than the intrinsic resolution of the CsI elements of the charged-particle detector system MICROBALL [7]. More information related to the ^{58}Cu proton decay can be found in Refs. [2,8] and Sec. III.

There are obvious conceptual differences between the 'conventional' direct proton and β-delayed proton emission processes and prompt discrete particle decays: The prompt particle decays compete with γ-radiation, while (ground-state) proton radioactivity competes with β^+ decay. The prompt proton decays carry an angular momentum of $l \sim 4\ \hbar$, while β-delayed protons are represented by s-waves and, to a much lesser extent, by p-waves. The time scale is different by several orders of magnitude. Proton emitters possess typical half-lives in the micro- to millisecond range. β-delayed protons are observable on a similar time scale but, of course, the intrinsic decay times are much faster, i.e., in the attosecond regime. The prompt particle decays, however, seem to lie in the picosecond range. There is a drastic change of nuclear shape in the course of the prompt particle decays – the initial states, situated at the bottom of rotational bands, have a deformation of $\beta_2 \sim 0.4$, while the final states are spherical shell-model states. For the other proton decays, the shapes associated with the initial and final nuclear states are essentially the same.

A second case of prompt proton emission was found in the decay-out of a rotational band in the doubly magic nucleus ^{56}Ni [9]. Moreover, a 4% decay-out branch from the second minimum in ^{58}Ni constitutes the first observation of a prompt monoenergetic α radiation [10] into the spherical 2949 keV 6^+ yrast state in ^{54}Fe [3,11,12]. At present it appears difficult to obtain quantitative theoretical descriptions of the process. In particular, the change of shape complicates the situation considerably, and that might link the process to time-dependent calculations as described in the contributions mentioned earlier [5,6]. The question on how long it actually takes to change the shape can probably be answered solely by such an approach.

In the following the experimental knowledge on specific features of the prompt proton decay of ^{58}Cu is revised and the three stages used to identify a prompt particle decay described. In Sec. III the new experiments are presented, and first, preliminary results are shown with respect to the proton decay in ^{58}Cu.

II PREVIOUS RESULTS AND METHODS

The data which provided the aforementioned results were collected at the GAMMASPHERE array [13] coupled to the 4π CsI ball MICROBALL [7] and to fifteen neutron detectors to measure the prompt γ radiation in coincidence with evaporated light particles. High-spin states in a number of residual nuclei were populated in the reaction ^{28}Si(^{36}Ar,xpynzα) at an effective beam energy of 136 MeV. The beam was provided by the 88-Inch Cyclotron at the Lawrence Berkeley National Laboratory.

The identification of a prompt particle decay proceeds in three steps. At first, unexpected γ-ray coincidences are noted. For example, the coincidences between ^{58}Cu band members (1197 and 1576 keV) and transitions in the low-energy regime of ^{57}Ni (1124 and 2577 keV), corresponding to the γ-ray spectrum in Fig. 1. Secondly, particle center-of-mass energy spectra are investigated, for which the proper decay path is selected by gating on proper γ-ray transitions in the parent and, if possible, the daughter nucleus (cf. top right of Fig. 1). For a final and decisive check $\gamma\gamma$ intensity ratios are considered. Fig. 1(b) shows the ratios of yields as a function of γ-ray energy for transitions in coincidence with one α particle, two protons, and one neutron, and γ rays associated with the proton decay ("^{58}Cu gated") as well as normal yrast transitions in the $1\alpha 2p1n$ reaction channel ^{57}Ni. An additional proton energy restriction of $E_{p,c.m.} < 3$ MeV for at least one of the two protons was demanded for the numerator. Since the energy of the proton peak amounts to 2.3 MeV, this ratio should be close to unity for the "^{58}Cu gated" transitions as it was confirmed experimentally. On the contrary, normal transitions within ^{57}Ni must be reduced significantly because the vast majority of protons from the fusion reaction have energies in excess of 3 MeV (cf. Fig. 1).

The results concerning the prompt proton decay of ^{58}Cu from this first experiment are summarized in Fig. 1(c). In the following, several dedicated experiments were planned and performed aiming at specific unknown or uncertain observables associated with this new decay mode.

III NEW EXPERIMENTS AND RESULTS

Because the experiment described in the previous Section was by no means optimized to detect prompt particle decays, a total of three experiments were performed subsequently to study (mainly) the prompt proton decay in ^{58}Cu and associated spectroscopic quantities in more detail.

A The Cologne experiment

The interest in the spin and parity of the 3701 keV state in ^{57}Ni is twofold. Firstly, it is the daughter state of the prompt proton decay in ^{58}Cu. Secondly, it is likely to reflect the neutron $1g_{9/2}$ single particle state with respect to the doubly-magic ^{56}Ni core [3]. A spin of $I = 9/2$ has been established but the parity has not been determined yet. Therefore, an experiment was performed at the Tandem accelerator of the University of Cologne specifically aiming at the measurement of the parity of the 3701 keV state in ^{57}Ni, and to confirm the spin assignment of that state. A ^{32}S beam of 90 MeV was impinging on a tantalum-backed ^{28}Si target, i.e., ^{57}Ni being produced following the evaporation of two protons and one neutron. The beam energy was chosen such that ^{57}Ni was the most intense reaction channel involving neutron evaporation. The set-up of the experiment is sketched in Fig. 2(a): A large-volume Ge-detector (AD) was used to map the angular distributions of γ rays in the region from 70° (0°) to 140° relative to the beam axis. Two other Ge detectors (M1 and M2, the latter is not shown) were placed at 140° on the opposite side of the beam to monitor the event rates, i.e., to normalize the events with respect to fluctuations in the beam intensity. A EUROBALL Cluster detector (P2) [14–16] and a six-fold segmented MINIBALL prototype detector (P1) [17] were positioned at 90° with respect to the beam. Their frontfaces are shown on the right hand side of Fig. 2(a), and the indicated 'horizontal' and 'vertical' Compton scattering events between capsules and segments, respectively, can be used to determine the electric or magnetic character of a γ ray due to their linear polarization [18]. In addition, $\gamma\gamma$ coincidences between the detectors at 90° and 140° can be used for the analysis of angular correlations. A large NE213 liquid-scintillator neutron detector was positioned at 0° to tag events in coincidence with evaporated neutrons. At the end of the experiment this detector was removed to allow for additional data points for the angular distribution part of the experiment at forward angles.

The analysis of this experiment provides the following information: (i) The neutron-gated angular distribution of the 1124 keV transition in ^{57}Ni (cf. Fig. 1) is consistent with a pure stretched dipole transition. (ii) The angular correlation between the 2577 and 1124 keV transitions supports this assignment. (iii) The

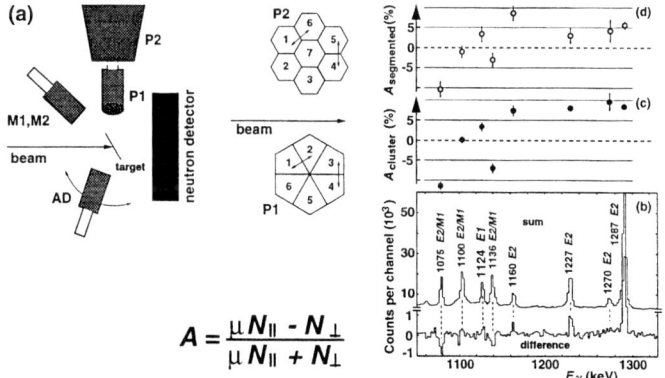

FIGURE 2. (a) Set-up of the Cologne experiment. See text for details. The γ-ray spectra in part (b) show the normalized difference (bottom) and sum (top) of events scattered parallel or perpendicular to the reaction plane in the Cluster or the segmented Ge-detector. The spectra were measured in coincidence with at least one detected neutron. The lines from pure charged-particle evaporation channels were carefully substracted using identical fractions of the spectra without the neutron coincidence. The peaks are labeled with their energies in keV and their (known) multipole character. The anisotropy measured for these transitions are shown for the Cluster (c) and the segmented detector (d).

neutron-gated linear polarization measured with the EUROBALL Cluster and the segmented MINIBALL detector clearly indicates electric character for the 1124 keV transition [19]. Neither the Cluster nor the segmented Ge-detector have the favoured orthogonal configuration for linear polarization measurements. Nevertheless, an anisotropy can be defined (cf. Fig. 2). Figure 2(b) shows the purified neutron-gated difference and sum spectrum according to the nominator and denominator of Eq. ?? using the statistics from P1 and P2. All labeled peaks in the spectra belong to ^{57}Ni [3]. The 1160, 1227, 1270, and 1287 keV lines are known to be stretched $E2$ transitions. They reveal the expected positive anisotropies for both P1 (d) and P2 (c). The 1075, 1100, and 1136 keV lines are mixed $\Delta I = 1$ transitions with $\delta(E2/M1) = -1.4(8), +0.94(^{18}_{13})$, and $-0.15(9)$ [3], respectively. The positive $E2$ admixture of the 1100 keV nicely coincides with an anisotropy close to zero, while those of the other two $E2/M1$ transitions are negative. Consequently, the 3701 keV daughter state of the prompt proton decay from ^{58}Cu has a spin $I = 9/2$ and positive parity, i.e., it corresponds to the neutron $1g_{9/2}$ single-particle state with respect to ^{56}Ni.

B The EUROBALL experiment

In fall 1998 there was a campaign of eight experiments combining the EUROBALL γ-ray spectrometer [21] with the 4π Si-ΔE-E-telescope ball ISIS [22] and the EUROBALL Neutron Wall [23] at the host laboratory in Legnaro, Italy. The 50-element Neutron Wall covered about 1π solid angle in the forward hemisphere.

One of the experiments used the reaction ^{40}Ca(^{24}Mg,$xpynz\alpha$) at 96 MeV beam energy. The compound nucleus was ^{64}Ge. This yields the same residual nuclei as the first GAMMASPHERE run, but at slightly higher excitation energies. The enriched 0.5 mg/cm^2 thin ^{40}Ca layer was backed by 7.0 mg/cm^2 gold and covered by an additional thin gold layer to prevent its oxidation. The main goals were electromagnetic decay properties associated with the prompt particle decays in ^{58}Cu and ^{56,58}Ni, i.e., (*i*) the lifetimes of the nuclear states in the decay-out regime of the deformed bands via Doppler-shift attenuation lineshape analyses of γ-rays detected in the Cluster section of EUROBALL, (*ii*) the spins via $\gamma\gamma$ angular correlations between the Clover and Cluster sections, and (*iii*) parities via linear polarization measurements in the Clover section.

Some of the first results from the experiment are the confirmation of the spin and parity assignment of the 3701 keV state in ^{57}Ni through angular correlation and linear polarization measurements. Another important issue related to the proton-decaying band is its deformation, especially in the decay-out regime. So far, only an average quadrupole moment could be measured [2]. Figure 3(b) provides a summed γ-ray spectrum in coincidence with the four members at the top of the proton-decaying band in ^{58}Cu (grey background) and one evaporated α-particle, one or two protons, and one neutron. The spectrum is projected on the Cluster section at backward angles and Doppler corrected with 95% of the speed of the

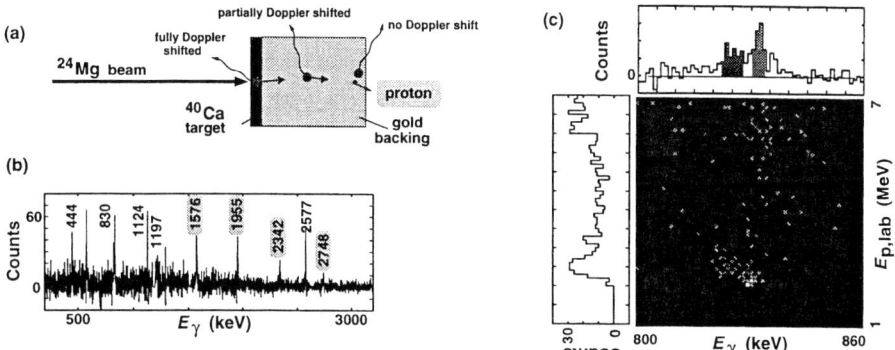

FIGURE 3. Part (a) shows a schematic drawing of the slowing down and possible decay pattern of the band in ^{58}Cu. Parts (b) and (c) are sums of spectra in coincidence with one of upper members of the proton-decaying band in ^{58}Cu and one evaporated α-particle, one or two protons, and one neutron. The γ-ray spectrum of part (b) is projected on the Cluster section of EUROBALL. Part (c) provides a two-dimensional E_γ=800-860 keV vs. $E_{p,\text{lab}}$=1-7 MeV spectrum and the corresponding singles projections.

compound nucleus for γ-ray energies 1.5 MeV$\leq E_\gamma \leq$ 2.4 MeV and $E_\gamma \geq$ 2.6 MeV. The Doppler shift is necessary to line up the energies measured in the Ge detectors at different angles. This implies that the states at the top of the band mostly decay already in the thin ^{40}Ca target layer. This clearly shows the very fast decay times along the band and, hence, confirms the relatively large deformation. It is, however, not possible to deduce individual lifetimes for the corresponding states. On the contrary, Doppler broadened lineshapes suitable for a lifetime analysis are visible for the two transitions at the bottom of the band (830 and 1197 keV). The lineshape of the 830 keV line has a stopped and a Doppler shifted component as can be seen in the expanded γ-ray projection of Fig. 3(c).

The two transitions in the daughter nucleus ^{57}Ni, 1124 and 2577 keV, reveal sharp peaks in Fig. 3(b), i.e., the ^{57}Ni residues were stopped when emitting them. This implies that either the band head in ^{58}Cu and/or the 3701 keV state in ^{57}Ni has a lifetime longer than some picoseconds. From γ-ray spectra in ^{57}Ni it becomes clear that at least the 3701 keV state has such a relatively long lifetime. This, unfortunately, prevents a conclusion on the lifetime of the band head in ^{58}Cu by solely looking into γ-ray spectra. It appears, however, as if there were an energy-shifted component in the proton peak seen with the first forward ring of ISIS, which indicates a sub-picosecond lifetime of the band head. The energy shift in the proton energy is visible in the two-dimensional E_γ=800-860 keV vs. $E_{p,\text{lab}}$=1-7 MeV correlation spectrum of Fig. 3(c). Clearly, the lower-energy part of the proton peak around 2.5 MeV is associated with the stopped component of the 830 keV line, while the higher-energy part is correlated with the shifted component of the 830 keV line. The average shift of $\Delta E \sim$ 400 keV in the proton energy nicely corresponds to the average shift of $\sim$ -7 keV in γ-ray energy, taking into account the respective detection angles relative to the beam direction.

C The GAMMASPHERE experiment

In the previous experiment at the GAMMASPHERE facility we observed the prompt monoenergetic protons only in rings one through four (40 CsI detectors) in the MICROBALL [7] charged-particle detector. This corresponds to 4° to 60° laboratory and 6° to 95° center-of-mass angles. Beyond ring four both the expected yield from the pronounced high-l angular distribution and the energy in the laboratory frame became too low to detect the decay-out protons [2]. The full width at half maximum (FWHM) of the proton peak turned out to be 700 keV, which is considerably higher than the intrinsic resolution of the CsI detectors ($\sim$ 200 keV). This worsening of energy resolution is due to the kinematic broadening because of the finite detector opening angles and, to a much lesser extend, due to scattering in the absorber foils in front of the MICROBALL elements. Therefore, only the first three rings (26 detector elements), which were considered vital for high-resolution particle spectroscopy, were replaced by four highly segmented ΔE-E-Si-strip detectors with a total of $4 \times 16 \times 16 = 1024$ pixels. The active pixels (due to corner shadowing) were around 800, and the opening angle per detector element was reduced by about a factor of eight. Together with an improved neutron array and a larger number of Ge-detectors we expected to be some 10 times more sensitive to prompt discrete-energy proton or α-lines.

The reaction was the same as in the first GAMMASPHERE experiment but at slightly higher beam energies to potentially pursue the bands closer to their point of termination. The data were preliminary analysed [24] and confirmed the existence of the proton peak. However, the FWHM turned out to be more than 500 keV in contrast to the expected < 200 keV. The source of this discrepancy is under investigation, with one explanation [24] being an isomeric ($\tau \sim$ 1 ns) band head. The isomerism would cause the proton to be emitted behind the target position such that the subtle emission angle determination would be off by several degrees. This explanation can be ruled out because of the result of the EUROBALL experiment, though other possibilities for slight but decisive angle differences, such as small beam offsets or detector misalignments, remain to be checked in more detail.

IV SUMMARY AND PERSPECTIVES

Clearly, the mass $A \sim 60$ region comprises a large variety of nuclear struture effects. First of all, the vicinity to the (soft) doubly-magic core ^{56}Ni provides an ideal testing ground for the spherical shell-model. Experimentally, a plethora of states in the first minimum are currently evolving, and the theoretical challenge seems to be not only the deduction of proper single-particle energies and effective two-body matrix-elements for the full fp model space but to include the $1g_{9/2}$ shell in one way or another.

At high spins ($I \sim 10-15$) collective structures dominate the experimental level schemes. Next to the strongly or superdeformed bands in the second minimum there are also a series of bands ranging from normally deformed $E2$ cascades over strongly coupled bands with about equally strong $M1$ and $E2$ transitions to plain $M1$ sequences. Experimentally, the knowledge is somewhat scarce as compared to the spherical shell-model states. On the theoretical side the HF(B) results on the strongly deformed bands indicate a significant effect on the choice of the effective interaction used [2]. More detailed investigations are underway [25]. In addition, the normally deformed and/or strongly coupled bands offer the unique opportunity to compare and relate conventional or Monte Carlo shell-model calculations to approaches based on mean field theory.

Most interestingly, however, prompt monoenergetic proton and alpha-decay lines were observed for the first time in the decay from high-spin states, which are associated with a deformed secondary minimum in the potential. At present, the drastic change of shape in the course of the decay (all the daughter states are spherical) seems to prevent a satisfactory theoretical description and, hence, prediction. Hopefully, time dependent calculations [5,6] can change the situation soon. The impact of the discrete decay-out transitions on models describing the tunneling process from the second to the first minimum remains to be determined. Experimentally, of course, there is the quest for more cases, in particular in other mass regions, and to collect a more comprehensive picture of the process. A steep yrast line of the first minimum may be important, as well as a closeby drip-line. The known decays also take place near or at the presumed band-head of the respective band. Hence, other favourable regimes are the $A \sim 80$ and $A \sim 90$ nuclei while, despite their relatively large excitation energy, the SD bands in the $A \sim 150$ region seem to be less good candidates. In the $A \sim 60$ region it is important not only to verify the decays in dedicated, independent experiments but to investigate the details of the decays such as precise branching ratios, decay times and, hence, strengths as well as potential fine structures. A combined analysis of the above mentioned EUROBALL and GAMMASPHERE experiments might answer the questions.

ACKNOWLEDGEMENTS

Let me first thank Jon Batchelder for the invitation to this successful workshop, and the UNIRIB consortium for the travel support. Many thanks to E. Caurier, D.J. Dean, J. Dobaczewski, L.S. Ferreira, P.-H. Heenen, R.J. Liotta, E. Maglione, W. Nazarewicz, F. Nowacki, A. Poves, and W. Satula for the effort they put into the theoretical description and understanding of the spherical and deformed states and their particle decays in the mass region. Without the perfect and persistent work of the collaborators from Washington University, namely R. Charity, M. Devlin, D.R. LaFosse, L. Sobotka, and last but by far not least D.G. Sarantites, this work would not have been possible at all. Thanks to A. Algora, C. Andreoiu, G. de Angelis, C. Baktash, D. Balamuth, M.J. Brinkman, R. Cardona, C. Chandler, R.M. Clark, F. Cristancho, J. Eberth, C. Fahlander, P. Fallon, E. Farnea, A. Gadea, A. Galindo-Uribarri, J. Garces Narro, P. Hausladen, O. Iordanov, H.-Q. Jin, R. Krücken, I.-Y. Lee, R. MacLeod, A.O. Macchiavelli, J. Nyberg, M. Palacz, Zs. Podolyak, L.L. Riedinger, D. Seweryniak, S. Skoda, Th. Steinhardt, Ch. Teich, O. Thelen, and C.-H. Yu, those not mentioned but having been involved in the set-up of the experiments, and the operating crews of the 88-Inch Cyclotron, the ATLAS facility, and the Legnaro and Cologne tandems for their assistance during the experiments. This research was supported in part by the Swedish Natural Science Research Councils.

REFERENCES

1. GAMMASPHERE, The Beginning ... 1993-1997, Science Highlights booklet, Ed. M.A. Riley, (1998); http://www-gam.lbl.gov.
2. D. Rudolph et al., Phys. Rev. Lett. **80**, 3018 (1998).
3. D. Rudolph, C. Baktash, M.J. Brinkman, M. Devlin, H.-Q. Jin, D.R. LaFosse, L.L. Riedinger, D.G. Sarantites, and C.-H. Yu, Eur. Phys. J. A4, 115 (1999).
4. A.V. Afanasjev, I. Ragnarsson, and P. Ring, Phys. Rev. **C59**, 3166 (1999).
5. N. Carjan, this proceeding.
6. P. Talou, this proceeding.
7. D.G. Sarantites et al., Nucl. Instrum. Meth. **A381**, 418 (1996).
8. D. Rudolph, in *Nuclear Structure '98*, Gatlingburg, TN, Ed. C. Baktash, AIP conference proceedings **481**, 192 (1999); *The Nucleus: New Physics for the new Millenium*, Faure, South Africa, Ed. J. Sharpey-Schafer, Plenum Press, Proceedings, 1999; *Achievements and Perspectives in Nuclear Structure*, Crete, Greece, Ed. S. Åberg, Physica Scripta, Proceedings, 1999.
9. D. Rudolph et al., Phys. Rev. Lett. **82**, 3763 (1999).
10. D. Rudolph et al., submitted to Phys. Rev. Lett.
11. J. Styczen et al., Nucl. Phys. **A327**, (1979) 295.
12. J. Huo, H. Sun, W. Zhao, and Q. Zhou, Nucl. Data Sheets **68**, 887 (1993).
13. I.-Y. Lee, Nucl.Phys. **A520**, 641c (1990).
14. J. Eberth, Phys. Blätter **49**, No. 11, 1016 (1993).
15. J. Eberth et al., Nucl. Instrum. Meth. **A369**, 135 (1996).
16. J. Eberth et al., Prog. Part. Nucl. Phys., Vol. **38**, 29 (1997).
17. D. Habs et al., Prog. Part. Nucl. Phys. **38**, 111 (1997).
18. D. Weisshaar, Diploma thesis, University of Cologne, unpublished.
19. D. Rudolph et al., Eur. Phys. J. A, in press.
20. L.M. Garcia-Raffi et al., Nucl. Instr. Meth. **A359**, 628 (1995).
21. EUROBALL III, A European γ-ray facility, Eds. J. Gerl and R.M. Lieder, GSI 1992.
22. E. Farnea et al., Nucl. Instrum. Meth. **A400**, 87 (1997).
23. Ö. Skeppstedt et al., Nucl. Instrum. Meth. **A421**, 531 (1999).
24. J.N. Wilson, private communication.
25. J. Dobaczewski et al., to be published.

Novel Experimental Techniques and New Directions

Prospects for Future Proton Studies at HRIBF

C. R. Bingham,[1,2] J. C. Batchelder,[3] T. N. Ginter,[4] C. J. Gross,[2,5] R. Grzywacz,[1,6] Z. Janas,[6,7] M. Karny,[1,6,7] J. W. McConnell,[2] K. Rykaczewski,[2,6] K. S. Toth,[2] and E. F. Zganjar[7]

[1] *Department of Physics, University of Tennessee, Knoxville, TN 37996, USA*
[2] *Physics Division, Oak Ridge National Laboratory, Oak Ridge, TN 37831, USA*
[3] *UNIRIB, Oak Ridge Associated Universities, Oak Ridge, TN 37831, USA*
[4] *Department of Physics, Vanderbilt University, Nashville, TN 37235, USA*
[5] *Oak Ridge Institute for Science and Education, Oak Ridge, TN 37831, USA*
[6] *IEP, Warsaw University, 00681 Warsaw, Hoza 69, POLAND*
[7] *Joint Institute for Heavy Ion Research, Oak Ridge, TN 37831, USA*
[8] *Department of Physics, Louisiana State University, Baton Rouge, LA 70803, USA*

Abstract. Great progress has been made in the last 20 years in the study of proton emission from unstable nuclei, but the prospects for additional strides in the next several years are bright. The present main limitations on the study of proton radioactivity are related to the inability to produce copious quantities of nuclides beyond the proton drip line, and the difficulty of measuring proton radioactivity of a mass-separated nucleus in the first few microseconds of its existence. At the Holifield Facility we will attack the second of these limitations by using new signal processing CAMAC modules DGF-4C. Digitizing of the preamplifier signals should enable the analysis of a proton decay occurring at times even less than 1 microsecond after an implant in a strip detector. In the same process, the threshold energy at which we can make measurements will be lowered. These two things will hopefully enable the measurement of lower-energy, but faster decays of isotopes in the ^{100}Sn region and below. For the latter region, the proton decays crucial for a rp-process scenario are of particular interest (e.g. ^{69}Br decay). Secondly, for very short-lived species, we plan to make measurements (without residue separation) at points much closer to the target, thus reducing the flight time between the target and detector. As more intense radioactive beams become available, eg. ^{56}Ni, we will utilize these to produce more neutron-deficient nuclides by use of colder reactions than is possible with stable beams. In some cases where delayed proton emitters are present in the same isobaric chain, the use of the cold reactions with radioactive beams can provide purer samples of the isotope of interest, with a reduction in background from the delayed proton emitters in the same mass chain.

INTRODUCTION

Proton radioactivity has been an active research area utilized in mapping the limits of nuclear stability for the last twenty years. The first ground state emitter, reported in 1981 [1], was ^{151}Lu which has a half-life of 80(2) ms and an $h_{11/2}$ configuration. Interestingly, a $d_{3/2}$ isomer of this well-studied proton emitter with a half-life of 16(1) μs was first discovered and reported [2] in 1999. Proton radioactivity studies have profited tremendously from the installation of on-line recoil mass separators at heavy ion accelerators, such as the DRS at Daresbury [3], the FMA at Argonne National Laboratory [4], and the RMS [5] at the Holifield Radioactive-Ion Beam Facility (HRIBF), and from the development of silicon strip detectors [6]. A strip detector facility was installed at HRIBF in recent years to pursue this field of study. Briefly, heavy ions from the HRIBF were directed onto targets at the entrance to the RMS at energies necessary to produce (HI,pxn) residues to assay for proton radioactivity. The recoiling residues moving in the forward direction are directed through the RMS, pass through a position-sensitive avalanche counter to determine the mass, and are implanted in a double-sided silicon strip detector behind the focal plane of the RMS. The signal produced by the implant passes through a preamplifier and is split to feed two amplifiers, one with low gain to treat the implant signal and another with high gain to view any subsequent signals from decay of the implant and its daughters. All implant (DSSD in coincidence with a PSAC signal) and decay events are time-stamped at the time of arrival, thus enabling the correlation of a proton decay signal with a particular implant and determination of the half-life. A summary of results obtained with this system is given in another article in these proceedings [7]. In this paper we discuss some of the ideas we will pursue in future studies of proton emission.

I THE FRONTIERS OF PROTON RADIOACTIVITY

To give some perspective on our directions it is instructive to review briefly the history of progress in the study of proton radioactivity. There are two features that emerge from viewing the list of discoveries: 1) in general the more recently discovered emitters have been made with increasingly energetic heavy ions, resulting in the evaporation of more neutrons (compare (HI,p2n) for first discoveries with (HI,p4n) and one (HI,p5n) reaction in recent times), and 2) the new half-lives are shorter on the average than the earlier measured ones, though there are some exceptions.

The production of nuclei farther from stability by fusion evaporation reactions becomes problematic due to the large decrease in the cross section for each additional neutron evaporated, largely due to the fact that proton evaporation becomes more competitive as one approaches the proton drip line. Perhaps the use of rare isotope beams will provide a means for production of isotopes beyond the drip line while utilizing lower beam energies that would increase the probability of evaporating just a few particles leaving the final residue in the isotope of interest. In addition to having a larger cross section for the isotope of interest, the number of other reaction products will be greatly diminished, thus reducing the background and making the measurements much cleaner.

The shortest half-life for proton radioactivity measured to date is 3.5 μs for ^{145}Tm [8]. This is at the short half-life limit of what we can measure using our current strip detector system and conventional electronics. There are two factors which impose this limit: the time required for the evaporation residue to pass through the RMS and the recovery time of our amplifiers after the deposit of a large signal from the implant. This paper will discuss some of the physics that we wish to pursue in further measurements, the plans we have to alter the electronics to enable measurement of sub-microsecond half-lives, and possible new experimental arrangements to alleviate the time-of-flight problem. One case where the use of a radioactive-ion beam may provide the cleanliness to do the measurement will be discussed.

II THE PHYSICS TO BE PURSUED

A portion of the chart of nuclides showing all previously reported proton emitters is shown in Figure 1. Also shown are the stable isotopes in this region and the expected position of the proton drip line. In addition various shades are used to indicate the predicted deformation of each nuclide [9]. It is noted that the proton drip line passes near ^{100}Sn, passes through a region of large deformation extending up to about mass 150, and then passes through a nearly spherical region up to the Pb region. Possibilities of study and related physics questions will be addressed for each region.

A The Spherical Region

In the region between Z = 64 and 82, the $s_{1/2}$, $d_{3/2}$, and $h_{11/2}$ proton orbitals are filling, and the ground states of the odd-Z nuclides in this range have been shown to have its valence proton in one of these orbitals. The half-life for emission of such an unbound proton in one of these orbitals is dependent on the dynamics of proton emission and the spectroscopic factor which is dependent on the nuclear structure of the proton emitter and the daughter nucleus. The dynamic part can be calculated by various techniques such as WKB, the Two- Potential Approch, or DWBA [See, eg. [10]]. By comparing the barrier penetration probability with the decay probability obtained from the experimental half-life, one can obtain an experimental value for the spectroscopic factor which can be used to test various shell model predictions. The precision of these experimentally determined spectroscopic factors is sensitive to the errors on the proton energy, but in many cases large error bars result from the inability to determine the branching ratio for proton

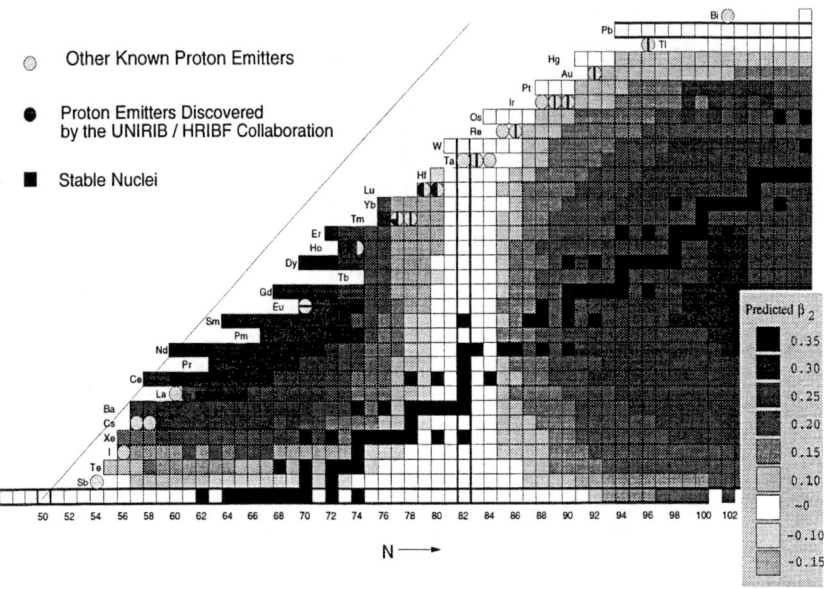

FIGURE 1. Chart of the nuclides showing the known proton emitters, the stable nuclides, the predicted location of the proton drip line and the predicted deformation.

emission, relative to beta decay. For the longer half-lives, the beta decay predominates and small errors in estimating the beta decay branchs result in huge errors for the proton decay branches. For half-lives in the microsecond region, the proton emission predominates and large errors on a nearly zero beta branch has little effect on the large proton branch. Thus, it is apparent that more precise spectroscopic factors for comparison with nuclear theory can be obtained by more measurements of short-lived proton emitters. As a practical matter, many of the proton emitters remaining to be discovered will have short (μs) half-lives. The half-life depends critically on the proton energy and the orbital angular momentum which it carries. As one proceeds farther from stability, the proton separation energy becomes smaller and the half-life falls precipitously. The observation of shorter-lived proton emitters will enable more extensive tests of the all-important mass formulas in the region beyond the proton drip line.

The proton separation energies of proton emitters observed in this region are plotted in Fig. 2, along with the predictions of the Liran-Zeldes mass formula [11]. While the agreement with this old formula is not extremely good, the predictions do exhibit a relationship with the experimental values. Thus, from the systematic trend, we have estimated the proton separation energy for a number of other $h_{11/2}$ emitters and these estimates are also plotted in the figure. It is apparent that there is potential for the observation of several new isotopes and isomers in this spherical region with energies about the same as that of ^{145}Tm ($t_{1/2}$=3.5 μs), and which also should have half-lives in the microsecond range. Similar projections for $d_{3/2}$ and $s_{1/2}$ proton emitters reveal several other cases which should be studied. We anticipate the study of a number of new isotopes in this region, and with the shorter half-lives, they will provide more stringent tests of mass formulas and shell model structure predictions.

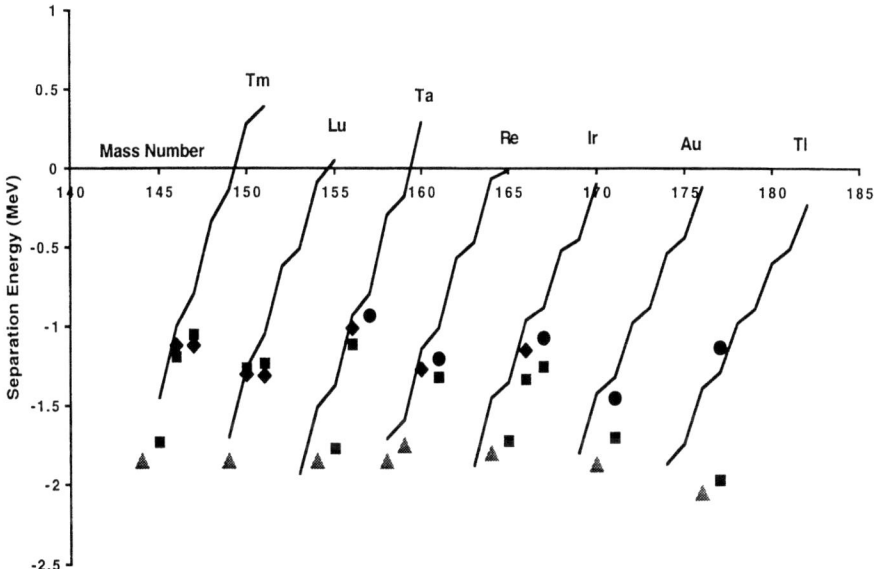

FIGURE 2. Proton separation energies of the known proton emitters for $s_{1/2}$ orbitals (circles), $d_{3/2}$ (diamonds), and $h_{11/2}$ (squares) in comparison with the predictions of Liran and Zeldes (lines). The triangles are locations of possible $h_{11/2}$ emitters which may decay with half-lives in the microsecond range.

The observation of fine structure in the proton spectrum from odd-odd nuclei in the spherical region may provide a means of identifying different single particle neutron states of odd-A nuclei near the proton drip line [12].

B The Deformed Region

The deformed region ($110 \leq A \leq 150$) is relatively less explored, and shows promise to provide the most significant new insights in proton radioactivity. The deformed states of course have a rather complicated single particle structure, which complicates the extraction of nuclear structure information from the proton decay half-lives. Nevertheless, a great deal of progress has already been made in understanding the decay process and in extraction of good spectroscopic information. Several contributions to these proceedings address the proton decay of deformed nuclei [13–16]. Since the single particle configurations of these deformed nuclei have many components, and since the first 2^+ states of daughter nuclei are at relatively low excitation energy, the probability for decay to the 2^+ state will be competitive with that to the ground state in some of the proton emitters in this region. When the fine structure line is observed, it provides data to extract information about a different piece of the deformed nucleus wave function. By examination of Fig. 1, it is obvious that there are many isotopes in this region requiring first or more extensive study.

C The Region Around ^{100}Sn and Below

The proton drip line passes very near to the doubly-magic nucleus ^{100}Sn. In order to understand the shell structure in the region, the study of the single particle orbitals is of great importance. Below ^{100}Sn spontaneous proton emission is of vital interest in understanding the details of the rp process. Very little information is available in this important region of the nuclear chart. The proton energies of the first possible emitters beyond the proton drip line will be small, and because of the lower Coulomb barrier, the half-lives will also be small. New experimental techniques may be required to study proton emitters in this region. The lightest proton emitter reported to date is ^{105}Sb with a proton energy of 478 keV [17]. While this result needs confirmation, other possible submicrosecond activities (e.g. ^{93}Ag, ^{103}Sb) will possibly provide valuable information regarding the valence orbitals and proton separation energies of resonance states beyond the proton drip line, yet close to a double-closed shell. The time scale and path of rp-process nucleosynthesis depends strongly on the two-proton capture rates necessary to create the elements above the unstable "termination points" like e.g. a ^{69}Br. The measurement of its low energy proton decay energy is a significant goal of our program.

D The Trans-Lead Region

This is a region of expected shape coexistence, observed already in α decay studies and one example of proton radioactivity [18]. The competition of proton emission and α decay in this region will provide information useful in discerning the role of the so-called intruder levels in driving the shape coexistence. Observation of decays from both the $h_{9/2}$ ground states and isomeric $s_{1/2}$ intruder states would establish the detailed locations and spectroscopic information on these important shape-coexistent levels.

III SIGNAL PROCESSING FOR SHORT-LIVED PROTON EMITTERS

In the use of DSSDs to correlate charged-particle decays from rapidly-decaying nuclei implanted in the detector, separation of the small decay signal from the larger implant signal coming from the detector is problematic. The correlated decay signals coming in the first few microseconds after an implant appear as a pile-up pulse. We propose to record such pile-up signals by digitizing the raw preamplifier signals to provide a trace of the pile-up signal with time bins of 25 ns. The input stage of new signal processing CAMAC modules DGF-4C available from X-ray Instrumentation Associates has been shown to accomplish this feat [19]. The traces can be analyzed with digital spectroscopy units, also on-board in the DGF-4C modules, to strip out the time and energy of a small decay signal riding upon a much larger implant signal. The implant times, decay times and energies are time stamped and buffered into the data stream to our data acquisition system. After some pre-analysis, these data can be analyzed with our standard sorting routines to study fast charged-particle decays. As an alternative, the digital traces of the piled-up implant-decay signals can be output to the host data acquisition computer. An example of this type of data is shown in Fig. 3. This trace was generated in a test run utilizing ^{113}Cs implants, and shows the peak shape when the proton is emitted approximately 300 ns after the implant. Our treatment of these data is still in a premature stage, but a fit by two curves, one to represent the implant and another to represent the decay, is shown also in the figure. It is our intent to use this mode of operation for the piled-up signals having starting times below 2 μs. We are in the process of developing programs to fit these traces to deduce the implant and decay times and the decay energies. These parameters will then be used to generate separate implant and decay events which will be fed to our normal sorting routines to observe correlations. The minimum decay times observable with this technique is of the order of 300 ns with our current preamplifiers, and the threshold energy is of the order of 200 keV. We believe this will enable the study of lighter nuclides which decay with lower proton energies and very short half-lives. The lower energy threshold will also be vital in the search for fine-structure peaks in the deformed region.

IV FLIGHT TIME LIMITATIONS

A schematic diagram of the Recoil Mass Spectrometer at HRIBF is shown in Fig. 4. It is comprised of a momentum analyzer followed by a mass separator. The instrument produces needed mass dispersion with a minimum of transmission of scattered beam [5]. The main disadvantage of this instrument for the study of short-lived species at the focal plane is its long length, ~25 m from the target to the focal plane. The time-of-flight for proton emitters through the RMS is dependent upon the reaction, but is typically 2-3 μs. This puts a rather severe limit on the minimum

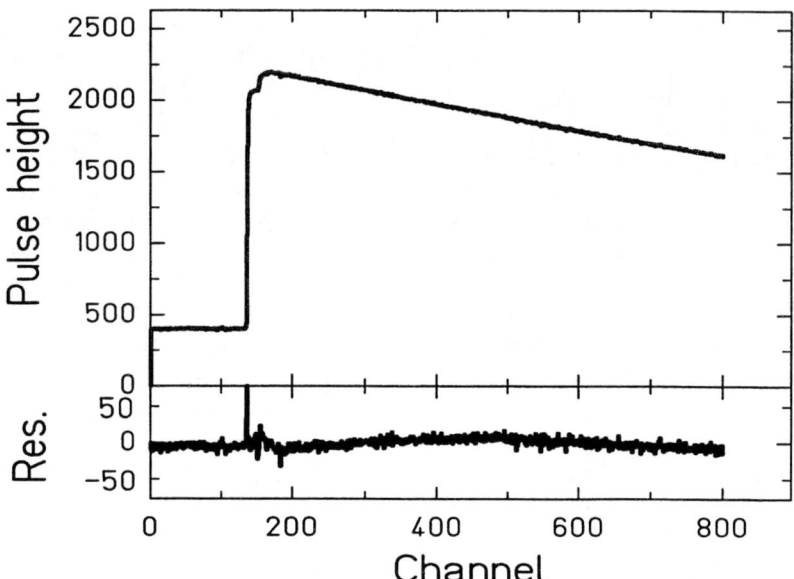

FIGURE 3. A trace of the pile-up pulse from a DGF-4C resulting from the emission of a proton 300 ns after the arrival of the implant in the DSSD. A fit incorporating two pulse shape curves is superimposed on the data and the difference in the data and the fit are plotted at the bottom.

half-life which can be studied at the focal plane. On the other hand it is possible that for the very short-lived species, mass separation will not be required to study

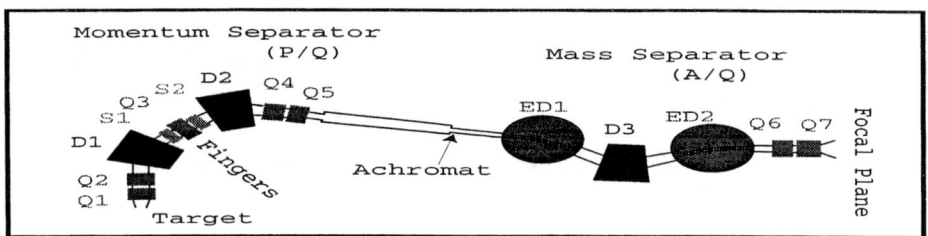

FIGURE 4. A schematic diagram of the Recoil Mass Spectrometer at HRIBF. In addition to doing experiments at the focal plane with full mass resolution, alternative counting stations can be introduced at the achromat and perhaps closer to the target in order to study decays of isotopes with half-lives in the μs range.

the correlated activity, since there is usually only one possible proton emitter with short half-life that would be produced with a particular beam and energy. Thus, we have prepared a chamber to make measurements at the achromatic focus of the RMS, which will shorten the time-of-flight by a factor of two. Preliminary plans are underway to investigate the possibility of obtaining appropriate focusing conditions at points even closer to the target.

As an alternative to the RMS, there is the possibility of utilizing the split-pole spectrometer online at HRIBF in a gas-filled mode in order to reduce the total flight path to less than 3 meters. Figure 5 illustrates the expected kinematics for the production of ^{149}Lu with ^{58}Ni and ^{56}Ni bombardments of ^{96}Ru at the appropriate energies and subsequent separation of the evaporation residues from the elastically scattered beam in the gas-filled spectrometer. The time of flight of the evaporation residues is seen to be about 300 ns, which should enable the observation of this proton emission if it has a half-life of at least 0.3 μs. For this particular case, it is noted that the use of the radioactive beam ^{56}Ni will lead to a cleaner separation of the evaporation residues from the scattered beam because of the lower-energy projectiles required for the p2n reaction.

ACKNOWLEDGEMENTS

This research is supported by the U. S. Department of Energy through contracts DE-AC05-96ER40983 with Lockheed-Martin Energy Research Corporation, DE-FG02-96ER40983(Tennessee), DE-FG05-88ER40407(Vanderbilt), and DE-FG02-96ER40978(Louisiana State). ZJ was partially supported by the Polish KBN2 P03B 086 17. We wish to thank Felix Liang for the Monte-Carlo calculations and valuable discussions.

REFERENCES

1. Hofmann, S., et al., *Z. Phys. A* **305**, 111 (1982).
2. Bingham, C. R., et al., *Phys. Rev. C* **59**, R2984 (1999).
3. James, A. N., et al., *Nucl. Instru. Meth. A* **267**, 144 (1988).
4. Davids, C. N., et al., *Nucl. Instru. Meth. B* **70**, 358 (1992).
5. Gross, C. J., et al., *Nucl. Instru. Meth. A*, Accepted for publication.
6. Sellin, P. J., et al., *Nucl. Instrum. Methods Phys. Res.* **A311**, 217 (1992).
7. Rykaczewski, K., et al., These Proceedings.
8. Batchelder, J. C., et al., *Phys. Rev. C* **57**, R1042 (1998).
9. Möller, P., Nix, J. R., Myers, W. D., and Swiatecki, J., *At. Data Nucl. Data Tables* **59**, 185 (1995).
10. Åberg, S., Semmes, P. B., and Nazarewicz, W., *Phys. Rev. C* **56**, 1762 (1997);**58**, 3011 (1998).
11. Liran, S. and Zeldes, N., *At. Data Nucl. Data Tables* **17**, 431 (1976).
12. Ginter, T. N., et al., These Proceedings.

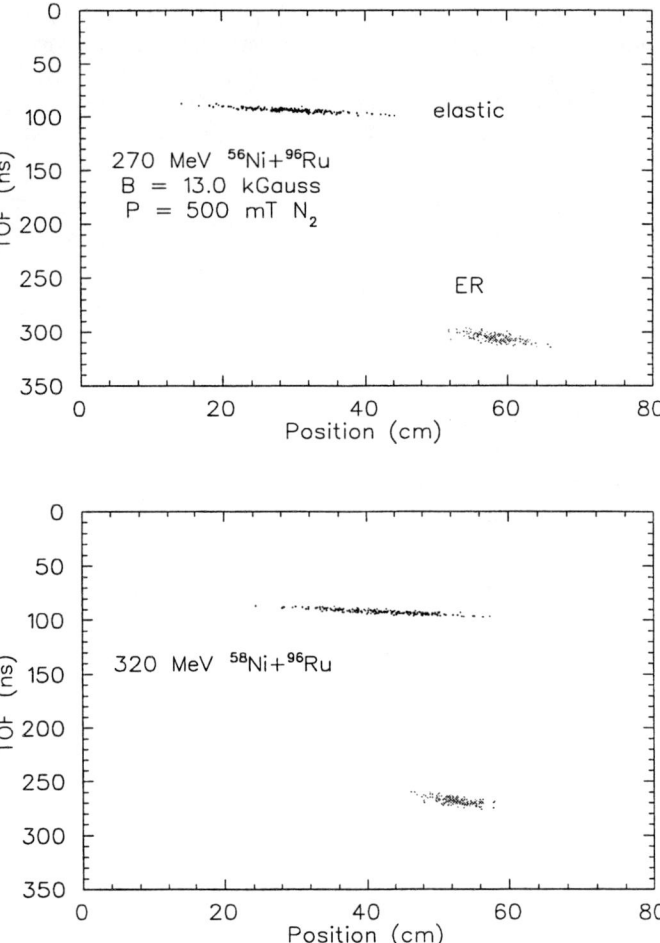

FIGURE 5. Monte-Carlo simulations of the time-of-flight vs. focal plane position of elastically scattered Ni ions and evaporation residues including the possible fast proton emitter ^{149}Lu. The lower(upper) panel illustrates the focussing properties for a reaction induced with ^{58}Ni(^{56}Ni) at the appropriate energy to produce ^{149}Lu.

13. Kruppa, A. T., Nazarewicz, W., and Semmes, P. B., These Proceedings.
14. Lalazissis, G., Vretenal, D., and Ring, P., These Proceedings.
15. Davids, C. N. and Esbensen, H., These Proceedings.
16. Maglione E. and Ferreira, L. S., These Proceedings.
17. Tighe, R. J., et al., *Phys. Rev. C* **49**, 2871 (1994).
18. Davids, C. N., et al., *Phys. Rev. C* **52**, 1807 (1995).
19. Hubbard-Nelson, B., Momayezi, M., and Warburton, W. K., *Nucl. Inst. Meth.* **A422**, 411 (1999).

Applications of real-time digital pulse processing in nuclear physics

Michael Momayezi[1], Peter Grudberg, Wojciech Skulski, William K. Warburton

X-Ray Instrumentation Associates, Mountain View, CA 94043, www.xia.com

Abstract. We present the core architecture of XIA's digital pulse processors and show how digital filtering and pile up inspection are performed in real time. XIA's pulse processors can simultaneously extract energy and pulse shape information from the detector pulses provided to its inputs. Two applications of this novel technology are discussed in the paper. High resolution for γ-rays was achieved at very short peaking times with digital ballistic deficit correction. Sequences of pulses characteristic of radioactive decay via proton emission can be recognized online with the real-time pileup inspector operated in the reversed logic mode.

INTRODUCTION

Digital pulse processors for γ-spectroscopy have been on the market for quite a while. The possible advantages of an all-digital system compared to an analog one are the following: (1) It is possible to operate the system remotely from a host computer without the need for any manual interaction, like adjusting potentiometers. This in turn allows to save and restore device settings by downloading a parameter file, as well as switching from one experiment to another with ease. (2) A digital system eliminates the need for a shaping amplifier because it can perform digital filtering of the waveforms obtained from an integrating preamplifier. (3) It is possible to extract more information from the detector signal than it has been possible using analog electronics. (4) The digital system functionality and performance are determined by its firmware and software. It is thus possible to change the functionality and to enhance the performance through the distribution of firmware and software upgrades.

The greatest challenge for any digital system is the enormous amount of computing that needs to be done to achieve reasonable throughput rates. For high throughput rates it is necessary to perform the computations in real time.

---

[1] Corresponding author, momayezi@xia.com.

THE CORE ARCHITECTURE

All of XIA's digital pulse processor models are built using a common core architecture, c.f. figure 1, which is geared to deliver precise energy and time information at the highest possible throughput rate. This architecture is outlined in figure 1. The core consists of an analog signal-conditioning unit (**ASC**), a real-time processing unit (**RTPU**) and a digital signal processor (**DSP**). The ASC accepts the output from an integrating preamplifier and sends it to a waveform digitizer (**ADC**). The ASC is used to adjust the gain and offset to make the signal range match the input range of the ADC. No shaping is performed. Indeed, great care is taken to not change or distort in any way the incoming waveform.

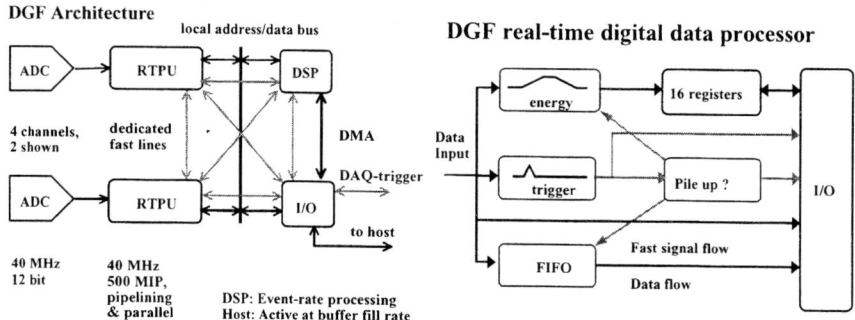

Figure 1: Left: schematic of the DGF-4C architecture, showing the waveform digitizer (ADC), the real time processor (RTPU), the digital signal processor (DSP), and the host interface (I/O). Two out of four device's channels are shown in the figure. The right-hand side shows the organization of the real-time processing units.

The ADC is a 40 MSPS digitizer with an accuracy of 12 bits. To avoid aliasing of high frequency noise or signal components an anti-aliasing filter is placed just before the ADC input. The filter is implemented as a fourth-order Gaussian filter with a 12dB corner frequency of 20 MHz.

The output from the ADC is a stream of 12-bit words clocked at 40 MHz, i.e. it constitutes a data rate of 60 Mbytes/s. This stream is sent to the RTPU, which shall be described in some detail in the next section. In the RTPU digital filters are applied in real time to extract energy and timing information, and to recognize pileup events, as follows. (1) A trapezoidal filter of programmable length and flattop time is used to replace the action of a conventional shaping amplifier. (2) A short trigger filter and comparator with programmable threshold is used to recognize the arrival of a pulse in the input signal. (3) A pileup inspector processes each pulse height step and recognizes those which occur too close together. Under most operating conditions, such pileup events are rejected. (A special case where pileup events are accepted rather than rejected is discussed below.) Once a well-isolated step has been found, the RTPU issues an interrupt request to the DSP, which in turn will read preprocessed data from the RTPU registers. The DSP quickly writes the data into a buffer area and exits the interrupt routine to minimize dead time. An event processing routine, which runs in background processing mode rather than interrupt mode, continuously finds events in the data buffer and processes these, i.e. reconstructs energies, increments spectra and formats list mode data complete with time stamps and pulse shape analysis results.

Through this division of tasks between the RTPU and the DSP, XIA's pulse processors can achieve very high throughput rates, well in excess of 500,000 c.p.s. Two key ingredients are (a) real-time processing of the raw waveform data in the RTPU, summoning the DSP only on an event-by-event basis to perform computations, and (b) buffering data in the DSP to accommodate statistical event rate fluctuations.

The trigger/filter processor

The real-time processing units (**RTPU**) form the core of XIA's pulse processor models. They employ a high degree of parallelism and pipelining to achieve the equivalent of 500 MIPS at a system clock rate of 40 MHz. Figure 1 shows a schematic view of the RTPU.

Each RTPU receives the continuous data stream of 60 Mbyte/s coming from the input channel's waveform digitizer and splits it into three parallel branches. The first branch is routed to a trapezoidal digital filter for energy measurement, as shall be described in section 3. This finite impulse response filter is operating continuously. The filter is implemented as two sliding sums of length L spaced apart by a gap of length G, measured in sampling clock cycles. The filter output is the algebraic difference between the leading and the trailing sum. The response of such a filter to a step input pulse is shown in figure 2. The rise time of the computed filter response is L cycles, the flattop time is G cycles, and the total filter pulse has length of $2*L+G$ cycles. The output of the filter is sampled at the end of the flat top portion and latched into a register from where it can be retrieved by the digital signal processor.

As figure 2 shows, two consecutive pulses need to be separated by only $L+G$ clock cycles. This separation is enough to ensure that at the sampling time of each event the filter output is caused only by the event in question. While this is valid for periodic pulses, the equivalent time to be used for computing the throughput for pulses occurring randomly in time is $2*L+G$ cycles. Let T be the time corresponding to $2*L+G$ clock cycles, ICR be the input count rate, and OCR the output count rate. The relationship between ICR and OCR is given by the following:

$$OCR = ICR*\exp(-T*ICR)$$

For a given T the maximum throughput rate is attained when ICR=1/T:

$$OCR_{max} = 1/(T*\exp(1)).$$

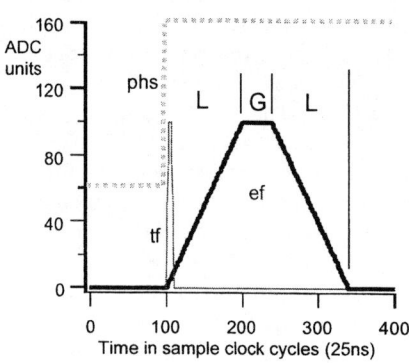 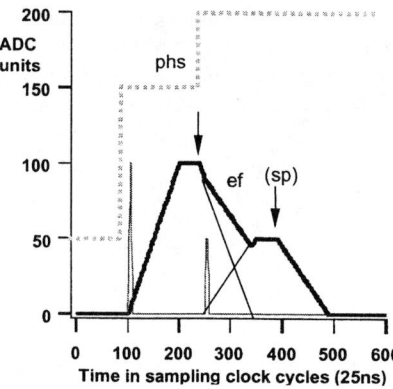

Figure 2: Left: A pulse (dotted line labeled "phs") arriving at the RTPU causes responses of the trigger filter (line labeled "tf") and the energy filter (line labeled "ef"). The output from the energy filter rises for L clock cycles, stays flat for G cycles and takes L cycles to return back to baseline. Right: Two consecutive pulses (dotted line labeled "phs") may cause overlapping energy filter responses (ef). If the first response has subsided at the sampling point (sp) of the second one, then both events are valid.

A second branch of the data stream entering the RTPU is connected to the pile up inspector. The events are detected by running the data stream through a trigger filter (TF) similar to the energy filter, but of shorter duration, comparable to the signal rise time. Each time the TF output exceeds a preset threshold, an internal trigger is generated within the RTPU. The pileup inspector will validate and latch the event data if after a time $T=L+G$ no second event has been detected. This case of pileup is demonstrated in figure 2 and is referred to as pileup in the energy channel, because the two signals are well resolved by the fast trigger filter. Two pulses separated by a very short time may cause pileup in the trigger channel, if the trigger filter cannot separate the two pulses. In such a case, the time spent by the TF over threshold is longer than it is the case for regular events. The pileup inspector is programmed to recognize and reject such events as well.

The third branch of the input data stream is directed into a FIFO memory. Whenever the trigger filter generates an internal trigger the FIFO is stopped when the pileup inspector validates the event, and its content can be read by the DSP.

The digital signal processor

The full power and flexibility of a digital system rests with the code running on its DSP. The code has to be organized such that it minimizes the system dead time, allows for complex on line data analysis, and can synchronize data acquisition across module boundaries.

System dead times can be minimized through event buffering, which helps to absorb bursts and to distribute the computational load evenly over time. The pileup inspector sends an interrupt request to the DSP, which faces two tasks: a) the DSP has to collect the preprocessed data from the RTPU, and b) the DSP has to process the event data. Task a) needs to be done immediately so that the RTPU can latch the next event. Task b) can be done later, if the intermediate data can be saved into a buffer for later processing. Therefore, the DSP is programmed to respond to the interrupt with just the necessary steps to save the data. If this can be achieved within *L+G* clock cycles then the DSP reading action does not cause any additional dead time beyond that dictated by the energy filter length and gap. This remarkable characteristic is due to the fact that the energy filter computation in the RTPU never stops. If the DSP data reading takes longer than the filter-induced dead time, then the readout time determines the system dead time. Thus the dead time is the maximum of DSP read time and filter dead time, but not the sum of both.

The DSP writes both the raw and the preprocessed data, including filter values, time stamps, and waveforms, into a buffer for further processing. Besides pulse height reconstruction and histogramming the DSP can also perform pulse shape analyses. One example will be discussed in section 3.

In addition to the event interrupt service, the modules have to respond in real time to system-wide commands that require synchronous action of all modules in a system. The most important examples are clock synchronization and simultaneous data acquisition START and STOP commands. It is not possible to implement these actions to within a 25ns time precision via regular CAMAC commands, or commands distributed over any other data bus such as Ethernet. In order to make module synchronization possible, all XIA modules use a hardware BUSY — SYNCH loop implemented in every module. In response to a START or a SYNCH command, all modules drive their BUSY output signal from logic 1 to logic 0. Right after doing that, the DSPs halt waiting for an interrupt, which will occur when the signal at the SYNCH input makes a 1→0 transition. The interrupt is used to zero the local timers and to start the run. This interrupt sequence is exactly the same in all modules.

In a system with many modules, all BUSY outputs are fed into an OR gate, the outputs of which are led back to the module SYNCH inputs. Thus, it is the last module, ready to start the data acquisition, which allows all other modules to start too.

When a module encounters an end-of-run condition it raises its BUSY output to logic 1. By the same mechanism as before, all modules will simultaneously stop the run. Through this mechanism it can be ensured that the life time of all modules in the system is exactly the same, and that at no point in time only a fraction of the system is ready to accept events.

PREAMPLIFIER BALLISTIC DEFICIT

Germanium detectors with large radii have long charge collection times, ranging from 100 ns to 500 ns. In addition, the charge collection time depends on the radial position at which the photon converts. Photons converting at either the inner or outer radius create charge carriers that have to travel the entire radial distance between anode and cathode. Charge carriers created about halfway between the inner and outer radius only have to travel half the distance and the charge collection time is shorter by a factor of two.

The output from a charge integrating preamplifier continues to rise during the charge collection time. It is well known that in a Gaussian shaper the finite preamplifier signal rise time will lead to a ballistic deficit, i.e. the measured output pulse height is smaller than that for the same charge had it been collected instantaneously. Using a trapezoidal filter, rather than a Gaussian shaper eliminates this problem and provides a much improved energy resolution, [2]. However, there still remains a ballistic deficit due to the fact that most low-noise preamplifiers use a differentiation stage with a 50 μs decay time. When using a trapezoidal energy filter, as shown in figure 1, one measures the amplitude of an exponentially falling signal after all charge has been collected. Consider a measurement of the exponential amplitude at a fixed time after the leading edge. For a fast rising signal, quick charge collection, there is more time for the resulting preamplifier signal to decay, than there is for a slow signal from slow charge collection. Thus, even with the use of a trapezoidal filter, there remains a contribution to the energy resolution from a kind of ballistic deficit. The effect is by no means negligible. In a small germanium detector operated at XIA, 2 inch diameter and 2 inch long, the rise times varied between 100 ns and 200 ns. The error due to ballistic deficit in the preamplifier is of the order of $\delta t/\tau = 0.1 \mu s/50 \mu s = 0.20\%$. Especially at higher γ-energies this is a significant contribution. Figure 3 shows the energy resolution for the 2614 keV line from ^{208}Tl as measured by the trapezoidal energy filter and with an applied ballistic deficit correction, computed by the DSP from the acquired waveforms.

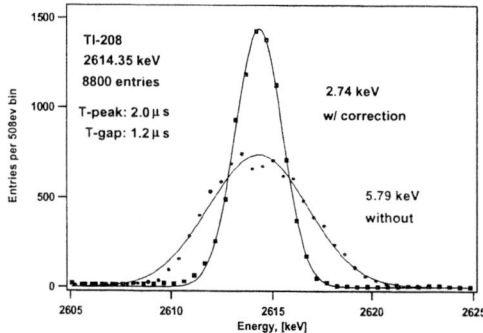

Figure 3: Energy resolution for the 2614 keV γ-line from ^{208}Tl. At 2 μs peaking time for the energy filter, an excellent energy resolution is achieved if the ballistic deficit correction is included.

It should be noted that an analog Gaussian shaper would in the limit of very long shaping times not have any difficulty with the signal rise time variations, while a trapezoidal shaper would. The analog trapezoidal shaper would remove only the ballistic deficit of a Gaussian shaper with short shaping times, but the preamplifier-induced errors would remain. The advantage of the digital system is that it can make the best use of the available information. Using pulse shape analysis it is possible to achieve excellent energy resolutions for very short filter times. For instance, the 2μs peaking time correspond to a 1μs shaping time in a Gaussian shaper.

USING PILEUP INSPECTOR TO SELECT PROTON EMISSION EVENTS

In most experimental situations signal pileup is something to be avoided and rejected, rather than to be called for. However, there are exceptions. In the study of very short-lived radioactive isotopes that decay by proton emission one seeks an unusual trigger condition.

At the Holifield radioactive beam facility, the radioactive isotopes are being produced in a collision of a heavy ion beam with a target. Via a mass separator the recoils are implanted into a double-sided silicon strip detector (**DSSD**). The vast majority of the implanted proton-rich recoil nuclei decay by the β^+-emission. A tiny fraction of them, however, are short-lived proton emitters. Their signature in the DSSD is a big implantation pulse (10..30 MeV) followed by a smaller 1 MeV pulse a few microseconds later at the same location. When looking for such events one can record the arrival times of all the implantation and decay pulse candidates and correlate these data offline. This technique tends to create fairly large, though still manageable data sets, [3].

A more convenient method to look for proton emission events is to record only the candidate events, based on the unique signature discussed above. It is possible to program the pileup inspector of the DGF-4C to recognize two pulses separated by less than 10μs as such candidate events. At the same time, the DGF-4C is programmed to record a 25μs trace for each such double event. We then invert the usual logic of the pileup inspector and validate only piled-up events. As a result the DGF-4C will now trigger on any pulse pair that has less than 10μs spacing, resulting in a large reduction of the amount of recorded data. Not all double events are due to proton emission, there will be electromagnetic interference and background radioactivity as well, but the data analysis is facilitated by a much smaller number of events, with more detailed information available for each event, than it was possible before.

We have tested these concepts with a radioactive source which also produces pileup events. Figure 4 shows such an event recorded in a pileup catcher mode. It was acquired with a ^{57}Co source illuminating a germanium detector. The captured trace shows two 122 keV γ-rays detected in short succession. Clearly, it is possible with this technique to capture and analyze even the most rapid decays, as long as the implant pulse and decay pulse can be distinguished at all.

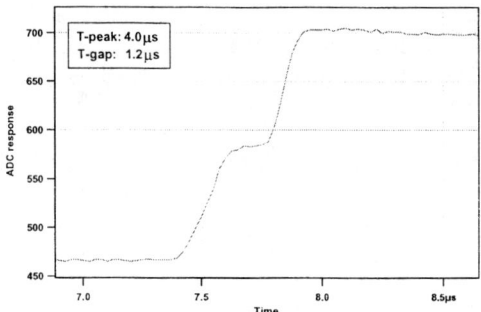

Figure 4: Two 122 keV γ-rays detected in short succession. Their separation is about 350ns. This waveform was acquired by deliberately triggering on pileup events.

CONCLUSION

We have discussed the architecture of an all-digital pulse processor. The device employs real-time processing for those tasks that need to run uninterrupted and at very high rate. Digital filtering fully replaces the conventional analog shaping amplifier, which makes it possible to directly process preamplifier signals. Event triggering, pile up inspection, and continuous waveform acquisition and storage, are all performed in real time. The on-board digital signal processor is programmed to ensure maximum data throughput and minimum system dead time. Clock synchronization, as well as run control, can be performed to 25ns precision with the help of a hardware BUSY-SYNCH loop.

Since the DSP has access to the waveforms it can perform sophisticated pulse shapes analyses on the fly, and it can make the best use of the available information. The dramatic improvement of energy resolution of a Ge-detector at short filter peaking time and for high γ-ray energies was achieved through online pulse shape analysis, taking into account ballistic deficit correction.

The real-time processing unit can be reprogrammed from the host computer to provide the user with unusual operating modes, for instance triggering only on pulses recognized as piled ups, in order to trigger preferably on rapidly decaying ions implanted into a silicon strip detector.

REFERENCES

1. T.W.Raudorf, M.O.Bedwell, T.J.Paulus, "Pulse shape and rise time distribution calculations for HPGe coaxial detectors", IEEE Trans. On Nucl. Sci. NS-29, No. 1, 764 (1982)

2. F.S.Goulding et al., "GAMMASPHERE – Elimination of Ballistic Deficit by Using A Quasi-Trapezoidal Pulse Shaper", IEEE Trans. On Nucl. Sci. NS-41, No. 4, 1140 (1994)
3. C.R. Bingham, these proceedings.

The statistical properties of the angular distribution of β-delayed protons from oriented nuclei

J. Rikovska[1,2], N. J. Stone[1] and A. Wöhr[1]

[1] *Department of Physics, Clarendon Laboratory*
University of Oxford, Parks Road, OX1 3PU, UK.
[2] *Department of Chemistry and Biochemistry, University of Maryland, College Park, MD 20742, USA.*

Abstract. Statistical model of β-delayed proton emission is briefly reviewed. Theory of angular distribution of β-delayed proton emission from nuclear states oriented at low temperatures is discussed and the design of the first trial experiment using ^{118}Cs oriented at low temperature at the ISOLDE/NICOLE facility at CERN is described.

I INTRODUCTION

The phenomenon of β-delayed proton emission occurs in wide range of elements in isotopes close to the proton drip-line. The process involves proton emission from highly excited ($\geq$ 2-3 MeV) states populated by β-decay of a precursor isotope. In the light elements the proton emitting states may have a large fraction of single-particle strength and hence large widths. Protons are emitted mostly from the lower part of the excitation spectrum with relatively large separation between states and individual peaks may occur in the proton spectrum in this case. In medium and heavy nuclei the proton emitting states have a narrow width and a complex structure with many contributing configurations. Protons originate mostly at the high energy part of the excitation spectrum where the level separation becomes small and thus the emitted proton spectra are effectively continuous [1].

Theoretical descriptions of the process have applied a statistical approach, assuming that the β-decay feeding of individual proton emitting states, and the amplitudes of contributions to the proton decay can be taken as randomly distributed. At some point near the proton drip-line the energies of the proton unstable states will fall low enough that the statistical distribution may fail. The experimental studies outlined in this paper are intended to probe certain of the statistical assumptions made in the current theories.

In the statistical model, (see e.g. [2] - [6]), the probability of emission of a β-delayed particle from a state $|J_i\rangle$, created by β-decay of a parent state $|J_0\rangle$ can be expressed as a product of three terms:

β-decay strength $J_0 \to J_i$ S_β	$\times$	weight of states with spin J_i $w_i(J_0, J_i)$	$\times$	Particle/photon emission ratio $f(\Gamma_p, \Gamma_\gamma)$

where (Γ_p, Γ_γ) are reduced widths for proton and γ-decay from level with spin J_i.

Determination of the β-decay strength and particle/photon emission ratio represents a rather difficult experimental problem. Theoretically, both quantities can be approximated within nuclear models [2,4,6, and refs. therein].

To date, in all analyses of β-delayed particle emission a random statistical approach has been taken to yield the quantity $w_i(J,J_i)$. Assuming pure GT β-decay of the parent state $|J_0\rangle$, intermediate states of $J_i = J_0 - 1$, J_0 and $J_0 + 1$ are populated. The number of states having each spin J_i is taken proportional to $(2J_i + 1)$, giving their relative weight $w_i(J_0, J_i)$ as $(2J_i+1)/3(2J_0+1)$. It should be remarked however that, from an initial state $|J_0\rangle$ with a dominant single particle component of well defined orbital angular momentum l, GT β-decay will preferentially populate states having the same l-value, thus favouring either $J_0 + 1$ or $J_0 - 1$. Hence nuclear structure effects can influence the β-decay process.

Statistical aspects of the theory also apply to level spacing and proton emission partial widths, leading to descriptions of the quasi-continuous proton energy spectrum which have been compared with experimentally observed spectra. To explore the validity of the weighting factor assumption further measurables are required. Through its sensitivity to the spins of the levels involved, measurement of the angular distribution of the emitted protons suggests itself. No such measurement have yet been made.

II β-DELAYED PROTONS FROM ORIENTED NUCLEI

Recent developments in technique for measurement of the angular distribution of charged particles (α,β) from oriented nuclei at ISOLDE, Oxford and Leuven, now permit their reliable determination [7-9]. The same techniques are applicable to the determination of the angular distribution of β-delayed protons.

An experiment has been proposed at ISOLDE to determine the angular distribution of these protons as a function of the proton energy and degree of orientation and compare the results with model predictions. As long as the proton emitting level widths are smaller then the level spacing, the total angular momentum of these states is a good quantum number. When the β-emitting precursor state is oriented, the intermediate proton emitting states inherit this orientation (reduced by deorientation associated with the β-decay). As a consequence, the angular distribution of emitted protons will show anisotropy, dependent on the spins J_0, J_i and J_f of the precursor, emitter and daughter states, respectively. This distribution can be calculated precisely as

$$W_i(\theta) = 1 + \sum_{k=2,4,\ldots} B_k U_k R_k P_k(\cos\theta) \quad (1)$$

where θ is the angle of emission with respect to the spatial orientation axis, index i is used to indicate that this distribution holds for a given value of J_i (see Eq. 4 below), $B_k(J_0)$ and $U_k(J_0, J_i, L_\beta)$ are the usual orientation and deorientation parameters [10], $P_k(\cos\theta)$ are Legendre polynomials and R_k are the proton angular distribution coefficients [11] given as

$$R_k(p) = \sum_{L_p L'_p S} g_{L_p S} g^*_{L'_p S} R_k(L_p, L'_p, J_i, S) \quad (2)$$

where

$$R_k(L_p, L'_p, J_i, S) = \hat{J}_i \hat{L}_p \hat{L}'_p (-)^{k+S-J_i} <L_p L'_p 00|k0> W(L_p L'_p J_i J_i; kS) \quad (3)$$

with W representing the Racah W-coefficient, the expression in brackets is the Clebsch-Gordan coefficient and $\hat{x} = (2x+1)^{1/2}$. Eq. 2 is expressed in the channel-spin representation. The transition between states $|J_i\rangle$ and $|J_f\rangle$ via emission of a proton with spin s is represented by a superposition of partial waves with angular momentum L_p and amplitude $g_{L_p S}$. S stands for a channel spin $S = J_f + s$; for protons $S = J_f \pm 1/2$. Parity conservation in the proton decay restricts L_p to either even or odd values depending upon the relative parity of the states $|J_i\rangle$ and $|J_f\rangle$. k is limited, for a parity conserving transition, to even values $|L_p - L'_p| \leq k \leq L_p + L'_p$. Coefficients R_k are normalised in such a way that $R_0(L_p, L'_p, J_i, S) = \delta_{L_p, L'_p}$.

For proton decay to a state with $J_f = 0$ the summation in Eq. 2 is avoided since only one value of L_p is allowed. Full evaluation of Eq. 2 for a general state spin is rather complicated, because amplitudes $g_{L_p S}$ are generally complex numbers containing phase factors arising from the presence of both the Coulomb and nuclear field.

To obtain the final expression for the proton angular distribution, individual contributions from all possible intermediate states $|J_i\rangle$ populated in β-decay can be added with appropriate weighting factors. Assuming a statistical distribution of states $|J_i\rangle$, we get

$$W(\theta) = \sum_i w_i(J_0, J_i) W_i(\theta). \quad (4)$$

Note that in this approximation the angular distribution is independent of proton energy.

To summarize, an accurate measurement of the angular distribution of β delayed protons from oriented precursor nuclei will be used to give experimental data for comparison with calculation given by Eq.4. The comparison may lead to a direct confirmation or denial of the assumption of random spin density of the proton emitting states:

- systematic deviation between the statistical model prediction and experimental data and/or energy dependence of the proton angular distribution could indicate presence of nuclear structure effects governing selectivity in the β-decay process;

- random deviation would indicate a need for a different statistical weighting approach;

- if no deviation is found the above statistical model used in Eq.4 would be confirmed.

III THE NUCLEAR ORIENTATION EXPERIMENT

The NICOLE ^{3}He/^{4}He dilution refrigerator on line to the general purpose isotope separator at the ISOLDE facility, CERN, will be used[1]. In the standard geometry, ORTEC/SILENA HP Ge detectors will be placed outside the cryostat at 0° and 90° with respect to the orientation axis. The protons will be detected by Si PIN diodes of 300μm thickness, with typical resolution 25keV for $\sim$8MeV α particles at 4K, mounted inside the cryostat at 0° and 90°. The performance of the detectors operating at 4K has been tested. The energy resolution has not been measured for protons, but can be estimated as 20keV. Assuming implantation energy 60keV, emitted proton energy spread can be estimated to be less than 5 keV due to scattering in the source. It may be necessary to use counter telescopes comprised of $\sim 50\mu$m thick ΔE transmission counter and 300μm thick E-counter to avoid β-emission background. The source of β-delayed protons will be parent nuclei of ^{118}Cs, produced in a spallation reaction of 2 GeV protons on a liquid lanthanum target coupled to tungsten surface ioniser. The expected yield is $\sim 5\times 10^6$ atoms per μC. β-delayed protons from ^{118}Cs have been observed before at room temperature [3,12,13]. The β-delayed proton branch was found to be $(4.2\pm0.6)\times 10^{-4}$ [3,13]. Low temperature nuclear orientation of ^{118}Cs in Fe has been studied through gamma-decay [14] and sizeable nuclear orientation has been achieved at 10-15 mK.

IV ACKNOWLEDGEMENT

The research was supported by UK EPSRC and US DOE grant no. DE-FG02-94ER40834.

---

[1] A trial experiment without counter telescopes has been completed in November 1999 and the data obtained are currently being analysed.

REFERENCES

1. Jonson, B., Nyman, G., *Nuclear Decay Modes*, ed. Ponearu, D. N., IOP Publishing Ltd, Bristol 1996, pp. 102-142.
2. Hansen, P. G., Jonson, B., and Richter, A. *Nucl.Phys.* **A518** 13 (1990).
3. Jonson, B., Hagberg, E., Hansen, P. G., Hornshoj, P., and Tidemand-Peterson, P., *Proceedings of the 3rd International Conference on Nuclei far from Stability*, Cargese, 1976, CERN 76-13, p.277.
4. Hardy, J. C., and Hagberg, E., *Particle Emission from Nuclei*, eds. Poenaru, I. and Ivascu, D. N., Boca Raton: CRC Press, 1989, Vol. III, chap.4, p.99.
5. Hornshoj, P., Wilsky, K., Hansen, P. G., Jonson, B., and Nielsen, O. B., *Nucl.Phys.* **A187** 609 (1972).
6. Hansen, P. G., *Hyp.Int.* **43** 381 (1988).
7. Wouters, J., Vandeplassche, D., Van Walle, E., Severijns, N., and Vanneste, L., *Phys.Rev.Lett* **56** 1901 (1986), Severijns, N., Wouters, J., Vanhaverbeke, J., Vanderpoorten, W., and Vanneste, L., *Hyp.Int.* **43** 415 (1988), Wouters, J., Vanderpoorten, W. De Moor, W. P., Schuurmans, P., Vanneste, L., and Eder, R., *Nucl.Instr.Meth.* **A313** 215 (1992).
8. Lindroos, M., Richards, P., Rikovska, J., Stone, N. J., Oliveira, I. S., Nishimura, K. and Booth, M. *Hyp.Int.* **75** 323 (1992).
9. Williams, D. A., Doran. D., Fogelberg, B., Jacobsson, L., Oliveira, I. S., Rikovska, J., Stone, N. J., Veskovic. M., *Hyp.Int.* **C1** 569 (1996)
10. Krane, K. S., in Low Temperature Nuclear Orientation, eds. Stone, N. J. and Postma, H., North Holland, Amsterdam (1986) Chap.2.
11. Satchler, G. R., Direct Nuclear Reactions, Clarendon Press, Oxford and Oxford University Press, London 1983 p. 375.
12. D'Auria, J. M., Grüter, J. W., Hagberg, E., Hansen, P. G., Hardy, J. C., Hornshoj, P., Jonson, B., Mattsson, S., Ravn, H. L., and Tidemand-Petersson, P., *Nucl.Phys.* **A301** 397 (1978).
13. Hagberg, E., Hansen, P. G., Hornshoj, P., Jonson, B., Mattsson, S., and Tidemand-Petersson, P., *Phys.Lett.* **73B** 397 (1978).
14. Shaw, T.L., Green, V. R., Ashworth, C. J., Rikovska, J., Stone, N. J., Walker, P. M., Grant, I. S., *Phys.Rev.* **C36** 413 (1987).

First observation of doubly-magic ^{48}Ni

J. Giovinazzo^a, B. Blank^a, C. Borcea^b, M. Chartier^a,
S. Czajkowski^a, A. Fleury^a, G. de France^c, R. Grzywacz^{d1},
Z. Janas^d, M. Lewitowicz^c, F. de Oliveira^c, M. Pfützner^d,
M.S. Pravikoff^a, J.C. Thomas^a

^{a} CEN Bordeaux-Gradignan, Le Haut-Vigneau, F-33175 Gradignan Cedex, France
^{b} Grand Accélérateur National d'Ions Lourds, B.P. 5027, F-14076 Caen Cedex, France
^{c} IAP, Bucharest-Margurele, P.O. Box MG6, Romania
^{d} Institute of Experimental Physics, University of Warsaw, PL-00-681 Warsaw, Hoza 69, Poland

Abstract. The doubly magic nucleus ^{48}Ni has been observed for the first time at GANIL. The experiment was performed using a high intensity ^{58}Ni beam at 74.5 MeV/A on a nickel target. This $T_Z = -4$ nucleus is the most proton-rich nucleus ever observed, and is a good candidate for two-proton radioactivity. In this experiment, we also implanted around 280 ^{42}Cr, 50 ^{45}Fe and 100 ^{49}Ni isotopes, for which no experimental information about the decay is available.

INTRODUCTION

^{48}Ni is of specific interest for mirror symmetry studies since it is the only case of a doubly-magic nucleus for which the mirror nucleus, ^{48}Ca, is bound. With an isospin projection of $T_Z = -4$, ^{48}Ni is also the most proton-rich nucleus. Together with ^{45}Fe, ^{48}Ni is also predicted by theory a good candidate for the yet unobserved two-proton radioactivity [1-4]. In addition, with ^{48}Ni, ^{56}Ni and ^{78}Ni, nickel is probably the only element with 3 doubly-magic isotopes, that makes them interesting candidates for a study of the evolution of shell structure with isospin quantum number.

The search for ^{48}Ni was possible due to recent develpements of projectile fragmentation facilities. ^{51}Ni ($T_Z = -5/2$) was observed for the first time in a GANIL experiment in 1986 [5], and ^{50}Ni ($T_Z = -3$) in 1993 at GSI [6]. A recent experiment at GSI, using the fragmentation of a 600 MeV/A ^{58}Ni beam, led to the observation of ^{42}Cr ($T_Z = -3$), as well as of ^{45}Fe and ^{49}Ni ($T_Z = -7/2$), with 10, 3 and 5 events, respectively [7].

---

[1]) Present address: Univ. of Tennessee, Knoxville, Tennessee, USA

We observed ^{48}Ni during an experiment at the LISE3 spectrometer in GANIL. This nucleus is probably the last doubly-magic nucleus accessible with present facilities.

EXPERIMENTAL SETUP

The exotic nuclei are produced by the fragmentation of a 74.5 MeV/A ^{58}Ni primary beam in a 230.6 mg/cm^2 thick nickel target followed by a 2.7 mg/cm^2 carbon stripper foil. This target is located between the two superconducting solenoids of the SISSI device in order to increase the acceptance of the spectrometer. The secondary beam goes through the alpha spectrometer to the LISE3 separator. A shaped beryllium degrader, located at the intermediate focal plane of LISE3, provided an achromatic tuning for a refined selection of projectile fragments. In addition, a final selection is performed with a Wien filter after the second dipole of LISE3.

The fragments of interest are implanted in a stack of five silicon detectors. Together with time of flight measurements, these detectors are used to identify the selected ions yielding their energy loss and residual energy. The first three silicon detectors are 300 μm thick, the forth one 500 μm and the last one is 6 mm thick. The first detector is used for a first energy-loss measurement and for time of flight determination with both the high frequency of the cyclotrons and a micro-channel plate (MCP) detector located before the velocity filter. The MCP had to withstand a few 10^5 particles per second, and its efficiency was only about 70%. The second detector, that is position-sensitive, measures another energy loss and a $X - Y$ position. The isotopes of interest are implanted in the third detector that gives the residual energy. This detector is also used, like the first one, to perform a time of flight measurement with the high frequency of the cyclotron. The last two detectors allow to veto on light particles that are not stopped in the third detector.

This set-up gives us a set of ten independant parameters for an event by event identification of implanted ions.

DATA ANALYSIS AND RESULTS

For each of those parameters, the mean value and the full width at half maximum (FWHM) are estimated for each isotope observed experimentally. We deduce from this the expected values of the parameters for unbound nuclei. The identification plot of energy loss in the first silicon detector versus the time of flight with the MCP is shown on figure 1. On this plot, events are selected when all parameters that are not represented lie in a window of two FWHM around the mean value, while no condition is required for the represented parameters. This plot shows 2 events of ^{48}Ni, 77 of ^{49}Ni, 29 ^{45}Fe and 164 of ^{42}Cr, that fullfil the crossed conditions on the parameters. There is no background in this plot, since no count appears where unbound isotopes should be located. This clearly show the unambiguous observation of ^{48}Ni.

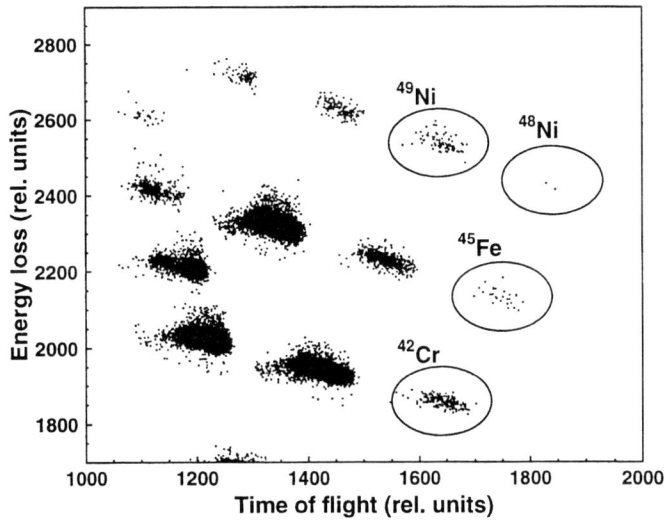

FIGURE 1. Identification plot representing time of flight between the MCP detector and the first silicon detector versus energy loss in this detector.

Due to the efficiency lack of the MCP, this plot suppresses around 30% of the total statistics. Another option is to perform an energy loss versus total energy identification plot. The condition on all other parameters remains to lie in a two FWHM window, except for the time of flight with the MCP, for which the cut occurs only when the signal is present. As a result of this analysis, shown in figure 2, we observe 4 counts of ^{48}Ni, 89 of ^{49}Ni, 48 ^{45}Fe and 264 of ^{42}Cr. On this plot, few background counts are present where unbound isotopes should be located. Nevertheless, it appears to be very unlikely that more than one count identified as ^{48}Ni comes from this background.

The transmission of ^{48}Ni isotopes through the separator is estimated to be $9.8 \pm 1.2\%$, from simulations performed with the LISE [8], LIESCHEN [9] and INTENSITY [10] codes. The total primary beam dose was 4.2×10^{17} particles and the dead time of the acquisition system 12.9%. Assuming the identification of four events of ^{48}Ni, this leads to an estimate of the production of cross-section $\sigma = 0.05 \pm 0.02$ pb, which is one of the lowest cross-sections ever observed in a nuclear physics experiment. The time of flight of ^{48}Ni in the beam line being 1.32 μs, we estimate a lower limit for its half-life of about 0.5 μs.

^{48}Ni is predicted to be a good candidate for two-proton emission from the ground state. Since it is an even-even nucleus, it is bound with respect to one proton emission. The barrier penetration half-life of this nucleus is shown on figure 3 as a function of the two-proton separation energy, in a simple picture of the tunnelling of a ^2He particle through the Coulomb barrier, assuming a spectroscopic factor of 1. The experimental lower limit of the half-life indicates that Q_{2P} may be lower than 1.5 MeV. This result is compared to half-lifes deduced from Q_{2P} values from different mass model [1–4,11–16].

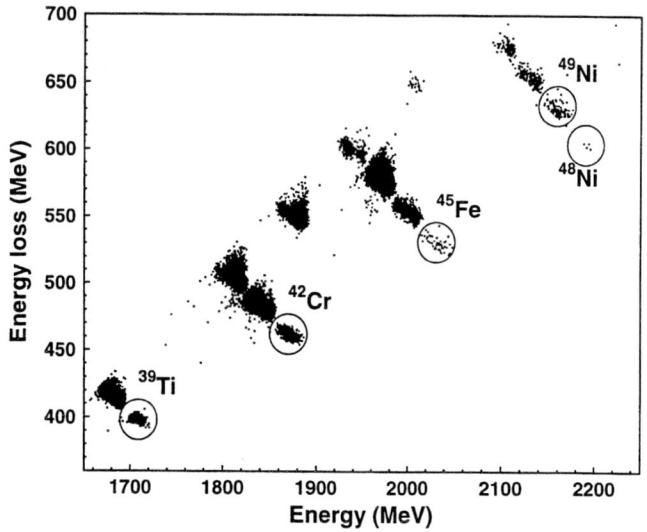

FIGURE 2. Identification plot representing the total energy versus energy loss in the first silicon detector.

CONCLUDING REMARKS

An unambiguous observation of doubly-magic ^{48}Ni was achieved at the GANIL/LISE3 facility from the projectile fragmentation of a ^{58}Ni beam. From this observation, we deduce a lower limit of its half-life about 0.5 μs and an estimate of the production cross-section around 0.05 pb. In addition, we confirm the existence of ^{49}Ni, ^{45}Fe and ^{42}Cr which were observed in a previous experiment at GSI with much lower statistics.

With the expected increase of primary beam intensity, further measurements should allow us to learn more about the mass and decay of ^{48}Ni.

REFERENCES

1. B.A. Brown *et al.*, Phys. Rev. C **43**, R1513, (1991).

2. W.E. Ormand et al., Phys. Rev. C **53**, 214, (1996).
3. W.E. Ormand et al., Phys. Rev. C **55**, 2407, (1997).
4. B.J. Cole et al., Phys. Rev. C **54**, 1240, (1996).
5. F. Pougheon et al., Z. Phys. A **327**, 17, (1987).
6. B. Blank et al., Phys. Rev. C **50**, 2398, (1994).
7. B. Blank et al., Phys. Rev. Lett. **77**, 2398, (1996).
8. D. Bazin, M. Lewitowicz, O. Sorlin, O. Tarasov, LISE simulation code, unpublished.
9. B. Blank, E. Hanelt, K.-H. Schmidt, LIESCHEN simulation code, unpublished.
10. J. Winger, B. Sherril, D. Morrissey, Nulc. Instr. Meth. **B70**, 380, (1992).
11. W. Nazarewicz et al., Phys. Rev. C **53**, 740, (1996).
12. W. Benenson et al., Nukleonika **20**, 775, (1975).
13. P.E. Haustein (ed.), At. Data Nucl. Data. Tab. **39**, 185, (1988).
14. P. Möller et al., At. Data Nucl. Data. Tab. **59**, 185, (1995).
15. J. Duflo, A. Zuker, Phys. Rev. C **52**, R23, (1995).
16. Y. Aboussir et al., At. Data Nucl. Data. Tab. **61**, 127, (1995).

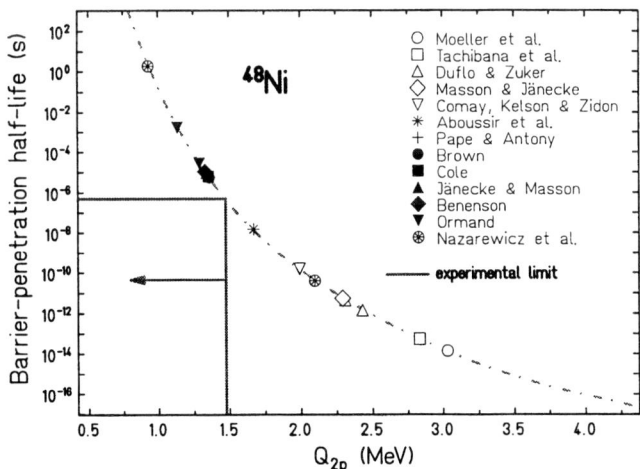

FIGURE 3. ^{48}Ni half-life calculated from the tunnelling of an ^2He particle through the Coulomb barrier as a function of the two-proton separation energy. The experimental lower limit is compared to several mass-model predictions.

Proton-Radioactivity Studies At the FRS After the SIS Intensity Upgrade

K. Schmidt

GSI, Planckstr.1, D-64291 Darmstadt, Germany

Abstract. The perspectives for the production of ground-state proton emitters via projectile-fragmentation reactions using beams from the heavy-ion synchrotron SIS at GSI are reviewed. Special focus is placed on the rare-earth isotopes and nuclei in the trans-lead region. In addition, estimates for cross sections and production rates will be given.

INTRODUCTION

Starting with the initial experiments on ground-state proton radioactivity, discovered at GSI and Munich in the early 1980 [1-3], fusion-evaporation reactions have proven to be an extremely powerful production mechanism for ground-state proton emitters, see [4,5] for recent reviews. Although novel experimental techniques, such as double-sided silicon strip detectors (DSSD) and recoil-decay tagging, have pushed the observation limit down to production cross sections of a few nb, see e.g. [6], a few regions along the proton drip line have remained inaccessible up-to-date. Especially an experimental investigation of the most neutron-deficient rare-earth isotopes between lanthanum and promethium, as well as of isotopes beyond lead would yield valuable new information. On the other hand, it has been shown that high-energy projectile fragmentation is a very useful tool for producing proton-rich exotic nuclei. After the completion of the intensity upgrade of the GSI accelerator complex, the production of hitherto unknown ground-state proton emitters via projectile fragmentation will become feasible.

In this contribution a summary of the intensity upgrade of the GSI facilities, along with intensities of selected beams will be given in the following section. Afterwards the perspectives for investigating proton decay at the projectile fragment separator FRS will be outlined.

THE SIS INTENSITY UPGRADE

In 1996, GSI started an intensity upgrade project [7] to reach the original design goals of the heavy-ion synchrotron SIS. The intensities available at that time amounted to about $1*10^8$ ions per pulse for the heaviest beams such as ^{238}U, while the incoherent space charge limit of the SIS is about a factor of 300 higher. Among the improvements foreseen were the installation of new high-current ion sources along with a suitable high-current injector and the installation of an electron cooler for the SIS. Details can be found in references [7] and [8], while Figure 1 gives an overview of the beam intensities available after the completion of the upgrade program in late 1999. All in all, the range of ion sources available at GSI comprises Mevva and Chordis sources for high-intensity beams, Penning sources for low-intensity experiments both at the SIS and the UNILAC accelerators, and, finally, ECR ion sources. Due to the efficient consumption of fuel, the ECR source is the tool of choice for producing beams from rare isotopes, such as e.g. ^{48}Ca. It should be noted, however, that the intensities for these beams usually do not exceed values of about $1*10^9$ ions per pulse for heavy beams.

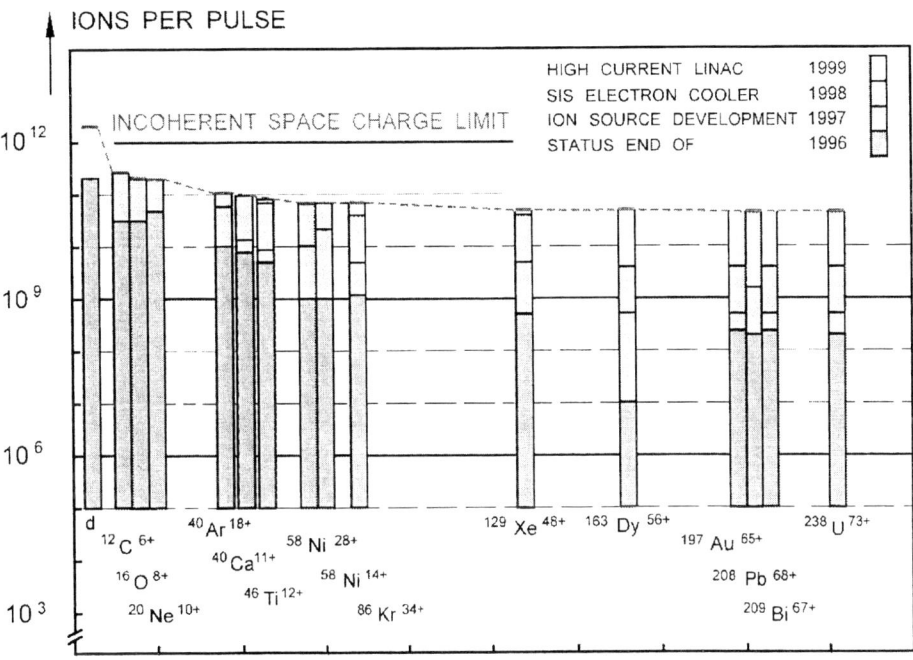

FIGURE 1. Intensities of selected beams from the heavy-ion synchrotron SIS after completion of the intensity upgrade [7,8].

PRODUCTION OF GROUND-STATE PROTON EMITTERS IN FRAGMENTATION REACTIONS

Detectable ground-state proton radioactivity is expected to be very rare in nuclei with Z < 50 owing to the low Coulomb barrier height and the concomitant high sensitivity of the tunnelling rate to the proton separation energy. Although heavy-ion fragmentation studies have largely mapped out the proton drip-line up to Z = 50, no examples of ground-state proton radioactivity have been discovered so far. Meanwhile, heavy-ion fusion reactions have produced all the known examples above Z = 50 via 1pxn evaporation channels, where x = 1-5. Two extended regions of the proton drip-line above Z = 50 remain for which no examples of proton radioactivity are known. These are (i) the region of highly deformed light rare-earth nuclei with Z = 57-61 (La, Pr, Pm) and (ii) heavy nuclei with Z > 83. The first region is difficult to access because of the lack of suitable beam-target combinations, requiring multi-neutron evaporation channels with very low cross-sections. There exists, however, a preliminary report on proton emission from ^{117}La [9]. In region (ii) the rapid increase in the fission probability is the main experimental obstacle.

In recent years, projectile fragmentation reactions at relativistic energies have proved to be a very powerful experimental tool for identifying exotic isotopes and rare decay events, see e.g. [10] for a survey. Although fusion-evaporation reactions usually have higher production cross-sections, the higher luminosity of experiments performed at relativistic energies, along with the possibility of identifying the reaction products on an event-by-event basis, offers excellent perspectives for an experimental investigation. The fragment separator FRS at GSI [11] possesses a large momentum acceptance of ±1 % and allows for a clean separation of the reaction products by using degraders at the intermediate focal planes. Moreover, the flight time of reaction products at typical beam energies is on the order of 200 ns, thus making the FRS the ideal tool to study extremely short-lived isotopes at the edge of stability.

In the following, estimates for production rates of interesting isotopes in region (i) and (ii) will be given.

Proton emitters in the rare-earth region

The main candidates for proton decay in region (i) are ^{116,117}La, ^{125,126}Pm, and ^{121}Pr, all of which are predicted to have strongly prolate-deformed ground state shapes [12]. Searches for the proton decay of other heavier isotopes of these elements using fusion evaporation reactions have been unsuccessful [13]. It is most likely that the proton drip-line has not been crossed far enough for proton emission to compete with β-decay in these isotopes. The identification of proton decay in the above elements will represent a major achievement, linking up the region of transitional nuclei above ^{100}Sn with the extended region of proton emitters continuing up through Z = 83. This will clearly establish the limits of particle stability over almost the entirety of the proton drip-line for odd Z nuclei. This in turn will provide a comprehensive test of the ability of mass models to extrapolate to these extreme limits and will enable the evolution of shapes and shell structure at the drip-line to be traced. Proton decay rates from different Nilsson configurations will provide a stringent test of the general validity of models of proton emission from highly deformed nuclei. Furthermore, it is likely that there will be additional examples of proton decay fine structure discovered in this region. This phenomenon, in which a proton decays to a low-lying excited state in the daughter nucleus, has recently been reported for the isotope ^{131}Eu [14]. The branch to the excited state is sensitive to different microscopic components of the Nilsson wavefunction in the parent nucleus.

The most neutron-deficient isotopes of the elements lanthanum, praseodymium, and promethium can be produced by fragmenting a 1 GeV*A beam of ^{144}Sm on a ^{9}Be target. The EPAX [15] predictions for the fragmentation cross section, along with the expected rates at the final focus of the FRS, can be found in Table 1. The rates were calculated assuming a primary beam intensity of $1*10^9$ ^{144}Sm ions per pulse, a repetition rate of 1 pulse every 6 s, a target thickness of $4 g/cm^2$, and a transmission of the fragments through the FRS of 80%. The losses due to secondary reactions were estimated to be 60%.

TABLE 1. Production Rates of Rare-Earth Isotopes.

Reaction	Reaction Product	Cross Section (b)	Estimated Rate at Final Focus (atoms/day)
^{144}Sm + ^{9}Be	^{116}La	$1.4*10^{-12}$	2
	^{117}La	$8.0*10^{-11}$	100
	^{121}Pr	$1.1*10^{-11}$	14
	^{125}Pm	$1.4*10^{-12}$	2
	^{126}Pm	$5.8*10^{-11}$	70

With the values listed in Table 1, experiments to search for proton emission from ^{117}La and ^{126}Pm appear to be feasible, while the other candidates look less promising. It should be noted, however, that a ^{144}Sm beam from the ECR source was assumed; using a high-intensity Mevva source, the production rates could be increased by at least one order of magnitude.

Proton emitters beyond lead

The only known proton emitter above the Z = 82 shell closure is ^{185}Bi [16]. In this case, proton decay was assigned to a low-spin isomeric state. The study of proton emission in region (ii) is therefore essentially unexplored. Mass measurements using the ESR at GSI [17,18] have established that the proton drip-line has been crossed for ^{197}At, ^{201}Fr, ^{207}Ac and ^{213}Pa (Z = 85-91). In order for proton decay to compete with short-lived α-decay branches, the Q_p-value must be larger than about 1 MeV. Extrapolation of the ESR mass measurements indicates that the nearest isotopes for which proton emission may occur with significant branches are ^{191}At, ^{197}Fr, ^{203}Ac and ^{209}Pa. In many cases it is expected that there will be a high-spin ground-state structure and a low-spin isomer which could be favourable to proton emission as was found for the case of ^{185}Bi. First generation proton and alpha decays will be correlated with extended alpha decay chains. These chains can be used to study the structure and proton binding energy of the daughter nuclei and can be linked to the mass surface established by the ESR measurements. Each new isotope therefore becomes a rich source of information on its descendants – this includes new alpha-decaying isotopes that will be produced along with the proton decay candidates. The macroscopic-microscopic model of Möller et al. [12] predicts that all of the above mentioned astatine-to-protactinium isotopes have significant ($\beta_2 \sim -0.2$) oblate deformations in their ground states. As no examples exist of the effect of oblate deformations on proton tunnelling, the measurement of proton emission from these isotopes will provide a distinctly new testing ground for the theory of proton emission from deformed nuclei.

The neutron-deficient isotopes of the trans-lead elements astatine, francium, actinium, and protactinium are produced best by fragmenting a 1 GeV•A ^{238}U beam on a ^{9}Be target. The cross sections for this reaction were obtained by extrapolating values measured for less exotic isotopes [19,20]. The production rates listed in Table 2 assume a primary beam intensity of $1*10^{10}$ ^{238}U ions per pulse, a repetition rate of 1 pulse every 6 s, and a target thickness of 4g/cm^2. The transmission of the fragments through the FRS was estimated to be 80%, while losses due to secondary reactions will amount to 80%.

TABLE 2. Production Rates of Trans-Lead Isotopes.

Reaction	Reaction Product	Cross Section (b)	Estimated Rate at Final Focus (atoms/day)
^{238}U+ ^{9}Be	^{189}At	$2*10^{-11}$	120
	^{191}At	$4*10^{-10}$	2400
	^{195}Fr	$1*10^{-12}$	6
	^{197}Fr	$4*10^{-11}$	240
	^{201}Ac	$1*10^{-10}$	620
	^{203}Ac	$1*10^{-9}$	6200
	^{207}Pa	$3*10^{-10}$	1900
	^{209}Pa	$1*10^{-9}$	6300

The predictions for the isotopes ^{209,207}Pa, ^{203,201}Ac, ^{197}Fr, and ^{191,189}At offer excellent perspectives for an experimental investigation at the FRS. In addition to the identification of new ground-state proton emitters, the measurement of proton emission from low-lying isomers and chains of alpha decays, which may either compete with or follow proton decay will be possible.

SUMMARY

After completion and commission of the upgrade of SIS accelerator at GSI, the following beam intensities will we available:
- several 10^{10} ions per pulse for heavy beams
- around 10^{11} ions per pulse for light particles with A < 40
- $10^8 - 10^9$ particles per pulse for rare-isotope beams

With these beam intensities, the investigation of ground state proton emission in new regions of the chart of nuclei will become feasible. Especially experiments focusing on the emission of protons from strongly deformed nuclei beyond the shell closure at Z = 82 and in the rare-earth region with Z = 57-61 look promising.

REFERENCES

1. Hofmann, S. et al., *Z.Phys.* **A305**, 111 (1982)
2. Klepper, O. et al., *Z.Phys.* **A305**, 125 (1982)
3. Faestermann, T. et al., *Phys.Lett.* **B137**, 23 (1984)
4. Woods, P.J. and Davids, C.N., *Annu. Rev. Nucl. Part. Sci.* **47**, 541 (1997)
5. Rykaczewski, K. et al., *Acta Physica Polonica* **B30**, 565 (1999)
6. Rykaczewski, K. et al., *Phys. Rev.* **C60**, 011301 (1999)
7. GSI Internal Report 95-05, 1995, ISSN 0171-4546
8. GSI WWW server at http://www.gsi.de/gsi.accelerator.html
9. Soramel, F., these proceedings
10. Geissel, H., Münzenberg, G. and Riisager, K. *Annu. Rev. Nucl. Sci.* **45**, 163 (1995)
11. Geissel, H. et al., *Nucl. Instr. and Meth.* **B70**, 286 (1992)

12. Möller, P. et al., *At. Data Nucl. Data Tab.* **59**, 189 (1995)
13. Woods, P.J., private communication, 1999
14. Sonzogni, A.A. et al., *Phys. Rev. Lett.* **83**, 116 (1999)
15. Sümmerer, K. et al., *Phys. Rev.* **C42**, 2546 (1990)
16. Davids, C.N. et al., *Phys. Rev. Lett.* **76**, 592 (1996)
17. Radon, T. et al., *Phys. Rev. Lett.* **78**, 4701 (1997)
18. Novikov, Yu. N. et al., in preparation
19. Junghans, A.R. et al., *Nucl. Phys.* **A629**, 635 (1998)
20. Taieb, J., private communication, 1999

International Symposium on Proton-Emitting Nuclei
Oak Ridge, Tennessee, USA
October 7-9, 1999

Yurdanur Akovali
Oak Ridge National Laboratory
P. O. Box 2008
Building 6000 MS 6371
Oak Ridge, TN 37831
USA
Telephone: 865-574-4695
Email: akovali@mail.phy.ornl.gov

Andrei Andreyev
IKS, KU Leuven
Celestijnenlaan, 200D
B-3001, Leuven
Belgium
Telephone: 32-16-327698
Email: andrei.andreyev@fys.kuleuven.ac.be

Cyrus Baktash
Oak Ridge National Laboratory
P. O. Box 2008
Building 6000 MS 6371
Oak Ridge, TN 37831
USA
Telephone: 865-576-7949
Email: baktash@mail.phy.ornl.gov

Dimiter Balabanski
University of Tennessee
Physics Department
1408 Circle Drive
Knoxville, TN 37996-1200
USA
Telephone: 865-974-7803
Email: dbalaban@utk.edu

Bryan Barmore
University of Tennessee
Oak Ridge National Laboratory
P. O. Box 2008
Building 6003 MS 6373
Oak Ridge, TN 37831
USA
Telephone: 865-574-4578
Email: barmore@mail.phy.ornl.gov

Jon Batchelder, Chair
UNIRIB/ORAU
P. O. Box 2008
Building 6008 MS 6374
Oak Ridge, TN 37831
USA
Telephone: 865-576-7656
Email: batcheld@mail.phy.ornl.gov

Jim Beene
Oak Ridge National Laboratory
P. O. Box 2008
Building 6000 MS 6368
Oak Ridge, TN 37831
USA
Telephone: 865-574-4622
Email: beene@mail.phy.ornl.gov

Fred Bertrand
Oak Ridge National Laboratory
P. O. Box 2008
Building 6000 MS 6369
Oak Ridge, TN 37831
USA
Telephone: 865-574-4737
Email: feb@ornl.gov

C. J. Beyer
Vanderbilt University
Department of Physics & Astronomy
P. O. Box 1807-B
Nashville, TN 37235
USA
Telephone: 615-322-2599
Email: cj@styx.phy.vanderbilt.edu

Carrol Bingham
University of Tennessee
Physics Department
1408 Circle Drive
Knoxville, TN 37996-1200
USA
Telephone: 865-974-7802
Email: cbingham@utk.edu

Maria Borge
IEM - CSIC
Serrano 113 bis
E-28006 Madrid
Spain
Telephone: 34-91-5901614
Email: borge@pinar2.csic.es

Richard Boyd
Ohio State University
Physics Department
174 W. 18th Avenue
Columbus, OH 43210-1106
USA
Telephone: 614-292-2875
Email: boyd@mps.ohio-state.edu

Nicolae Carjan
University of Bordeaux
CENBG
Le Haut Vigneau BP120
33175 Gradignan Cedex
France
Telephone: 33-557120772
Email: carjan@in2p3.fr

Ken Carter
UNIRIB/ORAU
P. O. Box 2008
Building 6008 MS 6374
Oak Ridge, TN 37831
USA
Telephone: 865-576-2642
Email: carter@mail.phy.ornl.gov

Joseph Cerny
University of California, Berkeley
119 California Hall MC 1500
Berkeley, California 94720
USA
Telephone: 510-642-7540
Email: jcerny@uclink4.berkeley.edu

Cary Davids
Argonne National Laboratory
9700 S. Cass Avenue
Argonne, IL 60439-4843
USA
Telephone: 630-252-4062
Email: davids@anl.gov

Thomas Davinson
University of Edinburgh
Kings Bldgs./James Clerk Maxwell Bldg.
Mayfield Road
Edinburgh EH9 3JZ
United Kingdom
Telephone: 44-131-650-5250
Email: td@np.ph.ed.ac.uk

Thomas Faestermann
Technische Universität München
Physik Department
James-Franck-Strasse
D-85748 Garching bei München
Germany
Telephone: 00-49-89-28912438
Email: thomas.faestermann@physik.tu-muenchen.de

Lidia Ferreira
CFIF, Instituto Superior Tecnico
Edificio Ciencia
Av Rovisco Pais
1096 Lisboa Codex
Portugal
Telephone: 00351-1-8419091-2
Email: flidia@beta.ist.utl.pt

Hubert Flocard
Institute Physique Nucleaire
IN2P3
Theory Group
91406 Orsay Cedex
France
Telephone: 33-1-6915-7946
Email: flocard@ipno.in2p3.fr

Alfredo Galindo-Uribarri
Oak Ridge National Laboratory
P. O. Box 2008
Building 6000 MS 6371
Oak Ridge, TN 37831
USA
Telephone: 865-574-6124
Email: uribarri@mail.phy.ornl.gov

Ermias Gete
University of British Columbia
4004 Wesbrook Mall
Vancouver, BC V6T 2A3
Canada
Telephone: 604-822-6576
Email: gete@alph04.triumf.ca

Thomas Ginter
Lawrence Berkeley National Laboratory
1 Cyclotron Road
Mailstop 88-234
Berkeley, CA 94720
USA
Telephone: 510-486-7356
Email: ginter@mail.phy.ornl.gov

Jerome Giovinazzo
CENBG-Bordeaux
IN2P3/CNRS
Impasse du Haut Vigneau
BP120, F-33175 Gradignan Cedex
France
Telephone: 33-557-12-08-52
Email: giovinaz@cenbg.in2p3.fr

Carl Gross
ORAU/ORNL
P. O. Box 2008
Building 6000 MS 6371
Oak Ridge, TN 37830
USA
Telephone: 865-576-7698
Email: cgross@mail.phy.ornl.gov

Robert Grzywacz
University of Tennessee
Physics Department
1408 Circle Drive
Knoxville, TN 37996-1200
USA
Telephone: 865-574-4498
Email: grzywacz@mail.phy.ornl.gov

Joseph Hamilton
Vanderbilt University
Department of Physics & Astronomy
P. O. Box 1807-B
Nashville, TN 37235
USA
Telephone: 615-322-2456
Email: hamiltj1@ctrvax.vanderbilt.edu

John Hardy
Texas A&M University
Cyclotron Institute
College Station, TX 77843
USA
Telephone: 409-845-1411
Email: hardy@comp.tamu.edu

Daryl Hartley
University of Tennessee
Physics Department
1408 Circle Drive
Knoxville, TN 37996-1200
USA
Telephone: 865-974-7803
Email: daryl@spinno.phys.utk.edu

Sigurd Hofmann
GSI
Planck Strasse 1
D-64220 Darmstadt
Germany
Telephone: 49-6159-71-2734
Email: s.hofmann.gsi.de

Zenon Janas
Warsaw University/GSI
Department of Physics
ul. Hoza 69
PL-00-681 Warsaw
Poland
Telephone: 48-22-823-18-96
Email: janas@zsjaxq.igf.fuw.edu.pl

Noah Johnson
Oak Ridge National Laboratory
P. O. Box 2008
Building 6000 MS 6371
Oak Ridge, TN 37831
USA
Telephone: 865-574-4105
Email: johnson@mail.phy.ornl.gov

Ari Jokinen
University of Jyväskylä
Department of Physics
P. O. Box 35 (Y5)
FIN-40351 Jyväskylä
Finland
Telephone: 358-14-602384
Email: ari.jokinen@phys.jyu.fi

Stanislav Georgievich Kadmensky
Voronezh State University
Plekhanovskaya Street, 22,360, p/b 24
Voronezh, 394030
Russia
Telephone: 0732-524654
Email: kadmensky@cd.vsu.ru

Marek Karny
University of Tennessee
Department of Physics
1408 Circle Drive
Knoxville, TN 37996-1200
USA
Telephone: 865-576-8763
Email: karny@mail.phy.ornl.gov

Jan Kormicki
Vanderbilt University
Department of Physics & Astronomy
P. O. Box 1807-B
Nashville, TN 37235
USA
Telephone: 615-322-0656
Email: kormicj0@ctrvax.vanderbilt.edu

Andras Kruppa
ATOMKI
Institute of Nuclear Research of
the Hungarian Academy of Sciences
Bem ter 18/c
Debrecen H-4026
Hungary
Telephone: 36-52-417266
Email: atk@chaos.atomki.hu

William David Kulp, III
Georgia Institute of Technology
School of Physics
Atlanta, GA 30332-0430
USA
Telephone: 404-894-9407
Email: david@nuclear.physics.gatech.edu

Georgios Lalazissis
Technical University of Munich
Physikdepartment T30
James Franck Str.
D-85747 Garching
Germany
Email: glalazis@physik.tu-muenchen.de

Marek Lewitowicz
GANIL
B. P. No. 5027
F-14076 Caen Cedex, 5
France
Telephone: 33-23-1454603
Email: lewitowicz@ganil.fr

Felix Liang
Oak Ridge National Laboratory
P. O. Box 2008
Building 6000 MS 6368
Oak Ridge, TN 37831
USA
Telephone: 865-574-4109
Email: liang@mail.phy.ornl.gov

Matej Lipoglavsek
Oak Ridge National Laboratory
P. O. Box 2008
Building 6000 MS 6371
Oak Ridge, TN 37831
USA
Telephone: 865-574-4708
Email: matej@mail.phy.ornl.gov

Enrico Maglione
University of Padova/INFN
Dipartimento di Fisica
Via marzolo 8
I-35131 Padova
Italy
Telephone: 0039-049-8277189
Email: maglione@pd.infn.it

Bogdan Mihaila
Coastal Carolina University
Chemistry-Physics Department
P. O. Box 261954
Conway, SC 29528-6054
USA
Telephone: 843-349-2251
Email: bogdan.mihaila@unh.edu

Michael Momayezi
X-Ray Instrumentation Associates
2513 Charleston Road #207
Mountain View, CA 94043
USA
Telephone: 650-903-9980
Email: momayezi@xia.com

Paul Mueller
Oak Ridge National Laboratory
P. O. Box 2008
Building 6000 MS 6368
Oak Ridge, TN 37831
USA
Telephone: 865-574-4725
Email: mueller@mail.phy.ornl.gov

Ivan Mukha
GSI
Kernphysik 2
Planckstr. 1
DE-64291 Darmstadt
Germany
Telephone: 49-6159-712887
Email: mukha@axp614.gsi.de

Witold Nazarewicz
University of Tennessee
Physics Department
1408 Circle Drive
Knoxville, TN 37996-1200
USA
Telephone: 865-574-4580
Email: witek-nazarewicz@utk.edu

Volker Oberacker
Vanderbilt University
Department of Physics & Astronomy
P. O. Box 1807, Station B
Nashville, TN 37235
USA
Telephone: 615-322-5035
Email: volker.e.oberacker@vanderbilt.edu

William Ormand
Lawrence Livermore National Laboratory
L-414
P. O. Box 808
Livermore, CA 94551
USA
Telephone: 925-422-8194
Email: ormand1@llnl.gov

Shashi Paul
Department of Nuclear & Atomic Physics
Tata Institute of Fundamental Research
Homi Bhabha Road
Colaba, Mumbai 400005
India
Telephone: 91-22-2152971, ext. 2333
Email: sdpaul@mail.phy.ornl.gov

Marek Pfützner
Warsaw University
Department of Physics
ul. Hoza 69
PL-00-681 Warsaw
Poland
Telephone: 48 22 823 18 96
Email: pfutzner@mimuw.edu.pl

Andras Piechaczek
Louisiana State University
Oak Ridge National Laboratory
P. O. Box 2008
Building 6000 MS 6371
Oak Ridge, TN 37831
USA
Telephone: 865-574-4723
Email: andreas@mail.phy.ornl.gov

Rodney Piercey
Mississippi State University
Physics & Astronomy Department
P. O. Box 5167
Mississippi State, MS 39762
USA
Telephone: 662-325-2806
Email: rbpl@ra.msstate.edu

David Radford
Oak Ridge National Laboratory
P. O. Box 2008
Building 6000 MS 6371
Oak Ridge, TN 37831
USA
Telephone: 865-241-5332
Email: radfordd@mail.phy.ornl.gov

Walter Reviol
University of Tennessee
Physics Department
1408 Circle Drive
Knoxville, TN 37996-1200
USA
Telephone: 865-974-7803
Email: reviol@utk.edu

Lee Riedinger
University of Tennessee
Physics Department
1408 Circle Drive
Knoxville, TN 37996-1200
USA
Telephone: 865-974-7805
Email: lrieding@utk.edu

Mike Rowe
Lawrence Berkeley National Laboratory
1 Cyclotron Road, Bldg. 88
Berkeley, CA 94720
USA
Telephone: 510-486-7310
Email: mwrowe@lbl.gov

Dirk Rudolph
Lund University
Department of Physics
Box 118, S-22100 Lund
Sweden
Telephone: 46-46-222-7633
Email: dirkr@alpha.kosufy.lu.se

Krzysztof Rykaczewski
Oak Ridge National Laboratory
P. O. Box 2008
Building 6000 MS 6371
Oak Ridge, TN 37831
USA
Telephone: 865-576-2636
Email: rykaczew@phy.ornl.gov

Karsten Schmidt
GSI
Kernphysik 2
Planckstrasse 1
DE-64291 Darmstadt
Germany
Telephone: 0049-6159-712743
Email: k.schmidt@gsi.de

Paul Semmes
Tennessee Technological University
Department of Physics
P. O. Box 5051
Cookeville, TN 38505
USA
Telephone: 615-372-3145
Email: psemmes@tntech.edu

Dariusz Seweryniak
Argonne National Laboratory
9700 S. Case Avenue
Argonne, IL 60439
USA
Telephone: 630-252-1514
Email: seweryn@anlphy.phy.anl.gov

Dan Shapira
Oak Ridge National Laboratory
P. O. Box 2008
Building 6000 MS 6368
Oak Ridge, TN 37831
USA
Telephone: 865-576-2648
Email: shapira@mail.phy.ornl.gov

Teemu Siiskonen
University of Jyväskylä
Department of Physics
P. O. Box 35
FIN-40351 Jyväskylä
Finland
Telephone: 358-14-602385
Email: teemu.siiskonen@phys.jyu.fi

Wojtek Skulski
X-Ray Instrumentation Associates
2513 Charleston Rd.,#207
Mountain View, CA 94043-1607
USA
Telephone: 650-903-9980
Email: skulski@xia.com

Alejandro Sonzogni
Brookhaven National Laboratory
P. O. Box 206
Upton, NY 11973
USA
Telephone: 516-344-5334
Email: sonzogni@bnl.gov

Francesca Soramel
University of Udine, INFN
c/o Dipartimento di Fisica
via delle Scienze, 208
I-33100 Udine
Italy
Telephone: 39-049-8277107
Email: soramel@pd.infn.it

Gene Spejewski
UNIRIB/ORAU
P. O. Box 2008
Building 6008 MS 6374
Oak Ridge, TN 37831
USA
Telephone: 865-576-7656
Email: gene@mail.phy.ornl.gov

Paola Spolaore
INFN
Laboratori Nazionali di Legnaro
Via Romea,4
35020 Legnaro (PD)
Italy
Telephone: 39-049-8068631
Email: spolaore@lnl.infn.it

Jirina Stone
Oxford University
Clarendon Laboratory
Department of Physics
Parks Road
Oxford OX1 3PU
United Kingdom
Telephone: 44-1865-272-243
Email: j.stone@physics.ox.ac.uk

Nicholas Stone
Oxford University
Clarendon Laboratory
Department of Physics
Parks Road
Oxford OX1 3PU
United Kingdom
Telephone: 44-1865-272-325
Email: n.stone@physics.ox.ac.uk

Patrick Talou
Los Alamos National Laboratory
Theoretical Division
Mail Stop B283
Los Alamos, NM 87545
USA
Telephone: 505-667-3821
Email: talou@lanl.gov

Michael Thoennessen
Michigan State University
NSCL Cyclotron Laboratory
East Lansing, MI 48824-1321
USA
Telephone: 517-333-6323
Email: thoennessen@nscl.msu.edu

Ken Toth
Oak Ridge National Laboratory
P. O. Box 2008
Building 6000 MS 6371
Oak Ridge, TN 37831
USA
Telephone: 865-574-4732
Email: toth@mail.phy.ornl.gov

Ron Townsend
ORAU
P. O. Box 117
Oak Ridge, TN 37831
USA
Telephone: 865-576-3302
Email: townsenr@orau.gov

Tamas Vertse
ATOMKI
Institute of Nuclear Research of
the Hungarian Academy of Sciences
Bem ter 18/C
Debrecen, H-4026
Hungary
Telephone: 36-52-417266
Email: vertse@tigris.klte.hu

William Walters
University of Maryland
Department of Chemistry
College Park, MD 20742
USA
Telephone: 301-405-1801
Email: ww3@umail.umd.edu

John Wood
Georgia Institute of Technology
School of Physics
Atlanta, GA 30332-0430
USA
Telephone: 404-894-5262
Email: jw20@prism.gatech.edu

Philip Woods
University of Edinburgh
Kings Bldgs./James Clerk Maxwell Bldg.
Mayfield Road
Edinburgh EH9 3JZ
United Kingdom
Telephone: 131 650 5283
Email: pjw@np.ph.ed.ac.uk

Chang-Hong Yu
Oak Ridge National Laboratory
P. O. Box 2008
Building 6000 MS 6371
Oak Ridge, TN 37831
USA
Telephone: 865-574-4493
Email: chy@mail.phy.ornl.gov

Edward Zganjar
Louisiana State University
Department of Physics & Astronomy
Baton Rouge, LA 70803-4001
USA
Telephone: 225-388-6842
Email: zganjar@rouge.phys.lsu.edu

Omar Zeidan
University of Tennessee
Physics Department
1408 Circle Drive
Knoxville, TN 37996-1200
USA
Telephone: 865-974-3342
Email: ozeidan@utk.edu

Jan Zylicz
Warsaw University
Institute of Experimental Physics
ul. Hoza 69
PL-00-681 Warsaw
Poland
Telephone: 48-22-621-38-10
Email: zylicz@fuw.edu.pl

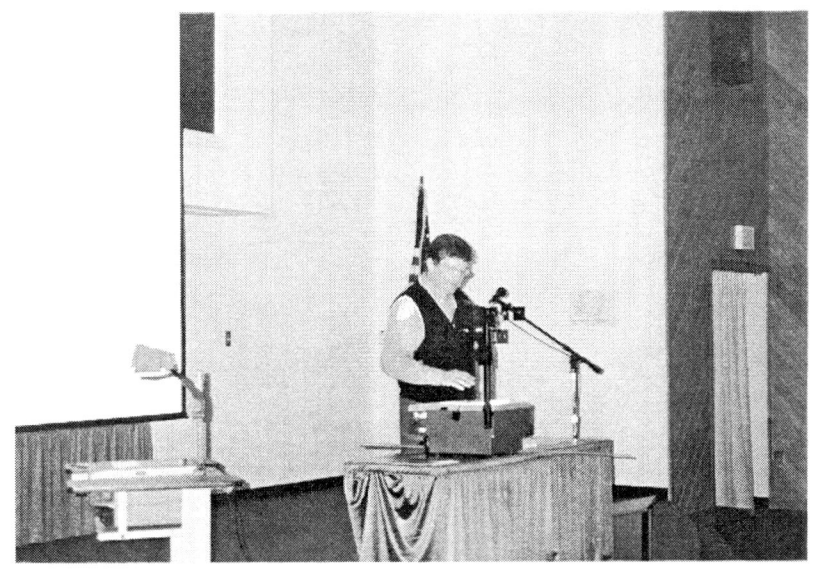

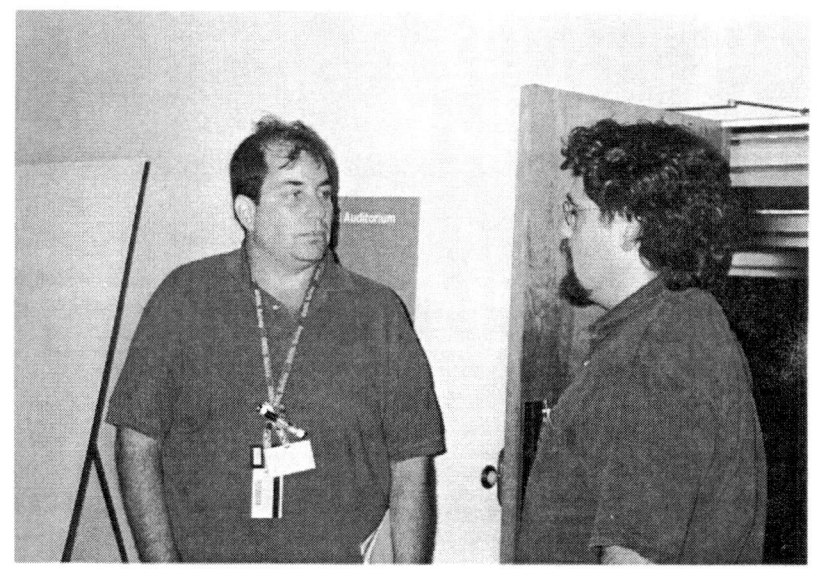

AUTHOR INDEX

A

Andrighetto, A., 68

B

Barmore, B., 184
Batchelder, J. C., 51, 83, 95, 297
Bednarczyk, P., 68
Bingham, C. R., 51, 83, 95, 297
Blank, B., 321
Boiano, C., 68
Bonetti, R., 68
Borcea, C., 321
Borge, M. J. G., 264
Boyd, R. N., 239
Broude, C., 68
Bryan, R. E., 51

C

Caggiano, J. A., 112
Carjan, N., 223
Carpenter, M. P., 59, 112
Cerny, J., 3, 95
Chartier, M., 321
Chromik, M. J., 105
Cizewski, J. A., 112
Czajkowski, S., 321

D

Dal Bello, A., 68
Davids, C. N., 59, 112, 200
Davinson, T., 51, 112
Dean, D. J., 217
de France, G., 321
de Oliveira, F., 321
Ding, K. Y., 112

E

Esbensen, H., 200

F

Faestermann, T., 24
Ferreira, L. S., 154
Fleury, A., 321
Fotiades, N., 112

G

Garg, U., 112
Ginter, T. N., 51, 83, 297
Giovinazzo, J., 321
Gregorich, K. E., 95
Gross, C. J., 51, 83, 297
Grudberg, P., 307
Grzywacz, R., 51, 83, 297, 321
Guglielmetti, A., 68
Gurvitz, S. A., 217

H

Hamilton, J. H., 51, 83
Hardy, J., 229
Heinz, A., 112
Hofmann, S., 14

I

Isocrate, R., 68
Ivascu, M., 68
Ixaru, L. Gr., 184

J

Janas, Z., 51, 83, 255, 297, 321
Janssens, R. V. F., 112
Joosten, R., 95

K

Kadmensky, S. G., 209
Karny, M., 51, 246, 297

Khoo, T.-L., 112
Kondev, F. G., 112
Kruppa, A. T., 173, 184

L

Lalazissis, G. A., 132
Lauritsen, T., 112
Lewitowicz, M., 321
Li, Z. C., 68
Lipas, P. O., 164
Lister, C. J., 112
Liu, Z. H., 68

M

MacDonald, B. D., 51
Maglione, E., 154
Malerba, F., 68
McConnell, J. W., 51, 297
Mihaila, B., 217
Momayezi, M., 307
Mukha, I. G., 144
Müller, L., 68

N

Nazarewicz, W., 173, 184, 217
Ninov, V., 95

O

Ormand, W. E., 275

P

Pfützner, M., 89, 321
Piechaczek, A., 51, 83, 95
Poli, G. L., 68
Powell, J., 95
Pravikoff, M. S., 321

R

Ramayya, A. V., 83
Reiter, P., 112
Ressler, J. J., 59, 112
Rikovska, J., 316
Ring, P., 132
Rizea, M., 184, 223
Rowe, M. W., 95
Ruan, M., 68
Rudolph, D., 285
Rykaczewski, K., 51, 83, 297

S

Scarlassara, F., 68
Schmidt, K., 326
Schwartz, J., 59
Semmes, P. B., 125, 173
Seweryniak, D., 59, 112
Shergur, J., 112
Signorini, C., 68
Siiskonen, T., 164
Skulski, W., 307
Sonzogni, A. A., 59, 112
Soramel, F., 68
Stone, N. J., 316
Stroe, L., 68
Strottman, D., 223
Szerypo, J., 51

T

Talou, P., 194, 223
Thirolf, P. G., 105
Thoennessen, M., 105
Thomas, J. C., 321
Toth, K. S., 51, 95, 297

U

Uusitalo, J., 59, 112

V

Vertse, T., 184
Vretenar, D., 132

W

Walters, W. B., 51, 59, 74, 83, 112
Warburton, W. K., 307
Wiedenhoever, I., 112
Wöhr, A., 316
Woods, P. J., 34, 51, 59, 112

X

Xu, X. J., 95

Z

Zganjar, E. F., 51, 83, 297